PLANTS AND PEOPLE
OF THE GOLDEN TRIANGLE

PLANTS AND PEOPLE OF THE GOLDEN TRIANGLE

Ethnobotany of the Hill Tribes of Northern Thailand

Edward F. Anderson, Ph.D., F.L.S.

Whitman College
and
Desert Botanical Garden

DIOSCORIDES PRESS
Portland, Oregon

Copyright © 1993 by Dioscorides Press
(an imprint of Timber Press, Inc.)
All rights reserved.

ISBN 0-931146-25-9
Printed in Hong Kong

DIOSCORIDES PRESS
9999 S.W. Wilshire, Suite 124
Portland, Oregon 97225

Library of Congress Cataloging-in-Publication Data

Anderson, Edward F., 1932-
    Plants and people of the Golden Triangle : ethnobotany of the hill
tribes of northern Thailand / Edward F. Anderson.
        p.   cm.
    Includes bibliographical references and indexes.
    ISBN 0-931146-25-9
    1. Ethnobotany--Thailand, Northern. 2. Ethnobotany--Golden
Triangle (Southeastern Asia) 3. Mountain peoples--Thailand,
Northern. 4. Thailand, Northern--Social life and customs. 5.
Golden Triangle (Southeastern Asia)--Social life and customs.
    I. Title.
    GN635.T4A53 1993
    581.6'1'09593--dc20   92-25857
                                                    CIP

Contents

Color plates precede page 145.

5

This book is dedicated to the people of the Golden Triangle:

To the friends I have made and greatly respect,

To the numerous hosts and hostesses who prepared meals and provided shelter during my field trips,

To medical practitioners who shared their vast knowledge of the plants around them,

To artisans who demonstrated their crafts,

To farmers who permitted me to visit their fields,

To headmen and elders who spent valuable time conversing with me,

For their hospitality, graciousness, and willingness to share their lives and experiences with an outsider.

May their love of life in the mountains, their respect for the forest, and their distinct ways of life continue.

Because of these many treasured experiences my life has been forever changed; it is richer by far, for I have met a truly noble and gentle people.

Preface

I first saw the people of the Golden Triangle in 1976. I had come to Thailand on a sabbatical leave from Whitman College to collect plants for the University of California, Berkeley. One day a friend who worked among the Akha and Lahu tribes asked if I would like to accompany him on a trip to the remote mountains of the northern region of Thailand. I eagerly accepted his invitation and began this trip, which changed my life!

Dr. Paul Lewis, Baptist missionary, linguist, and medical anthropologist, was the perfect person to introduce me to the hill tribes. He had spent nearly 30 years working among them, first in Burma and now in northern Thailand. At each village we visited, he gave out medical supplies, left bundles of donated clothing, and treated many of those who were ill or injured. I was immediately impressed by the love and respect he had for the people, and which they, in turn, showed him. I could not help being fascinated by the beauty of their way of life within the forested hills of Southeast Asia. What an amazing relationship they have with the surrounding environment, and yet how dependent they are upon it for simple survival.

Upon returning to Chiang Mai, Thailand, I searched for more information about these people, particularly about the plants with which they are so intimately involved. Clearly, plants are the source of their survival. Although several anthropologists have spent much time studying these mountain people, none have focused on the plant environment and the peoples' relationship to it.

My curiosity concerning these hill tribes led to extensive discussions with people who worked among them and knew the problems they were facing. All expressed a sense of urgency and impending disaster for these minorities who were being brought quickly and, at times, violently into the mainstream of Thai culture. I was shocked to realize that these cultures were rapidly disappearing through assimilation by the dominant Thai culture. What a legacy of plant lore would be lost if someone did not investigate these people and their plants.

As an ethnobotanist who specializes in the study of plants used by tribal people, the importance of this research became even more fascinating. I found the problems of the hill tribes as cultural minorities in Thailand to be immense. Whole villages of relatives immigrate from politically volatile regions in the north, east,

7

and west of Thailand. Because these tribal people continue to have high birth rates, thus creating ever larger families to feed and support, and because disease is an ever-present threat, poverty and isolation usually prevent them from receiving adequate medical and health care. There is a greater and greater shortage of land for cultivation as populations grow and hillsides are committed to reforestation. Also, discrimination occurs in many ways, often depriving these hill peoples of basic human rights and victimizing them in their dealings with Thai business people, soldiers, police, and government workers. What will happen to their knowledge of plants and their deep respect for the forest? I determined to learn what I could before it became too late. I have now been successful in obtaining considerable information about plants used by the six major hill tribes (Akha, Hmong, Karen, Lahu, Lisu, and Mien), as well as some of the smaller minorities.

The primary goal of this project was to record the plants that the tribal people use, relying primarily on firsthand observations and personal interviews. World Wildlife Fund–U.S. undertook financial support of my research in 1987, enabling me to purchase a field vehicle and make numerous extensive field trips.

As field work was of paramount importance and because uses of different plants varied seasonally, trips were planned so that villages could be visited throughout the year (see Appendix 4 for a list of the villages visited). However, this proved to be impossible on several occasions during the rainy season. Our group usually consisted of one or more tribal people, an interpreter, and at times student assistants. My companions and I traveled to villages by several methods of transportation, though the primary method was by my World Wildlife Fund vehicle. During the rainy season and when we knew the road would be particularly bad, we used a four-wheel drive vehicle. We also hiked to villages or areas where collecting was done and we even went to some areas by boat and elephant. A serious effort was made to visit villages that were isolated from major roads and tourist routes. Early in the study such villages were relatively easy to find; however, now it is almost impossible to locate areas uncontaminated by trekking groups, other types of tourists, or government and church development projects. We visited villages in many parts of northern Thailand at different elevations in an effort to learn the types of plants used wherever tribal people lived. If possible, we visited a village more than once, either in different seasons or with somewhat different objectives. Sometimes we concentrated solely on medicinal plants; at other times we looked for food plants in gardens, fields, and forests, and on the table. Often we first visited a village just to meet the headman and explain what we wanted to do. While we sat in his house drinking tea, we would make arrangements to return in a day or two either to accompany him or another informant to the forest or to have them bring in samples of plants they used.

Success depended on gaining the confidence and respect of the people with whom we were talking. In few instances did we go to a village without some type of prior contact; usually we asked someone to accompany us and introduce us to village leaders or those who had knowledge of plants. Sometimes this involved going with a Westerner who was known to the villagers because of prior visits or through development projects, but at other times we had tribal people make arrangements on preliminary visits and then return with us when we went to do the work.

One of an ethnobotanist's most important tasks is to collect and preserve material of the plants being used by the tribal people. These specimens, which are made into dry, flattened mounts on permanent herbarium paper, provide the all-important basis for the scientific identification of the plant, and serve as a formal voucher showing exactly what plant was described. The Latin scientific name enables taxonomists to catalog the material and indicates relationships. Moreover, these names are universally understood by scientists. These voucher specimens are then deposited in the herbarium of a major institution. I sent the main set of my voucher specimens to Harvard University's Economic Botany Herbarium, a second set to Arizona State University, and two additional sets to Payap and Chiang Mai universities in Thailand. In the future other botanists and pharmacologists can see the plant specimens with which I worked and perhaps gain further knowledge from them.

Sometimes I could not collect samples of a plant, either for practical or ethical reasons. Perhaps the foliage was out of reach, the plant was rare, or it was growing in someone's garden and I was not given permission to take any because of its great value. Occasionally I used photographs of plants to make sure that my tribal informant and I were talking about the same plant.

One of the major challenges of my field work was that of communication. I did not attempt to learn any of the tribal languages because field time in Thailand was at such a premium. Rarely did I even attempt to use my limited knowledge of Thai because, although some tribal people speak it (or the northern Thai dialect), it is often poorly pronounced and with limited vocabulary. It seemed apparent that their poor use of Thai and my limited ability to understand anything but some central Thai would result in a high degree of miscommunication. Therefore, I sought individuals who were fluent in one or more tribal languages, as well as in English, or who were fluent in northern Thai and English. Several Karen, Lahu, and Akha served capably in this capacity. During my first year's research I had an Akha interpreter who could speak English, Chinese, Akha, Lahu, central Thai, and northern Thai. For much of the later field work I relied extensively on William C. Young, a U.S. citizen who has lived his entire life in Southeast Asia and who speaks several languages fluently, including Lahu, which is the *lingua franca* among the hill tribes. Most frequently problems arose within a village when the informant spoke only his or her tribal language which I and my assistants did not understand. The informant's comments were first translated by another tribal person, usually into Lahu but sometimes northern Thai, which someone else then translated into English. Through this circuitous process there was considerable danger of meanings or descriptions being misunderstood, particularly when the informant, with almost no knowledge of medical terms, was talking about a plant used for the treatment of some injury or ailment.

In addition to obtaining information about how a particular plant or group of plants was used, I also attempted to record vernacular names. Sometimes a tribal person spoke into a tape recorder, which was later transcribed onto paper in a common phonetic spelling, but most often the informant pronounced the name several times and I wrote it using a common phonetic spelling. In some instances the informant was literate and he or she simply wrote the name or names in my field book. However, Mien names were often recorded in Chinese characters,

Karen in Burmese script, and other tribal languages in Thai script. Several have been difficult to change to standard phonetic spelling. Occasionally the common names helped me determine the identification of the plant and provided confirmation of another record.

Accuracy in recording the information was of great importance, but one could never be sure how much imagination or misrepresentation was involved. Though I never felt I was told a deliberate untruth, there was never a guarantee that what I was told was accurate. I was always pleased when a group of tribal people would gather around as I worked. One or two people would be the primary informant to whom I would direct questions. However, on numerous occasions bystanders would enter into the conversation to help an informant remember or better describe how a plant was used. At times they would nod in approval or correct an informant. On one occasion, a woman took a piece of bark from the plant a medical practitioner was describing for me and she put it in her mouth. Noticing this, my informant said that it reminded him of another use of the plant: to treat a sore throat. Rarely did the tribal people know correct medical terms for most ailments, except for common ones such as malaria, fever, colds, and itches. Therefore, questions were often directed specifically at the parts of the body affected, at symptoms, and then at how the plant was prepared to treat the problem. In some instances the informant even demonstrated preparation of the plant. Frequently I saw plants being prepared and used as fibers, as dyes, or for thatching. I then simply observed, photographed, and recorded what I saw and heard. These people willingly answered questions if I asked them.

Occasionally I was fortunate enough to be in a village for a special ceremony, such as a wedding, funeral, or spirit exorcism. I always asked permission to watch and to take photographs to avoid being disrespectful or possibly offending them (or the spirits) in any way.

In addition to field work, appropriate related publications were consulted as part of the comparison and verification process of the records I made. Reports in the literature had to be used with considerable caution because frequently plant descriptions were incomplete or identifications erroneous. In one report, for example, some medicinal plants supposedly used by a Thai tribal group were identified from a book on medicinal plants of Hong Kong.

A major source of information was NAPRALERT, a large computer database created by the University of Illinois, Chicago, of all medicinal plants reported in the literature. I obtained a print-out of all those from Thailand. Since these data were derived only from reports in the literature, they probably have some misidentifications.

I have created a large computer database from field records, herbarium labels, interviews, and the literature. This now enables me to sort and examine the data from various viewpoints and to extract information on numerous subjects in different combinations.

Several things became apparent soon after I began my field work. First, almost no one had investigated the culturally significant plants used by the hill tribes. Second, the tribal people, though often suspicious of the motives of Thai or Westerners, proved to be very cooperative once they understood that I respected their cultures and wished to share their knowledge of the natural surroundings.

They, too, wanted to know more about the plants that play such a significant role in their daily lives. Third, I rapidly came to appreciate that bamboo was of prime importance in their cultures, just as it is in cultures of the lowland people throughout Asia. However, our botanical knowledge of bamboos is amazingly incomplete. Fourth, I was alarmed to see that changes in the environment of the hill tribes, as well as in their cultures, are occurring at an unbelievable speed. Within the decade and a half that I have been studying these people, roads have been built into formerly isolated regions so that tourists now trek into once-undisturbed communities, and children attend Thai schools and speak the Thai language. For these reasons I have felt additional pressure to gather data as quickly as possible.

This book, therefore, presents to the interested reader and scientist a description of the people of the Golden Triangle, their environment, and their uses of plants. Theirs is a wonderful knowledge of the green world that surrounds them and upon which they are so dependent for survival.

Acknowledgments

The field work necessary for this project was made possible by two organizations that generously provided financial support and encouragement. The American Philosophical Society supported the research in its initial year, 1983–84, with a grant of $1000. World Wildlife Fund–U.S. (WWF–US) gave $5000 annually from 1987 through 1991. The WWF–US grants were given jointly to me and to my Thai colleague, Acharn Duangduen Poocharoen, of Payap University in Chiang Mai. Duangduen's assistance and that of other administrators and staff of Payap University rendered the necessary institutional support of the in-country research; the success of the project would not have been possible without their help.

Dr. Lamar Robert of the Payap University Research and Development Institute also provided valuable aid, both in the field and with computer expertise. Thanks are also expressed to Khun Pisit na Patalung and others of Wildlife Fund Thailand for their encouragement and cooperation. International air travel was financed through the Whitman College Aid to Faculty Scholarship Fund and the W. K. Kellogg Foundation.

I particularly want to thank Paul and Elaine Lewis, friends and colleagues of many years. Paul's recognition of the need for such a study and his knowledge of the people of northern Thailand were two essential ingredients in the early phases of the field work. He introduced me to several people who were knowledgeable of the plants used by the hill tribes.

A long-time resident of northern Thailand also proved essential to the project: William C. Young, with his language abilities, friendships among the hill tribes, great knowledge of life in the hills, and Land Rover made the last several years of work not only productive but also very enjoyable.

Rupert Nelson and Richard S. Mann introduced me to the tribal people with whom they were working and took me to various villages within their project areas. Their assistance was indispensable.

J. F. Maxwell of Chiang Mai University shared his vast knowledge of Thai plants and helped me obtain correct botanical identifications for most of the species. Others who assisted in plant classification were Dr. Tem Smitinand and Kuhn Chawalit Niyomdham of the Royal Forest Department Herbarium, Bangkok, Thailand, and Dr. P. F. Stevens of Harvard University. Drs. Soejatmi and John Dransfield of the Royal Botanical Garden, Kew, identified my bamboo and palm material. Katherine Bragg loaned me her collection of plants made during 1989 when she worked for the Mountain People's Culture and Development Education/Research Programme (MPCD) of Chiang Mai; these specimens are now deposited at Harvard University. Appreciation is extended to her and to Dr. Leo Alting von Geusau, director of MPCD.

Four individuals—Elizabeth A. Moore, Paul Lewis, Elaine Lewis, and Rupert Nelson—critically read all or parts of the manuscript and provided valuable comments and suggestions. The bamboo illustration was made by Naramit Xumsai na Ayudhya.

Thanks are also expressed to Keith Feldon of the International Rescue Committee, who was stationed at the large refugee camp in Chiang Kham. Twice he arranged for supervised field trips so that Hmong, H'tin, and Mien medicine men could leave the camp and collect plant materials for both themselves and for me. He then spent many hours with our team helping the tribal men describe the various uses of the plants.

Many people accompanied me in my field work. The following Whitman College students participated through the support of grants provided by the college: Katie Grew, Michelle Ladd, Tanya McKelvey, and Shannon Roys. Two other Whitman College students, Katie Cutler and Elizabeth Dolph, interpreted data as part of their senior research projects. Three of my children and my wife also worked with me on numerous field trips. Roy Miller, assisted me in the summer of 1989; in addition, he made a generous contribution to the project through World Wildlife Fund. Franklin Ott also accompanied me on several trips in 1991. Dr. Edwin A. Phillips joined me for field work in late 1987 and later allowed me to use his office at Pomona College, Claremont, California. Richard Wheeler, a U.S. Forest Service scientist assigned to the Royal Forest Department, took me on several trips and shared his expertise on Thai forests.

Several tribal people interpreted for me, introduced me, and guided me: Mark Laoko, Vinai "Ike" Pornsakulpaisarn, Aaron Pornsakulpaisarn, Pima Ana Cheu Meu, Ca Nu, Pratan Sakorn, Yua Tsong, Maw Phea, Law Jo, Maw La Baw, A Meh Jo, A Neh, Phaya Ca Nu, Chareon Pornchai, Abaw Baw Soe, and numerous others whose names were not recorded. Their knowledge and willingness to share it are greatly appreciated. Thai Tribal Crafts, a tribal-operated business in Chiang Mai, allowed me to study and photograph various artifacts in their store; their cooperation is much appreciated.

I also want to thank Kuhn Wanat Bhruksasri, Dr. John McKinnon (who worked with the TRI-ORSTOM Project), and Kuhn Chantaboon Sutthi of the Tribal Research Institute for their encouragement and assistance.

Finally, I want to express my sincere thanks to my wife, who prodded, encouraged, and inspired me throughout the field work and the writing of this book. It would have never been possible without her love and support.

Introduction

It began long before dawn: a rhythmic thumping from beneath the house, soon followed by the same sound from nearby houses. Slowly I realized that the women were already up. Some were busily working the large rice pounders as the first step in preparing the first meal of the day. Others were carrying water to their houses or starting the cooking fires. The Akha village was coming to life.

Sleep had not been easy for me, though the split bamboo floor had a degree of "give" and the temperature had dropped to a comfortable level. The night had been full of sounds, some familiar, others less so. Dogs barked, pigs grunted, and livestock directly beneath the floor on which I was trying to sleep almost constantly moved or moaned. Of course, there was also the odor from the cattle and water buffaloes! Long before dawn some ambitious rooster gave forth, followed by a chorus of his species throughout the village. During the night I had also been pelted from time to time by falling kernels of maize as rats ran along the rafters eating from the cobs that had been dried and hung for later use as animal food.

I had come to this remote site the day before by four-wheel drive vehicle. My companions were an Akha interpreter, who had come to Thailand from Burma several years earlier, and a student who would assist me in my field work. Only a handful of Westerners speak the tribal languages; perhaps less than ten speak Akha. Almost no Thai speak Akha, for they expect tribal people to speak either northern or central Thai. A few Akha who do speak Thai, speak it poorly. The Akha language, which is Tibeto-Burman, was unwritten until a U.S. missionary linguist committed it to writing about 30 years ago. Now there is even an English-Akha dictionary.

We had talked with the villagers long into the night about various things. The men sat on the floor smoking cigarettes and chewing betel nut. A few women stood along the walls or sat on some sacks of rice in a corner, some smoking pipes. By the door a cluster of curious children stared at the *farangs* (Westerners) and listened intently as we talked through an interpreter. The Akha are subsistence farmers living by slash-and-burn cultivation, and their big concern that night was whether the rice crop would be large enough for all to survive the coming year. They were also concerned that late rains would spoil the rice that had been cut in preparation for threshing, and they were very bothered by the rats. Last year one

15

of the most common species of bamboo had flowered. Because of the huge quantities of bamboo seeds available for food, the rat population had exploded, but this year the seeds were gone and the rats were not. Elaborate methods had been employed in the fields to keep the rats out: scarecrows, clappers made of bamboo, and long strings of cloth hung from cords. People watched the fields day and night to protect the precious crop. Still, much was being lost, and the rats had invaded the village to eat the harvested rice and maize. Once the cats in the village could have helped control the rat population, but cats were few because several years earlier a government program to eradicate malaria-carrying mosquitoes with DDT led to the death of most of the cats when they ate dead insects.

In "good" years these farmers had only to worry about the usual insect pests of their rice, maize, and peanut crops. Or perhaps there was not enough rainfall or too much. Fear of starvation was not new to these people, but today another factor is causing great concern: there is no longer enough land on which to grow the crops. While everyone knows the land must rest after a few years of cultivation, there is nowhere to move. There are too many people and much of the land has been set aside for reforestation by the Thai government.

The men talked wistfully about the great hunts they used to have in the forests. Old men could remember when game was always plentiful so that most of the time it was possible to have meat with rice and greens. The men still hunt, but now there are few birds, even fewer barking deer, and no monkeys. The patches of forest are becoming smaller and smaller because of illegal logging and the clearing of land for agriculture. Yet a trip to the forest can still be productive: the men gather fruits, rhizomes, seeds, wild greens, and perhaps some medicinal plants, which are always needed in the village. The forest is the source of food—and pleasure. It also provides many essential medicines, and as the forest disappears so do the plants that may make the difference between life and death. Unfortunately, reforested areas usually lack many of these important plants found in the natural vegetation. Times are hard and disease is always present.

I asked about the forests being planted by the Royal Forest Department. A few men laughed and commented that the trees were not being well cared for. Others expressed a quiet resentment that the land could no longer be used for farming. They love trees and understand the importance of forests, but they cannot help but think about the shortage of land for agriculture and the threat of starvation.

Some of the people in the village had recently come from Burma (now called Myanmar). Though the presence of these refugees aggravated the shortage of farmland, no one resented them because they knew of the death, danger, and disease that had caused them to leave their homeland. Although there is plenty of rich agricultural land in Burma, malaria is widespread and the Burmese army a constant threat. Many young men have been conscripted as porters and never seen again. Some have been shot, others starved. Others have been forced to walk through mine fields ahead of the soldiers. In some instances, entire villages have fled into a relatively safe Thailand.

One man spoke briefly of another concern. A few families still grow opium as a cash crop and to provide medicine for the village, even though they know the crop is illegal. Would government soldiers or police come to the fields and destroy the crop? Would anyone be arrested?

Leaving last night's conversation, my thoughts returned to the Akha village that was coming to life for another day of an almost-endless cycle of hard work, worry, and marginal existence in the mountains of northern Thailand. It was October and everyone was busy, for rice harvest had begun. Breakfast of rice, greens, fried peanuts, and a bit of meat was hurriedly prepared and eaten so that all able-bodied people could begin the long walk to the steep hillside fields of rice. We ate breakfast at a low, round, woven bamboo table.

From the split bamboo porch we watched the villagers going off to the fields. All the women had woven bamboo baskets with shoulder yokes and head bands, while many of the men were leading small ponies. The baskets were empty now, but on the return trip they would be filled with firewood, fodder for the livestock, and greens for the evening meal. The ponies would be laden with sacks of threshed rice. Many of the fields were several kilometers away, but the women did not waste a moment: as they walked and talked, their hands were busy spinning cotton thread using spindles rolled briskly on their thighs, pulling cotton from small bamboo containers tied onto their waists. Each person carried a midday meal of rice in a small, woven bamboo basket or simply in rolled-up banana leaves.

The older people and children stayed behind in the village. The elderly wove mats or baskets from split bamboo or rattan, and sewed articles of clothing to wear or sell for cash from foreigners in the handicraft markets. The children played with seeds, sticks, or even small coasters which they would ride down the steep roads within the village.

The morning view from the village was breathtaking. I could see a vast expanse of mountains and small valleys, and in the distance to the east was a large valley. The village site had been carefully selected on the saddle of a mountain where there was an almost-constant breeze. A small stream flowed nearby and the men had constructed an ingenious network of bamboo to bring water into the village. The night before I had watched a magnificent sunset from the porch of the headman's house. The village and its distinctive houses had a simple elegance in the practical design that has evolved over many generations.

Hundreds of tribal villages like this one are found in the rugged mountains of northern Thailand. The people who live in these remote settlements, numbering about half a million within the geographical boundaries of Thailand, have been culturally and geographically isolated from the dominant people of lowland Thailand, Burma, Laos, and China prior to the present. They are the ethnic minorities of Southeast Asia, and within their cultures lie a wealth of knowledge and a way of life very different from the main cultural groups in Asia.

As their cultures are assimilated into the dominant cultures, much is being lost. Traditional clothing is being set aside for more practical Thai and Western costumes, the Thai language is spoken by more and more tribal people, Western medicine is widely practiced, and market cloth and synthetic dyes are used in place of hand-woven cloth and vegetable dyes. More and more nonnative plants are being introduced into the gardens and fields. Finally, the traditional religious and ceremonial activities are being replaced by Buddhism and Christianity.

The traditional tribal cultures contain a wealth of knowledge of the natural world in which they live, for the natural environment is immediate to them. The people have not separated themselves from nature, in contrast to Westerners and many of the lowland Thai who have made for themselves an unnatural, technical environment. Tribal life is organized around ceremonies that protect the people from the spirits and placate them to allow the people to have successful crops, to hunt, to fish, and to be healthy. These people have developed an intimate knowledge of the forests and their products. Bamboo is one of the all-pervading plants within their lives, from the moment they are born until they are finally laid to rest. Rice, too, has a significance beyond merely being a food to eat. Because the spirits receive rice gifts, it (and liquor) are almost always part of the offerings. Opium also continues to play a role in the lives of many tribal people, though it is illegal to grow it.

The hill tribes have a knowledge of plants and animals that the world cannot afford to lose. The forests are disappearing, although many efforts are being made to save those that remain. We may not succeed, and if we lose the trees and other wildlife, we will also lose the knowledge about the forests and their contents. Thus, it has become crucial that the wisdom and experience of these tribal people, who for generations have lived in a close, balanced relationship with their ambient environment, be recorded. Otherwise, this knowledge will be lost forever. These people and their knowledge of the forest are a treasure of immense value to Thailand and to the rest of the world. The future of the forests may very well depend on these people; none of us can waste the wealth of their experiences and we must learn as much as we can from them. Indeed, I have come to believe that the tribal people themselves are an important tool for conserving the forests of the north and the treasures they contain, whether medicinals, fibers, foods, or bamboos.

Like many other concerned individuals around the world, the people of the hills do not want to see the treasures of their forests or tribal knowledge of them lost through the carelessness and greed of outsiders with short-term vision or a technology that promises to make the hill tribes independent of the natural world around them. The tribal people have willingly shared their "intellectual property" with me, hoping that it may help preserve the world they have known. This book therefore describes these people, the environment in which they live, the plants they use, and their farming practices and crops. Separate chapters deal with certain of the most important plants, such as rice, opium, and bamboo. Other chapters consider products from the forest, plants that yield fibers and dyes, medicinal plants, and construction materials. Two chapters portray other dimensions of tribal life: plants employed for pleasure and those used for dealing with the spirit world.

Rather than burden the reader with long lists of Latin binomials within each chapter, two detailed appendices are included at the end of the book. Appendix 1 lists all plants used by the hill tribes, the corresponding scientific name, family, use(s) of the plant, tribe(s) that use it, and voucher numbers of each. Appendix 2 lists all medicinal plants, the ailment treated, the plant part(s) used, and how the medicine is administered. The numerous illustrations and color plates should assist the reader in better understanding the plants and people of the Golden Triangle.

1

The People of the Hills

The area of Southeast Asia known as the Golden Triangle, where the countries of Thailand, Burma (Myanmar), and Laos meet, is infamous for the growing and trafficking of opium and heroin. Nestled in its upper valleys and on the slopes of the remote mountains of northern Thailand are more than 3000 villages (Plate 1). The inhabitants of these communities are called the "ethnic minorities" by some people, while others refer to them as "highlanders" or "hill tribes." I prefer the latter term, which in Thai is *chao khao*.

It is hard to believe that this is the last decade of the twentieth century in rapidly modernizing Thailand when one sees young women with elaborate silver-covered headpieces dressed in short skirts harvesting hill rice with sickles (Plate 55); or mothers sitting on porches, wearing heavily pleated, batik designed skirts made from hemp fibers and woven by hand, sewing elaborate tapestries representing their "picture literature and history" (Plate 7); or black-clothed hunters going to the forest with their hair cut in queues carrying crossbows and home-made black gunpowder rifles with barrels longer than the people are tall. Yet the cultures of these fascinating people with their long histories of independence from the main cultures of Southeast Asia still exist. During hundreds of years of isolation these people have evolved distinct and interesting strategies for surviving in the high, forested mountains.

Six tribes constitute the vast majority of these people in Thailand, with four smaller groups of Mon-Khmer origin making up the balance. My research has concentrated on the six larger groups: the Akha (Kaw, E-Kaw), Hmong (Miao, Meo), Karen, Lahu, Lisu, and Mien (Yao) (see map).

It is impossible to know the exact numbers of hill people because of their isolation and frequent migrations from region to region, but nearly all authorities claim that there are now at least half a million in Thailand (Lewis and Lewis 1984; Tribal Research Institute 1989). Thirteen of the fifteen provinces of northern Thailand contain 93 percent of the ethnic minorities, with the southwestern provinces containing nearly 5 percent (mostly Karen), and fewer than 3 percent in the central and northeastern parts of the country. The most current official population count of tribal people in Thailand estimates a total of 3408 villages with 551,144 inhabitants (Tribal Research Institute 1989). An unofficial census total for 1990

19

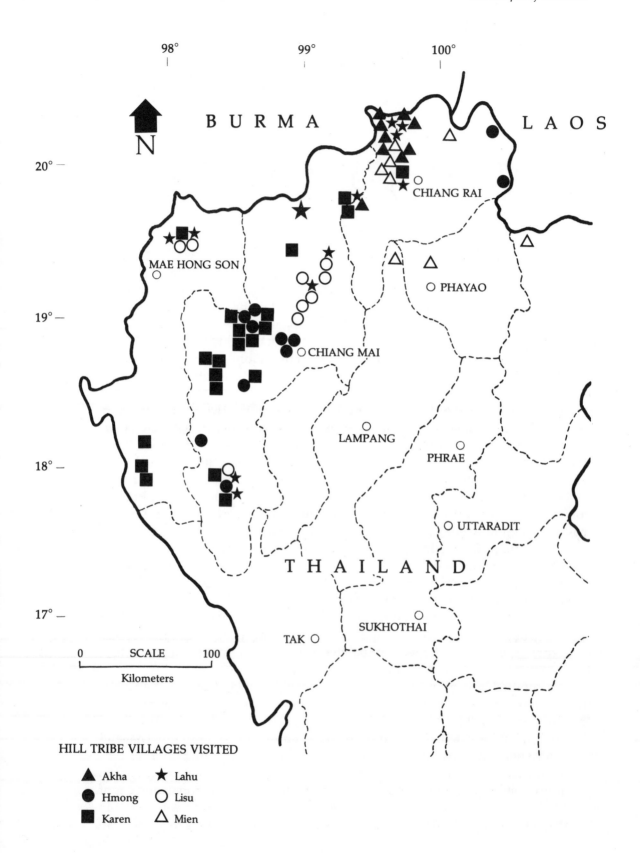

HILL TRIBE VILLAGES VISITED

▲ Akha   ★ Lahu

● Hmong   ○ Lisu

■ Karen   △ Mien

posted on a wall chart at the Tribal Research Institute states there are 554,171 tribal people in Thailand.

J. Edwin Hudspith (personal communication), a long-time missionary educator among the Karen tribal people, believes that both government and Tribal Research Institute figures are more than 50 percent below actual numbers. For example, the Tribal Research Institute summary of 1988 reports 270,803 Karen in Thailand (Tribal Research Institute 1989), but Hudspith claims there are at least 700,000 and that the government surveys ignored most Karen villages south of Mae Sariang in Mae Hong Son Province. Much of the census data were based on the reports of Thai malaria control teams, and one must assume that only those villages visited—and treated—were included in the count. Hudspith, who has visited about 250 villages in the region south of Mae Sariang, knows that most of the Karen villages were not visited by the medical teams. Rupert Nelson, another U.S. missionary who has spent several decades among the Karen, also believes the official figures are much too low; he states there are at least 450,000 Karen in Thailand (Rupert Nelson, personal communication). The oppressive political climate of neighboring Burma has forced thousands of Karen to enter Thailand, but because this region along the border is so dangerous, few Thai venture into it. If Hudspith's observations are correct, then probably there are more than a million tribal people now living in Thailand.

The origins of most of the tribal groups are relatively clear. Five of the six main tribes and all the minor groups, or closely related tribes, are living today in China, particularly in Yunnan Province. The Karen, the largest group, have an uncertain origin, though it is most likely the first Karen came to Burma long ago from southern Tibet or China (Lewis and Lewis 1984). The earliest reports of tribal people in northern Thailand are several centuries old. Particularly the Karen, but also the Hmong and Mien, were mentioned in two nineteenth-century writings, one edited by Mary Backus (1884) and the other titled *An Englishman's Siamese Journals: 1890–1893* (1895). However, the vast majority of tribal people have entered Thailand in this century, especially following the Second World War and during the almost-continuous political unrest and warfare in Southeast Asia over the past 45 years.

Linguists and anthropologists believe these hill people represent all three of the great linguistic groups of Southeast Asia (Matisoff 1983). The Karen, Akha, Lahu, and Lisu are of Tibeto-Burman origin, and thus are part of the Sino-Tibetian superstock. Their languages are monosyllabic and tone-prone (Matisoff 1983). The Mien and Hmong belong to the Miao-Yao group of the Austro-Thai superstock. Their languages are all monosyllabic and atonal (Matisoff 1983). Finally, the four minor tribes of the Mon-Khmer group belong to the Austro-Asiatic superstock, whose languages are described as "sesquisyllabic and registral" (Matisoff 1983).

The hill tribes tend to occupy somewhat distinct altitudinal areas within northern Thailand (Dessaint and Dessaint 1982). The Akha, Hmong, Lahu, Lisu, and Mien prefer to live in the high mountains above 1400 meters (4600 feet) elevation. These groups have traditionally cultivated opium, which grows best above 950 meters (3100 feet) elevation. They also grow hill rice and maize. The Akha, Lahu, and Mien also live in the lower mountains from 800 to 1400 meters

(2600 to 4600 feet) elevation. They grow opium at the upper levels of this zone, in addition to rice, maize, and a variety of cash crops. The Karen prefer the foothills from 400 to 800 meters (1300 to 2600 feet) elevation, often creating extensive areas of terraced wet rice, in addition to cultivating hill rice and maize. Few hill people, except the Karen, have migrated to the valleys because they are occupied by the dominant Thai.

Charles F. Keyes (1979:13) described the stereotyped view that the Thai government has of the tribal people:

> [They are] distinguished by their practice of upland swidden cultivation, by their production of opium, by their low level of economic development relative to the rest of the Thai population, and by their "alien" status as recent and illegal migrants into Thailand.

Unfortunately, this view, as Keyes so well pointed out, is inaccurate.

The rest of the chapter briefly describes each tribe, its origin, location and approximate number in Thailand, settlement pattern, organization, religion, economy, and clothing. An awareness of these features will permit the reader to better understand each tribe's relationship with the plants and their uses as described in the chapters that follow.

KAREN

Origin, Location, and Approximate Number. The Karen are the largest of the hill tribes of northern Thailand, the only tribal people who own and use elephants (Plate 2). Anthropologists have difficulty defining the Karen, arguing that they are so diverse culturally they may not be a single distinct ethnic group (Kunstadter 1967; Keyes 1979). In fact, perhaps the term *hill tribe* is also somewhat inappropriate for much of the group because so many of them no longer live in the hills.

The word *Karen* is from English, for the group itself has no generic term for all its members (Lewis and Lewis 1984). The central Thai refer to them as *Kariang*, whereas the northern Thai call them *Yang*. However, the term *Karen* has been used so widely and for so long that it is acceptable to virtually everyone, including the Karen themselves. Within Thailand there are two main groups of Karen, the *Sgaw* (*S'kaw*) and the *Pwo*; both groups are referred to as White Karen. Other minor subdivisions are the *Kayah* (*Bwe*) or Red Karen and the *Pa-O* (*Tanngthu*) (Technical Service Club 1987).

Many linguists believe the Karen represent a major linguistic subdivision of the Sino-Tibetian family of languages (Benedict 1972). Their ethnic origin is very unclear. Probably they are of Mongoloid stock (Department of the Army 1970), coming originally from the upper reaches of the Yangtse River, which they refer to as the "River of Golden Sand," near Tibet (Paul Lewis, personal communication). If so, their origins are much farther north than the other tribal groups, but unfortunately they left no easily identified ancestral stock or close relatives in China. Records show that the Karen have lived in Burma since the thirteenth century

(Rashid and Walker 1975a), beginning their migration into Thailand in the eighteenth century (Lewis and Lewis 1984). At present they probably make up 10 percent of Burma's population, but there, like in Thailand, they are very heterogeneous due to extensive acculturation with other cultural groups, such as the lowland Burmese and the Shan. Despite this heterogeneity, some of the Karen have long sought autonomy within modern Burma; they have fought with the Rangoon government over this matter for more than 40 years.

This tribe is widely dispersed throughout the northern and western parts of Thailand that are in proximity with Burma. As mentioned earlier, the official estimate of the number of Karen living in both hills and valleys of Thailand is approximately 271,000, but this figure may be very low (Tribal Research Institute 1989).

Settlement Pattern. The Karen typically live in small villages of about 25 houses, but often several villages are clustered together. They prefer to live in valleys and depressions within the hills near sources of water. Traditionally they have carried water to their houses in sections of bamboo placed in baskets on their backs (Plate 110), but recently a number of communities, through the assistance of mission projects or government development programs, have installed water systems using PVC pipe. Villages tend to be stable, although they may move at the death of the village priest (Lewis and Lewis 1984). Nowadays the Karen move only a short distance because there is no new land available and they have invested much time and effort in the development of fields near the old village site.

Organization. The basic unit for the Karen is the household, which consists of a nuclear family. Interestingly, the tribe is matrilocal in organization (Rashid and Walker 1975a). The village is governed by elders and the village priest, the latter position being hereditary. There is also a headman, but usually he has little authority other than as a representative to the Thai government. When newcomers arrive to settle in a village, they must get permission from the elders and village priest, not the headman (Rashid and Walker 1975a). As many Karen practice swidden or slash-and-burn agriculture, some land is always fallow; this reserve is controlled by the village as a whole, with the elders and village priest allocating it when appropriate. Marriage is quite stable among the Karen, with a low divorce rate and a prohibition of polygamy (Rashid and Walker 1975a).

Religion. Traditionally the Karen were animists, but they also have a long tradition of recognizing a Supreme Being. One very isolated group, the Talako, developed a very high moral and ethical system within their monotheistic religion. The Karen as a whole are constantly striving for harmony between themselves and the spirit realm (Lewis and Lewis 1984), so there are many rituals performed throughout the year. One of the most important spirits is the family or ancestral spirit (*bga*), but there are numerous spirits of nature as well. Many Karen, in fact, most of those in certain areas of Burma, are now Christians, this religion having been introduced into their monotheistic culture more than 150 years ago.

Economy. The vast majority of Karen are farmers, and those living in valleys have developed remarkably sophisticated terraced and irrigated fields (Plates 37–38). Such projects may be seen scattered throughout the hills of northern Thailand, including Doi Inthanon, Thailand's highest mountain. I examined one extensive series of terraced fields covering several hectares that were irrigated by a number of intricately dug channels coming from a small mountain stream only about a meter (one yard) wide (Plates 58–59). The Karen have never been opium growers, but in recent years some who live at the poverty level or are addicts have hired themselves out to other tribal groups as laborers in opium fields. Elephants and cash crops, such as maize and soybeans, are probably the two main additional sources of income, though handwork by the women has also become profitable the last few years. Until recently elephants were a major part of forest logging in northern Thailand. However, now the Karen find a much wider and more lucrative use of their elephants in the tourism industry.

Clothing. Karen women are remarkable weavers with the back-strap loom on which strips of fabric about 40 centimeters (16 inches) wide can be woven. Historically they have made all the family's clothing (Plate 169) and they also do a distinctive embroidery. Most clothing is made from two strips of cloth sewn together and differing only in length, color, and the presence of ornaments. Men wear a tunic or shirt with vertical red and white stripes and loose-fitting trousers or sarongs. Unmarried women in this tribe, in contrast to most of the hill tribes, wear the most simple women's clothing. In essence, theirs is a long white shift, with some color, usually red, woven into the waist where tassels are often added as well (Plate 3). Married women wear colorful blouses bearing distinctive designs and embellished with seeds, usually those of Job's tears (*Coix lachryma-jobi*) (Plate 4). They also wear sarongs with a horizontal pattern dominated by the color red. All women wear a head covering, but nowadays it is often a terry cloth towel that has been purchased in the market. Both men and women increasingly wear Western or lowland Thai clothing. Men are extensively tattooed, adding to their clothing's decoration (Goldrick 1989).

HMONG

The Hmong are the only tribal people who make clothing from hemp cloth upon which they dye intricate batik patterns (Plate 160). Many of them still grow hemp (*Cannabis sativa*) for its strong fibers which are stripped from the stem of the plant (Plate 155). Historically, the Hmong were also opium growers, and although opium was once an important cash crop for them, the opium poppy (*Papaver somniferum*) is now grown mostly for local medicinal use (Plates 111–113).

Origin, Location, and Approximate Number. These people, the second largest of the hill tribes in Thailand, number more than 82,000 (Tribal Research Institute 1989). Their population swelled rapidly at the end of the Vietnam conflict when many fled Laos because of terrible retribution by the Communist regime. Many of those who came to Thailand still remain in refugee camps, where they dream of once again returning to their homeland across the Mekong River. Others have immigrated to the United States and other Western countries. Hmong villages are now found in 10 of the 15 northern provinces of Thailand, as well as in Petchaboon, Loei, and Sukhothai provinces (Tribal Research Institute 1989).

Hmong roots are clearly in China, where there are at least two and a half million, possibly as many as four million, members of this minority (Department of the Army 1970). There they are called the *Miao*, a term they dislike because it means "rice shoot." They call themselves the *Hmong* or *Mong*. The Thai refer to them as *Meo*, following the Chinese, although the same word can mean "cat" in Thai if said on a mid tone. There are two main groups in Thailand, the *Hmong Deaw* or White Hmong and the *Mong Njua*, or Blue Hmong (Lewis and Lewis 1984). Probably the tribe arose in central Asia among Mongoloid stock, for there are records of them in the central Yangtse plain as early as 2200 B.C. (Jaafar 1975). They have long had conflicts and disputes with other ethnic people in China, and gradually the stronger and more numerous Han Chinese pushed them farther and farther south. There was almost continual hostility and warfare until 1870 when the Hmong were finally subjugated (Lewis and Lewis 1984). During the latter part of the nineteenth century they moved into Vietnam, Laos, Burma, and Thailand. Still characterized as having a strong desire for independence (Lewis and Lewis 1984), the Hmong continue to resist domination by other cultures, sometimes including the Thai government.

The Hmong language is part of the Miao-Yao family that some linguists place in the Austro-Thai supergroup. Although the origin of this language seems complicated, most experts generally agree that it, like the Mien (Yao) language, is Sino-Tibetian (Matisoff 1983).

There are about 230 Hmong villages in the 13 northern and central provinces of Thailand (Tribal Research Institute 1989). Because they continually search for land, they have tended to be the least stable tribal group (Jaafar 1975). Recently they were forced to settle at more permanent sites because no new land remains for them. They have been poor managers of the land in their slash-and-burn agriculture, usually farming a field until it was totally exhausted. The plot was then abandoned, and, in most cases, was able to recover its fertility only after lying fallow for many years. This practice has also changed not only because new land is no longer available, but development programs have taught them better ways of managing their fields so that cash crops do not exhaust the soil.

Settlement Pattern. Hmong villages vary greatly in size. Some have more than 100 houses while others may have as few as 15–20. The people prefer to live in the higher mountains and tend to isolate themselves from other tribes. Rarely does one find fences in a Hmong village, so livestock and children wander everywhere. A typical item in the center of many villages is a communal maize grinder (Plate 5). In recent years many communities have also developed water systems.

Organization. Two levels of organization exist: the family and the clan. The family is the basic cultural and political unit. It is strongly patriarchal and the oldest male has almost unlimited authority. Each household contains an extended family, and together the households compose the main units within the village. There are 12 clans but not all are in Thailand (Lewis and Lewis 1984).

These patrilineal groups, very important politically, culturally, and ceremonially, also play an extremely important role in both marriage and death. Marriage is usually monogamous, though sometimes a man may take more than one wife (Lewis and Lewis 1984). Marriages are seldom arranged and courting occurs particularly during the New Year's celebration. Once a young man and young woman decide they like each other, he then "captures" her and takes her to his home. Divorce is rare. In contrast to the Akha, the Hmong highly value twins.

Religion. The Hmong world is full of spirits. Some of the most important of these are the household spirits that protect the family from various types of calamity, such as disease, crop failure, and even death (Lewis and Lewis 1984). There are also good ancestral spirits (Jaafar 1975). Evil spirits reside mostly in the jungle and on the plains; each household performs numerous rituals to mollify these evil forces and to encourage the good spirits to protect them. Every village has at least one shaman, either male or female; these shamans have the special ability to go into a trance to exorcise evil spirits. Few Hmong have become Christian or Buddhist.

Economy. Most Hmong continue to be swidden farmers. Historically they have been active opium-growers, thus usually preferring fields above 1000 meters (3280 feet) elevation. They have been involved in the focus of numerous Thai, mission, and United Nations opium replacement programs. The Hmong are excellent traders, selling or bartering livestock, maize, and other cash crops.

Clothing. In addition to making hemp cloth and creating batik patterns, the Hmong also do wonderful needlework, creating tapestries, bedspreads, and other popular works of art (Plates 6–7). They weave cloth on looms that are a combination of the back-strap and pedal types (Lewis and Lewis 1984). Each of the two groups wears distinctive clothing. Blue Hmong women wear tightly pleated skirts bearing blue and white batik patterns to which top and bottom panels are attached. Their blouses are black with an embroidered large collar that is sewn on backwards so the beautiful design is on the inside. Each woman also wears an apron or "modesty panel" to cover her when she squats. Most women also wear leggings, but head dresses are seen only on special occasions. Usually their long black hair is gathered in back into a large bun.

White Hmong women wear white pleated skirts almost totally lacking ornamentation on special occasions; they wear black Chinese-style pants for every day. Their jackets are black, but often with blue borders. Their aprons for ceremonial occasions are very elaborately embroidered, but those worn most days are simply black. Turbans are worn on special occasions, but at other times their hair is gathered into a top bun. Silver jewelry is worn at all times by both groups and is an evidence of a person's wealth.

Men and boys of both groups wear loose-fitting black trousers held up with a leather belt. They have black jackets, often with some embroidery, and sometimes they wrap sashes around their waists (Plate 195). Traditionally, their heads were shaved, though often with a queue remaining. On special occasions men wear a skull cap with a red pompom.

LAHU

Origin, Location, and Approximate Number. About 60,000 Lahu now live in Thailand (Tribal Research Institute 1989), with more than 150,000 just across the border in Burma (Lewis and Lewis 1984) and about 270,000 to the north in Yunnan Province of China (Walker 1975). These people refer to themselves as *Lahu*, though the Thai call them *Mussur*, which means "hunter." The Lahu do not consider this derogatory because they have long prided themselves in being excellent hunters. The Chinese refer to this ethnic minority in Yunnan Province as the *Lo-hei* (*Lohei*) (Walker 1975).

Two main groups of Lahu live in Thailand (Lewis and Lewis 1984). The largest group, making up more than 75 percent of the population, is the Black Lahu. This group, in turn, is made up of three subgroups: the *Lahu Nyi* (Red Lahu), the *Lahu Na* (Black Lahu), and the *Lahu Sheh Leh*. The second main group is the Yellow Lahu (*Lahu Shi*), which comprise about 20 percent of the Lahu population in Thailand. The two subgroups of the Yellow Lahu are the *Ban Lan* and the *Ba Keo*. A few other small groups of Lahu, such as the *Ku Lao* and *La Ba*, make up the remainder.

The origin of the Lahu people is unclear, but Walker (1975) suggests they came from Tibet. By 1840 some had migrated southward and had settled in Burma (Lewis and Lewis 1984); by approximately 1880 a few had crossed over into Thailand and into Laos. Today the Lahu are located primarily in Chiang Mai and Chiang Rai provinces, but some also live in Mae Hong Son, Tak, Lampang, Nakhon Sawan, and Kamphaeng Phet provinces (Tribal Research Institute 1986). Their language clearly is Tibeto-Burman, specifically of the Yi or Lolo branch (Matisoff 1983). The Lahu Na dialect is the standard of those living in Thailand and has become an important *lingua franca* in much of Southeast Asia (Lewis and Lewis 1984).

Settlement Pattern. The Lahu traditionally prefer to live in the high mountains above 1200 meters (4000 feet) elevation. However, many have now been forced to move to lower elevations because of reforestation and resettlement programs. The number of houses in a Lahu village typically is low, though this will vary from group to group. Lahu Na villages may have up to 30 houses, whereas a Lahu Nyi community may have as few as 17 (Lewis and Lewis 1984). However, one of my favorite Lahu villages, the Lahu Na town of Goshen on Doi Tung at one time had 160 houses because of terrible political pressures in nearby Burma. The number of houses in this village is now 62 (Aaron Pornsakulpaisarn, personal communication).

The most important factors in choosing a village site are a nearby source of good water and arable land. A Lahu Nyi village will have a temple, the Lahu Sheh Leh have a sacred dancing area, and a Christian Lahu village will have a church.

Organization. In general, Lahu villages are loosely organized, though this has changed somewhat because of Thai government regulations. The Lahu do not have clans, so no one has a surname in the tribal language, although some have now taken Thai surnames (Lewis and Lewis 1984). The household is the basic unit, usually consisting of an extended family with two or three generations. The household heads, normally the oldest males (Walker 1975), elect a village leader, but generally community decisions are made by consensus (Goldrick 1989).

The village priest is in charge of religious matters and is second in rank to the village leader. The loose organization of a village means there are few strong ties; thus, small disputes often cause villages to fragment. Interestingly, the Lahu consider paternal and maternal sides to be of equal importance, thus making women equal to men in most day-to-day decision-making activities (Lewis and Lewis 1984). Marriage is rarely arranged and divorce is fairly common. Non-Christian communities sometimes practice polygamy (Lewis and Lewis 1984).

Religion. The Lahu believe in a Supreme Being, and in good and bad spirits and souls (Walker 1975). Over the years Buddhism has influenced some Lahu groups, but probably more than a third of the Lahu living in Thailand are now Christian, especially the Lahu Na and Lahu Shi (Lewis and Lewis 1984). Non-Christian villages usually have a village priest, whereas many Christian ones have a pastor. Some villages also have a shaman or seer, who specializes in trances. There may also be one to several medicine men or women who use supernatural means to cure illnesses, both natural and spirit-induced (Walker 1975).

Economy. The Lahu are excellent pioneer swiddeners producing a main crop of hill rice, but often a variety of cash crops as well. Traditionally, they grew opium, but now fields of it are rarely seen near their villages. Lahu Sheh Leh women have a reputation of being great basket weavers; these products, as well as the beautifully sewn articles made by the women, provide a good income for many families from the expanding tourist market. The men prepare the rattan and bamboo strips for the women, and sometimes they will make the open-weave burden baskets.

Clothing. The Lahu Nyi, Lahu Sheh Leh, and Lahu Shi women still wear traditional styles of clothing (Plate 8), but unfortunately the Lahu Na have widely adopted Thai and Western clothing; one must now go to their villages for special occasions, such as a wedding or Chinese New Year (Plate 9), to see these people in their ethnic clothes and jewelry. Some of their silver work is outstanding. Typically, Lahu Na women wear a long black tunic, and the various groups then embroider or trim this garment in distinctive but uniquely Lahu ways. The women are masters with both foot treadle and back-strap looms, now producing large numbers of articles, such as beautiful shoulder bags, for the tourist trade. Historically, some men were tattooed on their arms, back, chest, and thighs (Walker 1975).

MIEN

Origin, Location, and Approximate Number. Only about 36,000 Mien have migrated southward into Thailand from China (Tribal Research Institute 1989), where this minority numbers more than 1.3 million; in Vietnam perhaps there are as many as 200,000 (Lewis and Lewis 1984). These people call themselves *Mien* or *Men*, though others frequently refer to them as *Yao*, meaning "people" (Goldrick 1989). Some believe this latter term is a corruption of the first word of the whole name *Iu Mien*, which is commonly used by the Mien themselves, but rarely by others.

The Mien apparently originated in southern China about 2000 years ago (Lewis and Lewis 1984), slowly migrating southward due to pressures of the Han Chinese and other ethnic groups. They first entered Thailand in the mid-nineteenth century according to Lewis and Lewis (1984), though LeBar et al. (1964) claim they came at the turn of this century. Certainly the Vietnam conflict caused a great surge of immigration into Thailand. Most live in Chiang Rai and Phayao provinces of northern Thailand, with some other communities also in Lampang, Nan, Chiang Mai, Sukhothai, Kamphaeng Phet, and Tak provinces

(Tribal Research Institute 1989). The tribe is remarkably homogeneous, with all wearing the same style of clothing, speaking the same dialect, and following most of the same traditions.

Their language, in contrast to those of the other hill tribes, has been written for many centuries, and a considerable literature exists. It is of the Miao-Yao group of the Austro-Thai supergroup (Matisoff 1983), with many words borrowed from Chinese. In fact, the Mien continue to study the Chinese language within villages and traditionally use Chinese characters when writing.

Settlement Pattern. The Mien tend to live at the lower and middle elevations in northern Thailand, often in somewhat poorly defined villages. Typically there are 10–25 houses in a community, which lacks special structures, such as ceremonial gates or buildings (Lewis and Lewis 1984). Formerly this tribe was highly mobile but lack of available land and government regulations prohibiting them from going to many areas have made their villages much more permanent. The location of a house is very important to these people, for no other dwelling may obstruct the view from the front door or block a family's direct view of the local spirit shrines (Goldrick 1989).

Organization. The household with an extended family is the basic Mien organizational unit, but men have special status. Interestingly, these people readily adopt children, even from other tribes, into their families to provide more workers in the fields (Tan 1975). Each household works its own land. The most important unit of organization is the patrilineal clan in which members have a number of expected obligations to one another. Each village has a headman, who directs celebrations, presides at meetings of the elders, and helps maintain the security and welfare of the village (Lewis and Lewis 1984). He is also the village's representative to the Thai government.

Young people may choose whom they wish to marry, though there must be an affinity or harmony of their birth dates determined through astrological handbooks (Goldrick 1989). Weddings are major events within a village, taking many days, requiring the sacrifice of many pigs, and also involving elaborate and distinctive costumes worn by the couple.

Religion. The traditional Mien religion consists of two major systems that have become bonded together over the years (Lewis and Lewis 1984). The first is the belief that spirits and ancestors interact directly with human souls. Each person possesses many souls which reside in various parts of the body. The souls of children are easily frightened, so parents must be continually alert to prevent this. There are also spirits throughout the natural world (in streams, trees, fields, etc.). Because these spirits live in a world opposing that of humans, most of them are greatly feared. Ancestral spirits, on the other hand, are very powerful and can protect the household from the evil spirits of the "other world." A spirit priest is often called to perform rituals to ensure that the ancestral spirits are content and willing to protect the family.

The second major system is that of Taoism. Remarkably, this is the same form of Taoism that was practiced in China in the thirteenth and fourteenth centuries (Lewis and Lewis 1984). Religious specialists who can still read the ancient Taoist texts reside in most villages. Some amazing Taoist art work still persists, dating from several centuries ago; these paintings, which show the Taoist pantheon, are displayed during their most important ritual ceremonies.

In recent years some Mien have converted to Christianity.

Economy. The Mien are traditionally swidden farmers, though stability of villages, especially at lower elevations, has allowed some communities to develop irrigated rice fields. Until recently, some Mien grew opium; however, now they tend to grow very small plots of the plant only for local medicinal use. The Mien also have various cash crops, as well as much livestock, some of which are sold for cash income.

Clothing. Mien women wear a long black tunic with a large, bright red ruff around the neckline (Plate 10). Beneath this they wear loose-fitting pants with beautiful cross-stitch embroidery on the front of each leg. To this is added a large sash at the waist and a large black turban covered with embroidery on the head. On special occasions the women don elaborate and large pieces of silver jewelry (Plate 11). Men typically wear a loose-fitting black or indigo-colored jacket and Chinese-style trousers. Neither item has many decorations. Men in some areas have adopted Thai or Western clothing. Both sexes carry large shoulder bags, and babies are never seen without skull caps.

AKHA

The Akha people are cooperative, friendly, and gracious. They are also the most distinctively dressed, wearing elaborate, silver-colored, pointed headpieces, brief black miniskirts, and colorfully decorated tunics (Plate 12).

Origin, Location, and Approximate Number. One of the smaller of the six main tribal groups in northern Thailand, the Akha number possibly 33,000 people in approximately 200 villages, primarily in Chiang Rai Province. A few villages are also found in Kamphaeng Phet, Chiang Mai, Tak, Phrae, and Lampang provinces (Tribal Research Institute 1989). It is estimated that there are at least 180,000 Akha in Burma, as well as a large number in Yunnan Province, China. They are believed to be of Hani stock (Lewis and Lewis 1984).

The Akha refer to themselves by this name, though the Thai call them *Kaw* or *E-kaw,* terms they do not like. There are several subgroups of Akha, which differ somewhat in dialect, but the main group in Thailand is the *Jeu G'oe.* The Akha language is of the Yi or Lolo branch of Tibeto-Burman, and is similar to Lisu and Lahu, although the three languages are not mutually intelligible. The Akha language was not written until the middle of this century (Lewis and Lewis 1984).

It seems likely that the Akha arose in Yunnan Province, coming into Burma about the middle of the nineteenth century (Lewis and Lewis 1984). It was not until the beginning of the twentieth century that they entered Thailand, migrating about the same time into Laos and Vietnam. The headman of the village of Paya Phai Kao just a few kilometers into Thailand from Burma in Chiang Rai Province claims his is the oldest Akha village in Thailand, having been there since 1887. Prior to World War II only a few thousand Akha lived in Thailand, but in recent years there has been heavy migration to escape the terrible persecution by the Burmese military.

Settlement Pattern. Akha villages vary greatly in size, from as few as 10 to over 200 households. The people prefer to live above 1200 meters (3900 feet) elevation, often locating themselves on the saddles of hills to take advantage of breezes (Plate 1). The organization of an Akha village is unique. Normally there is an elaborate gate on the main trail entering the village. On one side of the gate are two large, carved, wooden fertility figures. Under no circumstances should a person touch the gate. Nearby there is often a high swing constructed from four long tree trunks, which is used only at certain special times during the year (Plates 165–166). There is also a village courting area, frequently placed in the vicinity of the swing and gate. At night young men and women come to this area to sing and dance together; some "visit the jungle" if they find each other attractive.

Organization. The Akha household is ruled by the oldest male, and Akha society consists of patrilineal clans (Jaafar and Walker 1975). Several leaders share responsibility within the village. The headman is more or less the political leader and is chosen by the elders, but the village priest is more powerful. The priest is responsible for the sacred sites in and around the village; he performs the most important ceremonies and makes sure that "the Akha way" is followed (Lewis and Lewis 1984). Finally, there are spirit specialists.

Marriage involves a period of courting in the evenings, and there is considerable sexual freedom for young people (Jaafar and Walker 1975). In contrast to the Lisu, the Akha have no bride price or dowry. On the other hand, polygamy is practiced. Divorce is a relatively simple procedure, and one frequent reason given for it is if the woman fails to bear children, who are very important within their society. However, one of the greatest tragedies that can befall a family, as well as the village, is the birth of "human rejects." These are deformed children and twins. Traditionally, these babies are immediately suffocated by the father; the bodies are then taken to a remote area of the jungle and buried. Both family and village must then spend considerable time and effort being cleansed. In fact, the parents must destroy all their belongings and live outside the village for a year.

Religion. The Akha believe in a supreme deity called *A poe mi yeh*, and they go to great efforts not to offend him. Lewis (1970) states there are four categories of Akha spirits: (1) great spirits, such as of the sun and moon, which are not harmful; (2) owner spirits, which are in charge of the fields, the livestock, and even the people themselves, and which are continually asked for help; (3) afflicting spirits, which bring sickness if offended; and (4) spirits who enter people; these are demonical in nature and must be driven out immediately or they will kill the victim.

Each person has a soul, but it may leave the body if frightened. Rituals are performed in this case to bring the soul back into the body to prevent death.

To placate these many spirits, the Akha make offerings frequently and for a variety of occasions throughout the year. In addition, they have many festivals, which are religious in nature.

Finally, ancestors are of great importance to the Akha household, and there must be an ancestral altar on the women's side of the house (Plate 82). The ancestors are fed several times a year; they in turn will feed or care for the family (Lewis and Lewis 1984).

Economy. Swidden farming is the primary occupation of the Akha, and historically they were major growers of opium. They also have great prowess as hunters, an activity much relished by the men. The Akha have no permanent land rights, and have moved fairly frequently until recent years. They raise many dogs (which are eaten) and have much livestock that are used for food, ceremonies, and to sell for cash. Some now grow cash crops in addition to the rice, most of which they consume.

Clothing. The woman's headpiece is certainly the most notable item of Akha clothing. It is made of woven bamboo which is covered with dyed feathers and gibbon fur, buttons, coins, silver, and aluminum (Plate 13). These headpieces are works of beauty, with each of the Akha subgroups having a distinct headpiece. The headpiece is never removed, even at night. This has caused some conflict when women are taken to lowland Thai hospitals where they are ordered to remove them for sanitary reasons. Young girls are not allowed to wear headpieces while attending government schools where they must also wear the Thai school uniform: a white blouse and dark blue skirt. Women's clothing is made of dark blue or black homespun material. There is a short jacket, often with beautiful stitchery on the sleeves and back, a breast band, a short skirt, leggings, and a shoulder bag (Plate 12). Men typically wear a loose black jacket, baggy trousers, and a turban on special occasions, but when they travel to the lowlands they usually wear modern clothes to be less conspicuous and to avoid having to show their identity cards to prove they belong in Thailand (Grunfeld 1982).

LISU

Origin, Location, and Approximate Number. The Lisu, described by some as highly competitive and even aggressive, number approximately 25,000 in about 130 villages (Tribal Research Institute 1989). They live primarily in Chiang Mai and Chiang Rai provinces, but with scattered villages in Kamphaeng Phet, Tak, Petchaboon, Phayao, Mae Hong Son, Lampang, and Sukhothai provinces. According to current estimates, there are an additional quarter of a million Lisu in Burma and half a million in western Yunnan Province, China (Lewis and Lewis 1984).

These people call themselves *Lisu*, though the meaning of the word is unclear. The Thai refer to them as *Lisaw* and the Burmese call them *Yaw-yen* (Department of the Army 1970). Some ethnologists separate the Lisu into three groups: the White (*Pai*), the Flowery (*Hua*), and the Black (*He*) (LeBar et al. 1964). All those living in Thailand are from the Flowery or *Hua* group. The tribe originated in China and each group differs somewhat in dress and dialect. The Lisu themselves believe they arose near the headwaters of the Salween River in Tibet (Lewis and Lewis 1984), and then over the years migrated southward along the river, reaching Burma and Thailand about the turn of the century (Rashid and Walker 1975b). However, even in this short period of time the Lisu in Thailand have developed several differences from those in Burma.

The Lisu language is clearly of the Yi or Lolo branch of the Tibeto-Burman family of languages, with many similarities to Lahu and Akha. Numerous words have also been borrowed from Yunnanese. All the Lisu in Thailand speak the same dialect (Lewis and Lewis 1984).

Settlement Pattern. The Lisu face an interesting dilemma: they want to be separate but at the same time they like to be near lowland markets to sell their crops and handicrafts. Traditionally, however, they have preferred ridges in the high mountains for their village sites, but, like other tribal groups, many communities have been forced to move to lower elevations and even into lowland valleys (Plate 14). A good village site should be near water and the forest, and it should have good soil nearby for farming. Water is especially respected by the Lisu; they believe an almost mystical power resides in it. Therefore, they often spend considerable time and effort developing elaborate aqueducts for their villages. Their communities also require abundant patches of *Artemisia atrovirens*, known as *lu khwa*, growing nearby (Plate 15), for this weedy plant is important in many of their ceremonies (Lewis and Lewis 1984). Every village must have a shrine of the guardian spirit; this center, always located above all the houses in the village, is tended by the village priest. Females may not enter the area of the shrine, and anyone not wishing to offend the guardian spirit, who protects the village from all sorts of disasters and problems, carefully respects this rule (Goldrick 1989). Clearly, the guardian spirit is an important unifying factor for each community (Lewis and Lewis 1984). In contrast to some tribes, the Lisu tend to have relatively stable villages, but, like most tribes, these villages vary greatly in size, some having more than 150 houses (Rashid and Walker 1975b).

Organization. The household is the basic Lisu unit of organization, and it contains a nuclear family. The village is the next highest unit, but there is no structured political system; the elder males simply rule by consensus (Rashid and Walker 1975b). The patrilineal clan is the organizational level above the village. Six clans are traditionally Lisu and have Lisu names, but there are an additional nine clans that have evolved from inter-marriage with the Yunnanese (Lewis and Lewis 1984). There is considerable rivalry and conflict between clans, with feuds often leading to death. Marriage, which is usually monogamous, must be within the clan and preferably between cousins (Rashid and Walker 1975b). The bride price is very high and often becomes very competitive. Following marriage, the woman becomes her husband's property.

Religion. The Lisu recognize many spirits, but the all-powerful spirit is called *wa sa* (Lewis and Lewis 1984). The various spirits include ancestral spirits, forest spirits, "tame" household spirits, and bad ghosts of dead people. Every hill has a spirit, as does the sun, moon, trees, and even a rifle or crossbow. The Lisu also firmly believe in the presence of weretigers and vampires (Lewis and Lewis 1984). As mentioned earlier, a village priest is responsible for the shrine of the guardian spirit. In addition, there are usually several spirit men or shamans who are capable of making a link between the human and spirit worlds.

Economy. Traditionally, the Lisu have been swidden farmers. Opium was a major cash crop for many families until recent years; now they grow a variety of other cash crops in addition to both hill and wet rice. They try to use a piece of swidden land for only two years, then abandon it for new land if it is available.

Clothing. Lisu women wear a very distinct, blue or green tunic, which is knee-length in front and to the middle of the calves in back (Plates 15–16). It has wide red sleeves and fastens under the right arm. The tunic's yoke and upper sleeves are then covered with many narrow strips of cloth in contrasting colors. The lower sleeves are often red and the lower part of the tunic bright blue. Under the tunic they wear knee-length loose black trousers and sometimes red leggings. For special occasions women also don a wide black sash around the waist, to which are added numerous multicolored tassels. There is considerable competition among young women as to who can have the most tassels. Lisu women sometimes cover their heads with a black turban, with many multicolored wool yarn tassels, and they wear large quantities of elaborate, heavy silver jewelry.

Men usually wear a simple black jacket, blue or green trousers, and a red sash around the waist. At times they wear a turban or bath towel on their heads. For ceremonial occasions the men wear velvet or satin jackets studded with silver buttons.

These six tribal groups, though living near one another in the mountains of northern Thailand, tend to remain distinct despite the pressures for acculturation. They have also responded somewhat differently to the plants of their environment as will be seen in the chapters that follow.

2

The Land of the Hill Tribes

I gripped the wheel firmly and carefully applied the brakes to keep the pick-up truck on the road. Lamar, my passenger, tried to appear relaxed but I sensed this was clearly a "white knuckle" experience for him, too. The small truck slithered and momentarily headed for the steep cliff at the side of the road; the chains on the rear wheels seemed worthless, for the thick mud was like grease and the poorly constructed road was badly banked and rutted. We were less than 10 kilometers (6 miles) from the main road, but that distance seemed almost impossible to cover following an unusually heavy rain during the night. Just a few days earlier this steep mountain road in northern Thailand had been completely dry and covered with several centimeters of extremely fine dust that formed dense "clouds" when disturbed, covering everything within many meters of the road. We were caught in the start of the monsoon rains, a period of several months of clouds, heavy rains, and unbelievable mud, that replaces a season of totally dry weather.

This is the environment in which the hill tribes live and to which they must continually adapt. I once asked a Lahu which season of the year the people of his tribe liked most; his answer was an affirmation of my feelings as well: the wet season. Although the rainy months bring many problems with mud, mold, and difficulty of travel, they also produce a magnificent growth of plants in the fields and in the forests. The world becomes a beautiful green in just a few weeks and gives those who live in it a feeling of great fertility. The dry season, on the other hand, feels very oppressive; much of it is hot, and the surrounding environment steadily becomes more and more brown—and dead. Dust is everywhere, even in the most remote village; then there is the smoke, sometimes so dense that the sun merely appears as a red ball in the sky. Thousands of fires, both large and small, are set throughout northern Thailand to clear fields of debris and to provide a "quick shot" of nourishment for the crops which will soon be planted. Some of these fires escape into the nearby forests, often passing over vast areas of land and removing most of the ground cover. Stands of bamboo burn and the culms explode like gunfire, leaving behind darkened, bare slopes. Indeed, though the mountain roads may be passable during the dry season, it is not a pretty time of year. These are the monsoon forests of northern Thailand and the home of the hill tribes.

GEOGRAPHY

Northern Thailand lies between 17° and 20°30′ north latitude, and 97°20′ and 101°20′ east longitude, making up a total area of 97,695 square kilometers (37,720 square miles). This region is usually referred to as tropical, lying mostly between the Tropics of Cancer and Capricorn, and is characterized by considerable solar radiation, minimal seasonal fluctuations of temperature, and no winter season (that is, no freezing temperatures) (National Research Council 1982). More significant in defining a tropical region, however, is the area's biotemperature. Biotemperature is a measurement of the heat that is effective for plant growth. Many plant ecologists believe that plants best grow and reproduce in the temperature range of 0–30°C (32–86°F), although of course, numerous plants can survive both higher and lower temperatures (Holdridge 1967). Therefore, the mean annual biotemperature for any given region is a yearly average of the temperature, a number that falls somewhere between 0 and 30°C (32 and 86°F). Most biologists define the tropics as any region in which the mean annual biotemperature at sea level is greater than 24°C (75°F). Typically the mean annual biotemperature decreases 5–6°C (9–11°F) for every 1000 meter (3280 feet) increase in elevation.

Holdridge et al. (1971) extensively studied various regions of Thailand, calculating that the mean annual biotemperature of 24°C (75°F) is found in Thailand at sea level only as far north as 12° north latitude. Therefore, considering both the latitude and the elevation at which the tribal people live in northern Thailand (300–1500 meters or 980–5000 feet), they actually live in a subtropical environment rather than a tropical one. Specifically, northern Thailand is classified as the Subtropical Moist Zone below 1000 meters (3300 feet) elevation and the Subtropical Lower Montane Belt, including Wet and Rain Forest zones, above that altitude (Holdridge et al. 1971).

PHYSIOGRAPHY

Thailand is typically divided into five physiographic areas: the Northwest Highlands, Chao Phraya Plains, Korat Plateau, Chanta Buri region, and the Peninsular region. The Northwest Highlands region contains the vast majority of the hill tribes and is widely called northern Thailand (Ogawa et al. 1961). This area is actually an extension of the Shan highlands of Burma and consists of a number of more or less parallel mountain ranges oriented in a north-south direction. A relatively massive mountain range lies astride the international boundary between Burma and Thailand, extending northward into China and southward essentially along the entire Malay Peninsula. The highest mountain of Thailand, Doi Inthanon, lies in Chiang Mai Province and reaches an elevation of 2576 meters (8450 feet). Other major mountain peaks in the north include Doi Chiang Dao, a massive limestone landmark of 2185 meters (7170 feet) elevation, and several other peaks in northern Thailand exceeding 2000 meters (6560 feet) elevation; all are the homes of one or more hill tribe group.

Five major rivers drain these mountain ranges within northern Thailand; each runs more or less southward. They are the Chao Phraya (into which the other four flow and which ultimately passes through Bangkok and into the Gulf of Thailand), the Mae Ping, the Mae Wang, the Mae Yom, and the Mae Nam Nan. The gigantic Mekong (Maekong) River forms part of the eastern boundary of northern Thailand; many of the same groups of tribal people live in the hills of Laos to the east of this river. The valleys through which these tributaries flow are occupied primarily by northern Thai, who have developed the valley bottoms for wet rice cultivation, thus restricting the hill tribes mainly to the hill slopes and mountains.

These more or less continuous mountains extending far to the north into China have been a natural migration route for most of these tribal people. Because there is a poorly defined and remote boundary between Burma and Thailand, highlanders continue to move easily throughout their mountainous domain.

GEOLOGY

Northern Thailand consists of mountains separated by wide, remarkably fertile alluvial-filled valleys. The mountains of the north have mainly developed as north-south folds in the sedimentary rocks due to the intrusion of batholiths. Though most of the area is covered by forest and has become greatly eroded with time, one can still see that the main outcrops consist of granite of early Mesozoic origin; hard, very pure, nearly vertically oriented limestone of probable Ordovician time (such as near Doi Chiang Dao); and some quartzites (Pendleton 1962). Northern Thailand has had no significant volcanic activity in recent geologic history.

SOILS

Many of the soils of northern Thailand are still unclassified and our understanding of them limited (Pendleton 1962; Phillips et al. 1967). The soils of the steeper mountains are mostly red-yellow podzols that are derived from acidic parent rocks. These tend to be deposited in the upland plateaus and thinly on the hill slopes. They are fairly acidic (pH 4.5–5.5), poor in water retention and fertility, and usually have very little organic content. Unfortunately, these soils are easily eroded, so cultivating slopes with this soil type is hazardous and usually yields little success. There are also reddish-brown laterite soils at lower elevations; these are more deeply weathered and only slightly acidic. Though not very fertile, they are stable and capable of retaining considerable water; they tend to resist erosion, thus permitting swidden agriculture.

The reddish brown soils that have developed from the weathering of limestone are more susceptible to erosion as they are less porous; however, they are also less acidic. Some reddish brown latosols are derived from the limited outcrops of basalt and andesite in the north; these soils are much less fertile than most other types. Finally, there are the much richer alluvial soils of the valleys, but few tribal people occupy the areas in which these occur.

CLIMATE

Northern Thailand, like most of Southeast Asia, is strongly influenced by the monsoons. This term refers to the distinctive annual weather pattern of seasonal changes in which the wind blows from the southwest during one half of the year and from the northeast during the other half. The influence of the Indian Ocean causes that half of the year with winds coming from the southwest to be wet, whereas the season having winds from the northeast is dry. The two seasons are interrupted by a dormant or transition phase in which the weather is hot, dry, and quite stable (Webster 1981). Some fluctuations in this pattern occur, though agricultural practices have long been adapted to the almost certain arrival of rains in northern Thailand in May.

There are four distinct seasons in Thailand due to the effects of the monsoon winds and the intervening periods (Ogawa et al. 1961; Phillips et al. 1967). The first is the season of the northeast monsoon, which is from November to February. It is the most pleasant time of year, with relatively low temperatures, dry air, and little or no rain. The coolest month of the year is January; in the north the mean temperature during this time is 19–24.5°C (66–76°F), and sometimes temperatures will dip below freezing at night in villages above 1000 meters (3300 feet). Occasional "mango showers" occur during these months caused by temporary fronts forming between the ocean winds from the east and south and the northeast monsoon wind.

The second season, a transition period, occurs from March to April. The most unpleasant time of year, this pre-monsoon season is very hot and dry. In addition, swidden fields are burned during this season, making most of the north extremely smoky and dusty. Mango showers may occasionally occur during this season too.

The third season is that of the southwest monsoon; it is the rainy season and occurs from May into early October. The equatorial low pressure belt moves northward with the sun in the late spring to cause this sudden change in climate. The hottest month of the year is often May, just prior to the beginning of the rains; during this time the mean temperature in the northwest mountains is 27.5–31°C (81.5–88°F). Massive cloud systems are carried by winds from the south, southwest, and west, resulting in heavy rainfall over all Thailand during most of the season. Average precipitation during this time in Chiang Rai Province, the northernmost region of Thailand at approximately 380 meters (1250 feet) elevation, is 14,380 millimeters (56.6 inches) (Nuttonson 1963), but the higher mountains receive even more rainfall. Mean precipitation for Chiang Mai, located about 150 kilometers (93 miles) to the south and at 314 meters (1030 feet) elevation, is 11,070 millimeters (43.6 inches). The peak of the rainy season is August, in which more than 500 millimeters (19.7 inches) may fall.

The fourth season is referred to as both the retreating or post-monsoon season; it occurs in October and is characterized by a gradual change in wind direction from the south and southwest to the west and then to the north. Mean precipitation drops by a third from that of September, followed by a major decrease and then cessation of rain in November. However, temperatures remain much the same as in the previous season. The weather can be quite unpredictable during this season, sometimes being quite dry but at other times producing heavy rains.

VEGETATION ZONES

Considerable disagreement exists concerning the names of the several vegetation zones of northern Thailand. Most scientists agree that the general term *monsoon forest* used first by Schimper (1903) is appropriate. However, subdivisions within this broad category vary widely. The Royal Forest Department (1962) has established a classification of the forests of Thailand that has been widely followed by many geographers, foresters, and botanists. However, Pendleton (1962), Ogawa et al. (1961), Richards (1966) (for Burma), Holdridge et al. (1971), Whitmore (1984), and Maxwell (1988) have proposed modifications or entirely different zones from the system of the Royal Forest Department, which subdivides the forests of Thailand into two major groups: deciduous and evergreen.

Deciduous Forests

Deciduous forests are of two types. Both types are found widely in northern Thailand and occur up to about 1000 meters (3280 feet) elevation (Plate 17). The **Mixed Deciduous Forest** is widespread throughout the north and is best characterized by the presence of teak (*Tectona grandis*). Ogawa et al. (1961) refer to this forest as the Tall Deciduous or Monsoon Forest. Maxwell (1988) calls it "Mixed" Deciduous Forest. In general, the trees are relatively tall, reaching a height of 30 meters (98 feet) or more, and with a more or less closed canopy. Some of the taller trees are *Vitex peduncularis, Terminalia mucronata, Xylia xylocarpa, Holarrhena pubescens*, and *Engelhardia serrata*. A second stratum of smaller trees and shrubs occurs at 7–10 meters (23–33 feet) in height; frequently there is much bamboo and lianas are fairly common. Fire is frequent in this zone, occurring almost every dry season, but causing relatively little damage.

A second type of deciduous forest is the **Deciduous Dipterocarp Forest,** referred to by Ogawa et al. (1961) as a Savanna Forest, and by Maxwell (1988) as a Deciduous Dipterocarp–Oak Association. This forest type can be described as having an open or discontinuous canopy at about 15 meters (49 feet) height and an undergrowth consisting primarily of grasses. Occurring up to 1000 meters (3300 feet) elevation in some areas of the north, this vegetation type is generally found on poorer soils. Most of the trees totally drop their leaves during the dry season and bamboos are not common. Again, fire occurs frequently in this region during the dry season, removing fallen leaves and other organic debris. The dominant trees are of the Philippine mahogany family, Dipterocarpaceae, the most common being *Dipterocarpus obtusifolius, D. tuberculatus, Shorea roxburghii,* and *S. siamensis.* Other significant trees are *Gluta usitata, Quercus aliena, Q. kerrii,* and *Aporusa wallichii.* Some workers suggest that this forest type should be subdivided into a xeric or dry phase and a moist or mesic phase. Ogawa et al. (1961) refer to these phases as a Dipterocarp Savanna Forest and a Mixed Savanna Forest, respectively.

Evergreen Forests

The Royal Forest Department lists four types of evergreen forests in Thailand, but one type, the Mangrove Forest, does not occur in the north.

The first type is called the **Tropical Evergreen Forest.** It occurs in the north

in damp, low-lying situations near streams and rivers and on the hills up to an elevation of about 1,000 metres, beyond which they are replaced by hill evergreen forests (Royal Forest Department 1962:1).

It is a bit difficult relating this forest type to those described by others. It probably corresponds to the Evergreen Gallery Forest of Ogawa et al. (1961) and to part of the Subtropical Lower Montane Wet Forest of Holdridge et al. (1971); it would still be part of the Mixed or Closed Deciduous Forest of Maxwell (1988). This forest receives a fairly continuous supply of underground water throughout the year and the completely closed canopy is at least 30 meters (98 feet) high. There are abundant epiphytes and lianas on the larger trees, but the ground plants are mostly herbaceous. Some bamboos and palms occur in this vegetation, including the rattans. Again, members of the Dipterocarpaceae, including *Dipterocarpus alatus, D. costatus,* and *Hopea odorata,* are important trees in this type of forest, which the hill tribes tend to leave undisturbed.

The second evergreen forest is the **Coniferous Forest,** but my observations, as well as those of others, are that these rarely, if ever, consist only of conifers. The Royal Forest Department (1962) states that these forests occur from 700 to 1000 meters (2300 to 3300 feet) elevation. Perhaps this forest type is best described as a transition zone between the deciduous forests at lower elevations and the temperate evergreen forests at higher elevations. Only two pines are native to northern Thailand, *Pinus kesiya* and *P. merkusii,* and both occur in this forest type, along with *Castanopsis, Lithocarpus,* and *Quercus* species of the Fagaceae. Because of its distinctive combination of trees, Pendleton (1962) referred to this forest type as the Northern Thailand Pine and Oak Mixed Forest. Serious degradation has occurred in some of these forests through the tapping of the trees for resin, as well as the scorching and then removal of pitch-filled wood from their bases for sale as kindling to start charcoal fires. The hill tribes often occupy this zone of vegetation, so much of it has also been cleared for agriculture.

The third type of evergreen forest listed by the Royal Forest Department (1962) as occurring in northern Thailand is the **Hill Evergreen Forest.** Lying mostly above 1000 meters (3300 feet) elevation, this vegetation type is clearly temperate in its characteristics. It is one of the zones most heavily affected by the hill tribes (Plates 18–19). Where it still exists in an unaltered state, this forest type consists of very dense, extremely rich stands of evergreen trees, making a solid canopy that is up to 50 meters (164 feet) high. There are usually two or three strata of woody plants and abundant epiphytes, lianas, and ground cover. This vegetation zone is the same as the Temperate Evergreen Forest of Ogawa et al. (1961), is part of the Subtropical Lower Montane Wet Forest of Holdridge et al. (1971), and is the Primary Evergreen Forest of Maxwell (1988). Ogawa et al. (1961) comment that the tropical components of this zone's vegetation totally drop out at about 1300 meters (4265 feet) elevation. The effects of the northeast monsoon are less in this zone because wet fogs and some rainfall provide sufficient moisture during the dry season so that growth may occur in the forests all year. The conifers and most dipterocarps (with the notable exception of *Dipterocarpus costatus*) are absent in this area, with members of the Fagaceae, Theaceae, Lauraceae, Euphorbiaceae, and Magnoliaceae all present.

The vegetation zones of northern Thailand are similar to those of neighboring Laos and Burma. Despite the notable differences in the lowland vegetation between Burma and Thailand (Ogawa et al. 1961), there appear to be few major differences in the mountains, although eastern and northeast Burma are poorly known botanically. I have noticed many different species of plants in the eastern portion of northern Thailand; apparently these are more closely related to the plants of Laos, which has a considerably different flora from most parts of northern Thailand. Clearly, the flora of the north is quite different from the flora of the other parts of the country.

MODIFICATIONS OF THE VEGETATIONAL ZONES

Forests have been and continue to be removed from areas of northern Thailand (Plate 20). Two different things may occur once an area has been cleared. If surrounding vegetation and soil are not badly disturbed and the region is allowed to regenerate naturally, a successional series will quickly begin and within a few decades much of the climax vegetation will reappear. If the area has been badly disturbed but is then left to recover, a completely different assemblage of plants will form into a secondary vegetation. These include fast-growing, short-lived trees such as *Macaranga denticulata*, *Ficus hispida*, *Trema orientalis*, and *Callicarpa arborea* (Maxwell 1988). In time these trees may eventually be replaced by those of the climax vegetation at that elevation and soil condition.

Areas within the mountains of northern Thailand that continue to be disturbed (along roads and trails and near villages) are often heavily infested with both native and exotic species, some of which create serious problems. The introduced plants include *Mimosa invisa*, *M. pigra*, *Bidens pilosa*, *Plantago major*, *Tithonia diversifolia*, *Eupatorium odoratum*, and *Passiflora foetida*. Several of these weedy species are used by the tribal people. For example, *Plantago major*, *Tithonia diversifolia*, and *Eupatorium odoratum* are used medicinally. *Thysanolaena latifolia*, a native but weedy roadside grass, is harvested for broom makings and sold to the lowland Thai. Another grass, *Imperata cylindrica*, is very widespread and probably naturalized, especially in swidden areas, and it is used medicinally and for thatch.

Another consequence of the modification of a vegetation zone by humans is the creation of savannas. All of these open areas, which are dominated by perennial grasses and largely lacking trees, are the result of human interference with the natural forest vegetation. The usual cause of such savannas is the continued use of a hill side for slash-and-burn agriculture until the soil has been exhausted and serious erosion has occurred. Moreover, the continued burning of these hill sides gradually kills the native plants that have persisted through an extensive period of cultivation. The main plant that invades devastated hill sides is the cogon grass, *Imperata cylindrica*. Though valuable as a source of thatch for houses, *I. cylindrica* essentially ruins the area of land it comes to occupy because of its almost solid, impenetrable mass of rhizomes at and just below ground level. Vast areas of these savannas created by human activities may be seen in Doi Inthanon National Park (Plate 21).

DEFORESTATION

Northern Thailand comprises about one-fifth of the area of the country, with a total area of nearly 100,000 square kilometers (38,000 square miles). It is estimated that more than 75 percent of northern Thailand was forested at the beginning of the twentieth century; the rest of the country probably never had more than about 70 percent forests (Myers 1980). Removal of forests for agriculture occurred at an estimated rate of about 3 percent per year during parts of the first half of the twentieth century (Feeny 1988), and, with the advent of remote sensing in the 1960s, more accurate figures of deforestation could be seen, as well as its causes. The picture is not encouraging. From 1961 to 1965 the remaining forests in the north were reduced by more than 8 percent. The next seven years showed a further reduction of more than 10 percent, or a total of 10,978 square kilometers (4239 square miles). Later figures are even more shocking: from 1972 through 1978 almost 28.5 percent of the remaining forests disappeared (Myers 1980)! A further calculation made in 1988 indicated that only 47.39 percent of the forests of any type, both virgin and regrowth, remain in northern Thailand (Mountain People's Culture and Development Project 1989). Perhaps more significant is the fact that only 30 percent of the forests remain in the mountains of the north at elevations above 800 meters (2625 feet); this region comprises some of the most important watershed in the entire country.

An important question is what impact the hill tribes have had on this rapid removal of forests. To put the question in proper perspective, figures from east and northeast Thailand, where no hill tribes occur, need to be compared to figures from the north. In 1961 nearly 61 percent of eastern and nearly 41 percent of northeastern Thailand were forested. By 1988 only 14 percent of the forests in the northeast and 21.5 percent of those in the east remained (Mountain People's Culture and Development Project 1989). Clearly, the forests of some other parts of Thailand are disappearing at a much more rapid rate than in the north where the hill tribes are primarily found.

Several reasons have been given for the rapid removal of forests in Thailand. The primary cause almost certainly is the development of new agricultural land for the steadily growing population. Only a fourth of the total land area of Thailand is suitable for agriculture, so very early in its history the forests were removed from most of the lowland areas for the development of wet rice cultivation. As the prime lowland areas were occupied and farmed, gradually an ever-increasing population cleared and cultivated more marginal land in the foothills and mountains. Many lowland Thai continue to migrate into the foothills, cut away the forest edge, and plant crops of maize or cassava. The land is poorly maintained, so in a year or two the Thai hack away more of the foothill vegetation and plant on new land. These squatters are often followed by farmers who operate on a larger scale, growing sugar cane and other crops that quickly deplete the soil. No one attempts to rotate crops or allow the land to be fallow, as do swidden farmers; thus, within a few years the land is essentially destroyed. The few trees that do eventually return are often cut down by the poorest people to make charcoal.

A second, and very significant, cause of deforestation is legal and illegal logging. Logging has long been a highly profitable business in Thailand because of many species of valuable hardwoods, including teak; however, all mature teak had been removed by the late 1970s and the government was forced to prohibit the export of this valuable tree. Trees are cut not only for lumber, but also for firewood, for domestic cooking and heating, for tobacco-curing sheds, and for the making of charcoal. Logging is also done for other forest products, such as lacquer, oil, rattan, resin, bamboo, and pulp. Deforestation from logging was so extensive that many people were killed by serious flooding and mud slides in the south in the winter of 1988; the devastated hillsides were so easily linked to careless logging practices without reforestation that the Thai government banned logging of any kind throughout the country early in 1989.

According to many observers, illegal logging accounts for more forest removal than does legal cutting. Some people estimate that this illegal harvesting of logs may be more than three times that which has been legally allowed. Despite the current country-wide prohibition of timber-cutting, much still goes on even in national parks and reserves. Numerous loop-holes exist in the various government regulations attempting to control the cutting of trees. One example is the domestic use of teak. The law prohibited the cutting of teak trees for commercial purposes in 1980; however, the Thai were permitted to continue to use teak for the construction of houses. Therefore, they simply cut down some teak trees and built simple houses having especially large posts; a few months later the house was torn down and sold for the making of furniture and veneer, or for some other commercial use.

A third reason for deforestation is the clearing of land for slash-and-burn or shifting agriculture, which is practiced by all the hill tribes. However, it is interesting to note that more lowland Thai now practice this form of hill cultivation than do all the tribal people combined. Estimates are that this form of agriculture accounts for the clearing of about 500,000 hectares (1,235,500 acres) annually throughout Thailand (Feeny 1988). Shifting cultivation will be dealt with extensively in the next chapter.

Attempts by the Thai government to control deforestation began with the

Forest Protection Act of 1897, which attempted to regulate commercial logging. Numerous modifications and new acts were established over the following years, one of the most important being the Forest Reservation Act of 1936–1937, which designated protected and reserved forest areas. This Act was later replaced by the National Reserved Forest Act of 1964. At first the government set the lofty goal of keeping 50 percent of the country's land in forest. However, continued deforestation has forced the government to revise its target downward. For example, the Fourth Plan of 1977–1981 said that 37 percent of the country was to remain in forests (Feeny 1988). In light of the continued removal of trees, even this level has proven impossible to maintain.

Deforestation, or the "despoiling of land," meaning the removal of forests for shifting agriculture, was one of three main factors that led the central government to revise its long-term official policy of non-interference with the hill tribes to one of centralized control over them. The other two factors were their growing of the opium poppy, which the government wishes to suppress for the welfare of the people and because of international pressure, and for security, which meant keeping the loyalty of the hill tribes so that they formed an effective buffer against Communism (Bhruksasri 1989a).

WATERSHED CONSERVATION AND REFORESTATION

Alarmed by the rapid removal of forests, extensive soil erosion, serious sedimentation of waterways, and occasional catastrophic flooding in urban areas as far away as Bangkok, Thai government agencies have attempted to respond by creating various regulations and programs.

The Royal Forest Department has also set aside vast additional areas of protected areas in Thailand within the last 25 years. Since 1961 the government has created 52 parks, 28 wildlife sanctuaries, and 41 nonhunting areas, many of which are in the north (Ewins and Bazely 1989). Clearly of great importance for the conservation of forests and watersheds, these protected areas, amounting to more than 40,000 square kilometers (15,444 square miles) in 1985, have exacerbated the problems of the hill tribes. With the continued immigration from neighboring countries of tribal people who require land for agriculture and the setting aside of additional land for conservation, the pinch has become quite severe; there simply is not enough land upon which to practice responsible shifting cultivation with fallow fields being allowed sufficient time to recover.

Forest destruction and the consequent land degradation have been blamed for changes in climate, both in the north and throughout Thailand. Little hard evidence supports this claim, but the problem of erosion cannot be contested, for it is indeed serious in many areas. Not only does poor management of cultivated fields lead to serious loss of soil, but other human activities such as road building and logging may have even more profound effects. However, it is important to note that only about 4 percent of the mountains are actually under active cultivation at any one time, the remaining area being in some state of regrowth and covered with vegetation (Hoare 1985). The tribal people are well aware of the dangers of erosion, and many do all they can to prevent it. Various development programs are also helping educate them about other techniques, such as permanent cropping and contoured grass or pigeon pea strips, which have proven to be very effective in stabilizing (and enriching) the soil.

Sedimentation of waterways throughout Thailand as a result of deforestation has caused serious problems. However, the reasons for this are not well-documented and possibly the hill tribes have been blamed for causing sedimentation when others are at fault. Some research carried out in northern Thailand indicates that road building may cause far more sedimentation problems than does shifting agriculture (McKinnon 1989).

The Land Development Department of the Ministry of Agriculture is reclassifying all the land in Thailand, including the watershed regions and areas where the hill tribes live in northern Thailand. Alarmed that 1.3 million rai (208,000 hectares; 513,968 acres) of forest are currently being destroyed annually in the north, including what is believed to be critical watershed, the government is attempting to regulate various human activities within the mountainous areas (Prapun Phonpunpoa, personal communication). The most critical zone, Class A1, consists of naturally forested and undisturbed areas on the steepest slopes; no development of any kind is allowed in this zone. Unfortunately, the designation of the zones sometimes has been decided only with the use of topographic maps and without careful consideration of actual terrain and human activities. Consequently, some tribal groups that have lived for many years in areas that are now designated Zone A1 are being relocated. Zone B1 is also set aside for watershed protection and still has at least some forests. The other zones of the land classification system deal with terrain that may be part of a critical watershed but has no forest remaining or a badly disturbed forest.

One of the main thrusts of this program is to move the tribal people slowly out of the high mountains to lower elevations, but such a relocation policy is fraught with problems and potential conflicts. In effect, the government wants to move the tribal people to the so-called middle zone of the mountains, which is from 600 to 1500 meters (1970 to 4920 feet) elevation. Some of this zone is to be reforested, whereas other parts can continue to be used for agriculture. The government has arbitrarily decided that each family can have 15 rai (2.4 hectares; 5.93 acres) and that all other land is to be reforested. Many highlanders welcome conservation activities, believing that they are long overdue. However, for many it is too much too late, and serious problems are being caused for the tribal people. Some also complain that the Royal Forest Department takes the fallow land for reforestation, thus eliminating the possibility of land rotation.

Reforestation is a major program of the Royal Forest Department, and extensive plantations of teak (*Tectona grandis*) have been established at lower elevations throughout northern Thailand. The government has stated that a teak tree can be cut when its circumference is 1.9 meters (6.2 feet) at a height of 1.3 meters (4.3 feet); unfortunately, it takes approximately 120 years for a tree to grow to this size. Some plantings are now about 80 years old, but the vast majority are only about 30 years old. Thus, these slow-growing teak forests will not be commercially available for many decades. The Royal Forest Department chooses the types of trees to be used in reforestation; the vast majority are species of pine, mostly of the two native species, *Pinus kesiya* and *P. merkusii* (Plate 22). These fast-growing trees may produce a harvestable crop in about 15 years, but they must be 120 centimeters (47 inches) in circumference to be cut (Prapun Phonpunpoa, personal communication). Unfortunately, the resultant pine forests at elevations where they do not naturally occur, not only alter the soil conditions, but create forests that are very susceptible to fires which commonly sweep through the mountains. Several species of the introduced tree, *Eucalyptus,* are also planted in reforestation projects (Plate 23). Like the pines, these species are fast-growing but change the soil conditions. They are also of limited commercial value.

The Royal Forest Department permits tribal people to use the reforested areas to get firewood, medicinal plants, and vegetables; however, they are not allowed to cut the trees except for personal use, such as house construction. They are also allowed to hunt in these new forests (Prapun Phonpunpoa, personal communication).

Unfortunately, the government policy is not understood by many tribal people. One Lahu village leader believes that reforested areas cannot be used by tribal people. An Akha headman told me that cutting trees for any purpose was strictly prohibited and that military rangers would arrest anyone if he did so. Thus, there were no new house posts available in his village. He also stated that part of the land set aside for use by his village had to be reclaimed (i.e., planted to trees) and would no longer be available for agriculture. He was very worried that there would no longer be enough land for his village, especially in those years when they did not get good crops. Moreover, he was also worried that the fields would have to be rotated too frequently.

A Lahu headman expressed similar concerns, stating that each household needed at least 10 rai (1.6 hectares; 3.95 acres) of land to cultivate. Being required to reclaim some of his village's land would cause a serious shortage of fields for growing rice; however, he did say that the government had given his village permission to plant tea as some of the "forest trees" in the reclaimed area, which would provide a cash income in a few years. He thought they might even try planting some rubber trees. Ironically, the reforestation of former agricultural land has meant that some villages of tribal people cannot grow enough rice to meet their needs, so they have returned to growing opium to earn cash to buy the needed rice.

The government is also attempting to keep bamboo from growing in the newly forested areas because it is such a serious fire problem. Thus, the tribal people must use valuable agricultural land to grow bamboo for house construction and the numerous other uses to which they put it (Chapter 6), or they must go great distances to find stands of it in forested areas that are not newly planted.

The mountains of northern Thailand, formerly covered by rich, diverse forests, now provide far less security to the hill tribes, who live there and who are so dependent upon the plants. Only rigorous measures will protect the remaining vegetation and thus provide the resources upon which the highlanders so depend.

3

Farming in the Hills

The charred tree trunk still smoldered at the side of the road amidst the remains of bamboo and shrubs that had been nearly devoured a few minutes earlier by flames sweeping through the area. My companions and I continued a few hundred meters to a scene of utter desolation and destruction (Plate 26). Only a few blackened tree trunks broke the skyline; all else on the hill in the distance had been burned to a blackened smoothness, and all life appeared to be gone. The smoke-filled air made it hard to breathe. Another field nearby was covered with dry, dead shrubs, bamboo, and young trees that had been cut several weeks previously; this debris would soon become a conflagration, resulting in another blackened hill side (Plates 29–30). Certainly nothing could grow on the hill for years. Yet only a few short weeks later a miraculous change occurred. With the start of the rainy season everything burst into life; what appeared to be dead bushes now showed new buds and leaves. Even the blackened hillside changed to a bright vital green (Plate 27). The mountain came to life with a beauty that was hard to comprehend, especially when the memory of the dry season and its fires was still so vivid. We were witnessing the slash-and-burn farming technique practiced by most of the hill tribes, a technique that involves clearing, burning, and then cultivating a piece of mountain land.

SHIFTING CULTIVATION

The mountains of northern Thailand are a patchwork of vegetation in various stages of succession (Plate 28). Here and there fields are under active cultivation. Most of the hill tribes practice **shifting cultivation,** also called **slash-and-burn agriculture.** The Thai refer to this method as *tham rai.* It is a form of cultivation that has been practiced for thousands of years in many parts of the world, though many people think it is carried out only in the tropics by primitive peoples. Historically, this form of subsistence farming has been practiced by peoples living in hilly terrain throughout the world. One term for shifting cultivation is **swidden farming,** which comes from an Old English dialect word, *swithe,* and refers to the burned

47

clearings that were created in the heathlands of the British Isles hundreds of years ago. Some also refer to this method of farming as **rotational agriculture.** This volume uses most of these terms interchangeably, though the term rotational agriculture refers to only one type of farming in which fire is used in preparing the field.

Shifting cultivation has been extensively studied and described. Some writers have been highly critical of this method of farming because of its ecological effects, such as the destruction of forests and subsequent erosion. Others, however, argue that this form of farming, properly practiced, is both appropriate (because of the type of terrain) and ecologically sound. One of the major problems in tropical and subtropical regions of the world is the rapid loss of nutrients from the upper levels of the soil; they are leached to such depths by the heavy rains that plants cannot absorb them. The use of synthetic fertilizers or heavy manuring can alleviate this problem in some areas, but fertilizing is essentially impossible for subsistence farmers in mountainous regions.

Generations ago these people learned that a piece of hilly terrain could be farmed for brief periods and then allowed to lie fallow, thus permitting the successional plant life to restore the original level of fertility and organic matter, improve soil structure, and prevent erosion. Though virgin forests were removed when the plots of land were originally cleared, much of the vegetation was allowed to return during the fallow periods.

Slash-and-burn techniques vary somewhat in various parts of the world, depending on terrain, soil types, crops grown, and outside pressures. However, the method has several common features and is still the most widespread method of cultivation in hilly, rain-fed tropical and subtropical regions. No matter where it is practiced, this method involves clearing the land of forest, brushland, or savanna; burning the cut debris; planting crops on the ash-rich soils for periods of one to several years; and abandoning the plot to allow it to regenerate during a fallow period of a few to many years (Plates 29–33). During the rest period the farmers hope that fertility will return to the soil through the natural vegetation. J. E. Spencer (1966:43), in an extensive and balanced study of shifting cultivation in Southeast Asia, comments that this method of agriculture has been developed because there is

> no other system of greater efficiency, effectively suited to the rather poor physical environments and specific ecologic situations in which shifting cultivation is still employed.

For generations tribal people have practiced slash-and-burn agriculture; it has become a significant or integral part of their tradition, and techniques have been passed down from their ancestors. Many even believe that if they were to abandon this form of agriculture, they would be punished by the spirits of their ancestors. Thus, an important part of the agriculture calendar is dedicated to working in harmony with the ancestors, as well as the spirits of the mountains and the crops. As Terry B. Grandstaff states (1980:3):

> Swiddening is their primary occupation, receives the majority of their labour, and is the basis for countless discussions, legends, and stories, and

problem-solving. Such groups do not simply *do* swiddening, they *are* swiddeners.

Most of the nutrients of tropical and subtropical forests are contained within the plant biomass rather than in the soil where they would be quickly leached away by the heavy rains. The shifting cultivator takes advantage of this nutrient source by cutting the plants covering the land and burning them during the dry season, thereby releasing the nutrients in the form of readily soluble ash. A crop is planted directly in this ash layer, thus receiving a quick fix of nutrients with the first moisture of the rainy season. The agricultural calendar is carefully tuned to take advantage of this brief period of high soil fertility.

Another characteristic of shifting cultivation is that only simple hand tools are used by the farmers, including long-handled hoes, axes, long-bladed knives or machetes, dibble sticks for planting seeds, a reaping knife or sickle, and a small L-shaped hoelike tool for weeding (Plates 34–35). Animals are rarely used, except for carrying harvested products from the field, nor are plows; thus, the soil is never turned over deeply.

Shifting cultivation is widely practiced in northern Thailand, with probably more than one million people dependent upon it for subsistence and another four to five million practicing it to some extent; the hill tribes make up only about 15 percent of this group. As many as 40,000–50,000 hectares (98,850–123,550 acres) are under this form of agriculture (Srisawas and Suwan 1985).

Tribal people in the north practice two forms of shifting cultivation: pioneer and cyclical.

Pioneer Swiddening

This method, also referred to as **primary forest cultivation,** accounted for about 40 percent of the land cultivated in the mountains of northern Thailand in the past. Practiced primarily by the Hmong, Mien, Lisu, Akha, and Lahu, this method traditionally involved clearing a piece of virgin forest and farming it for a period of time (perhaps only three or four years, but sometimes up to 20 years) until the soil fertility was completely exhausted. This form of slash-and-burn agriculture is still practiced to a limited extent at higher elevations of the Hill Evergreen Forest, a vegetational zone that begins at 1000 meters (3300 feet) elevation and has considerable nutrient richness, both in the plant biomass and in a humus layer on the ground. Thus, although much labor is required to fell the trees to clear a plot of land, there is assurance that nutrient levels will stay high for several years and serious weed problems will not occur for some time. Eventually, however, diseases and weeds become uncontrollable, and the field's fertility is exhausted so that it becomes savanna and does not regenerate into secondary forest. The land is then abandoned, which, historically has meant that the village would move, and a new piece of forest is cleared. Kunstadter and Chapman (1978) refer to this form of shifting cultivation as **"long cultivation–very long fallow"**; usually it is practiced at higher elevations in the mountains where opium, which requires fairly deep hoeing of the soil, is grown. Criticism has been directed at the hill tribes for practicing this form of shifting cultivation, not only because they have grown opium, but also because of the destruction of forests and valuable watershed. The other adverse ecological impact that can be seen in areas farmed under this method is the artificial savannas that have been created, consisting mainly of the grasses *Imperata cylindrica* and *Saccharum spontaneum*; forests almost never return to these areas.

Although five of the six hill tribes have practiced this form of slash-and-burn farming, it is becoming less and less significant, with fewer than 5 percent of the highlanders still practicing it, mainly because there are few virgin forests to be cut. Thai government restrictions prohibit the destruction of these forests; therefore, highland farmers are, in effect, being forced to adopt cyclical swiddening and other farming practices, even on heavily degraded fields, where fertility and production of crops is dangerously low. Unfortunately, the tribal groups that have practiced pioneer swiddening, such as the Hmong, use a piece of land too long before allowing it to lie fallow, with the result that all stumps and shrubs are removed or killed, thus preventing a rapid return to the successional sequence. Rather, aggressive grasses, such as cogon (*Imperata cylindrica*), move into the site and effectively prevent a normal succession from ever occurring.

Cyclical or Established Swiddening

The Karen and Mon-Khmer groups have long practiced this form of slash-and-burn agriculture, which is usually done at middle elevations in the mountains. It is also referred to as **land rotation, bush fallowing, secondary forest swiddening,** and **short cultivation–long fallow agriculture.** It is now by far the most common type of slash-and-burn farming, with nearly all the hill tribes practicing it in one form or another.

Traditionally, the Karen use a cleared piece of land for only one year, allowing it to then lie fallow for several years; during this rest period many shrubs and trees return to the site and fertility is rapidly and almost fully restored. One significant difference between this form of swiddening and pioneer slash-and-burn agriculture is that the fields are not cleared so extensively (Plate 36); recovery during the fallow period is therefore much more rapid, but it also makes weeds a more serious problem. With cyclical swiddening, extensive areas of nearby secondary forest remain relatively undisturbed, as well as a few pockets of primary forest, especially on hill tops and along water courses; these facilitate the regeneration of the cleared land. In fact, most established swidden farmers comment that they could not use a field more than a year or two simply because of the immense amount of labor that would be required to combat the weeds. This agricultural practice is ecologically sound when fallow periods are lengthy; moreover, the rotational system of fields involving a predetermined sequence of land use also permits villages to remain permanently at a particular location. A typical successional sequence in a well-managed, cyclical swidden field would be one or two years of intensive cultivation, with a crop such as rice, followed by a fallow period of six or more years. During the rest period a rapid regrowth of vegetation occurs, with herbs and shrubs dominant during the first two or three years, followed by a regrowth of trees from coppices and stumps. In effect, there is only a very short period of time that the land is completely devoid of vegetation and susceptible to heavy leaching and erosion. The fallow field does not remain unchanged during this succession, but each year new species of plants become evident, while others attain dominance; still others slowly decrease in abundance or disappear. The peak of plant diversity is usually during the middle years of the succession, with as many as 75 different species occurring in a given area (Kunstadter 1978a); many of these plants are utilized by the highlanders for food, fiber, and medicine. In other words, even the fallow fields surrounding a village play a significant role in the survival of the hill tribes.

Highlanders who could be called "established" swidden farmers now probably number more than half a million. The Northern Thai also widely practice slash-and-burn agriculture at lower elevations. Kunstadter and Chapman (1978) describe this method as **"short cultivation–short fallow,"** and it is rare that the land ever returns to a forested condition. Severe land degradation thus occurs over a period of years, creating areas of scrubland bearing scars of heavy erosion. Few tribal people are practicing this form of swidden agriculture.

PERMANENT FIELDS

More and more tribal villages are seeking permanent locations, not only because the government is restricting their movements, but also because there are few areas that are now unoccupied and into which they could move. Unfortunately, there is little incentive to devote labor and capital in developing a piece of permanent farmland when there is no assurance of ownership. Though land ownership is still a critical problem for many highlanders, some are gaining titles and therefore legally own the land they are working. This is especially true of the Karen and some other tribes that have moved into mountainous valleys where irrigated fields and orchards can be established. However, there is little sense in constructing terraced fields if there is insufficient water to irrigate them; without water they are nothing more than "flat swidden" (Cooper 1984).

The Karen have developed some impressive terraced wet rice fields, using simple engineering techniques and only hand tools, with the result that many hectares can be irrigated using the water from only a small mountain stream (Plates 37–38). In an area west of Chiang Mai the village men dug a canal along the side of the valley to allow them to irrigate an area of more than 100 hectares (250 acres). Their tools consisted of a piece of plastic hose filled with water (which served as a level), and digging hoes. In two or three places they had to resort to using a bit of explosive to cut through solid rock. This canal, more than three kilometers (1.9 miles) long, was fed by a small stream no more than two meters (6.5 feet) wide. A nearby village had purchased a ram pump to raise water from a stream that the villagers had dammed; this raised water would then be dispersed over a series of terraces constructed downstream from the dam.

In contrast to shifting cultivation on the hillsides, permanent irrigated fields require a greater investment in implements and animals. Wet land farmers must have a water buffalo to pull a plow and harrow. However, total labor investment in these fields is less than in swidden areas because much less time needs to be spent weeding the fields.

If permanent fields have a year-round water source, a farmer may adopt a multiple cropping scheme. Often the first crop in a field is rice, followed by a second (or third) of peanuts, soybeans, tobacco, or beans.

Most hill tribe farmers will develop irrigated fields if water and appropriate terrain are available; probably a fourth of all hill tribe farming is now of this type. Both Lahu and Akha villages on Doi Tung, for example, have created extensive terraced fields, for such pieces of land allow them to grow more than one crop a year and thus harvest greater yields (Plate 60).

Sometimes hillside areas without irrigation are developed into permanent plantations of trees, such as tea (*Camellia sinensis*), peaches (*Prunus persica*), and, more recently, coffee (*Coffea* spp.).

GARDENS

All tribal groups establish gardens and back yard plantings within and very near their villages, usually with longer-living plants than those grown in the swidden fields (Plate 39). Often gardens are carefully tended and protected by bamboo fences, but in other villages they may simply be areas in which several plants are placed and allowed to survive. An amazing number of different plants are grown in and around villages; these serve many uses, including food, fibers, dyes, medicines, and ornamentation. As an example of the diversity of useful plants within a village, I observed the following plants in the Lisu village of Huay Kong: balsam (*Impatiens balsamina*), banana (*Musa* × *paradisiaca*), basil (*Ocimum basilicum*), bottle gourd (*Lagenaria siceraria*), cabbage (*Brassica oleracea* var. *capitata*), canna (*Canna indica*), castor bean (*Ricinus communis*), celosia (*Celosia argentea*), chili (*Capsicum annuum*), coffee (*Coffea arabica*), day lily (*Hemerocallis lilioasphodelus*), eggplant (*Solanum melongena*), garlic (*Allium sativum*), ginger (*Zingiber officinale*), green bean (*Phaseolus vulgaris*), guava (*Psidium guavaja*), jackfruit (*Artocarpus heterophyllus*), kapok (*Ceiba pentandra*), kidney bean (*Phaseolus vulgaris*), lemon grass (*Cymbopogon citratus*), lettuce (*Lactuca sativa*), lime (*Citrus aurantifolia*), litchi (*Litchi chinensis*), mango (*Mangifera indica*), marigold (*Tagetes patula*), papaya (*Caryca papaya*), passion vine (*Passiflora edulis*), peach (*Prunus persica*), pineapple (*Ananas comosus*), poinsettia (*Euphorbia pulcherrima*), pumpkin (*Cucurbita maxima*), roselle (*Hibiscus sabdariffa*), sorghum (*Sorghum bicolor*), sugar cane (*Saccharum officinarum*), sweet orange (*Citrus sinensis*), sweet potato (*Ipomoea batatas*), tamarind (*Tamarindus indica*), taro (*Colocasia esculenta*), tomato (*Lycopersicon esculentum*), tree cotton (*Gossypium arboreum*), watermelon (*Citrullus lanatus*), and yard long bean (*Vigna unguiculata*). I saw most of the same plants in the Akha village of Ba Go Akha, but with the addition of caryota palm (*Caryota mitis*), coriander (*Coriandrum sativum*), *Eucalyptus*, green pea (*Pisum sativum*), mint (*Mentha arvensis*), mustard (*Brassica rapa*), onion (*Allium cepa*), peanut (*Arachis hypogaea*), and pomelo (*Citrus maxima*).

ORIGINS OF AGRICULTURE AND DOMESTICATION OF PLANTS

Spirit Cave in northern Thailand is a remarkable archeological site that was occupied by hunters and collectors between 9500 and 5500 B.C. A number of important preserved plant remains were found at this site, some of which probably were in the process of becoming domesticated; these include cucumber (*Cucumis sativus*), bottle gourd (*Lagenaria siceraria*), soybean (*Glycine max*), and castor bean (*Ricinus communis*) (Hutterer 1984), all of which are grown by the hill tribes today. On the other hand, there was no evidence from Spirit Cave that rice was cultivated. Exciting as these discoveries may be, the early hunter-gatherers were not related to the hill tribes that live in the same mountains today. Rather, as stated in Chapter 1, we must look farther to the north in southern China to find the roots of these people; within this area many other plants were brought into domestication and true agriculture was developed. This region, called the Southern Asia Belt (Li 1970), is a warm tropical to subtropical region with abundant rainfall and a rich flora. Three major food plants—rice (*Oryza sativa*), taro (*Colocasia esculenta*), and yams (*Dioscorea alata* and *D. esculenta*)—were cultivated in this belt in very early times. Another cereal, *Coix lachryma-jobi* (Job's tears), was also brought into cultivation in this zone. With these and the wealth of other plants surrounding them, the hill tribes adopted a wide variety of cultivated plants, and as they slowly migrated southward they carried many of their plants with them. Most are still found in their gardens and fields today, as well as numerous others that they have encountered in either Burma, Laos, or Thailand. The highlanders have not been hesitant to adopt new plants as they have been introduced from other tribal groups; the lowland Thai; or through government, church, and United Nations development programs.

THE AGRICULTURAL CALENDAR

The hill tribes have developed an agricultural calendar that corresponds intimately to the distinctive climate and seasons within Southeast Asia. There are some variations from tribe to tribe with regard to types of plants grown and seasonal ceremonies, but all tribes follow nearly identical activities in preparing and planting the hillside and valley fields.

January is a month of little agricultural activity, except for the harvesting of opium latex (see Chapter 7). Many village activities are carried out at this time, including house construction, weaving, and a variety of celebrations.

During February, families and village leaders select the fields to be cleared and planted in the coming season. Several factors are considered, such as slope, elevation, soil type, exposure to the wind and sunlight, and the presence or absence of indicator plants. Farmers are pleased to see the wild banana, *Musa acuminata*, for it indicates good soil for poppies and fair conditions for rice. On the other hand, large stands of field bamboo (*Oxytenanthera albo-ciliata*) indicate the land is poor for opium poppies but good for rice; farmers know from experience that the soil is too hot and has too much clay for poppies, but is fine for several varieties of hill

rice. If *Dendrocalamus hamiltonii* is present, farmers realize the soil has more loam content and is therefore good for both rice and poppies, especially if above 1000 meters (3280 feet) elevation. Farmers looking for good fields at moderate to high elevations are happy if they find *Boehmeria sidaefolia*, a plant that indicates good soil for rice and temperatures probably cool enough for poppies. Historically, this plant also provided fibers with which the Hmong made paper money for various spirit ceremonies. The shrub, *Debregeasia velutina*, is an excellent indicator of possible poppy sites, as are some species of *Litsea*. If the new site has grasses and sedges, such as *Thysanolaena latifolia*, *Themeda arundinacea*, or *Scleria terrestris*, then the soil is quite likely hard and dry, and therefore poor for either rice or opium poppies (Geddes 1976). Many farmers are particularly pleased to see *Mimosa invisa* growing in a new site as it does not need to be cut down but actually dies back and is completely dry about the time they are ready to burn; when the fire moves over the field, it completely consumes this plant, which is a good source of nitrogen and greatly enriches the soil. High elevation plants that indicate good land for poppies are *Polygonum chinense* and *Pterospermum grande*. During February poppy seeds are also harvested for the next year's planting.

In late February the cutting of trees and clearing of the fields begins, with this difficult phase extending into March. The men do the heavy work of felling the trees, but the women assist in cutting the brush and smaller plants, including bamboo. Much of the wood is carried to the edge of the field and carefully stacked to be used later for firewood.

Once the trees and brush are cut, the debris is left to dry (Plates 24, 29). Most larger trees are left standing or their branches simply lopped off as too much work is required to cut them down and they will not greatly interfere with the growing of crops. Other trees are cut at about a meter's height or slightly less; sprouts will soon grow from these, greatly accelerating the return of plant life as the land goes into fallow. The vegetation along water courses and on ridge tops is usually left uncut.

Fires are set as soon as farmers determine that the materials are sufficiently dry and will burn completely (Plate 30). If the fires are set too soon, not all the debris is burned; the family must then return to the area, stack what remains, and reburn the field. Several men participate in the burning activity as they wish to contain the fire within the selected field; often, however, they are unable to control the blaze and it moves into nearby forested areas or fallow swidden fields.

Field burning continues into April, the most unpleasant month in the highlands of the north. Dust is deep, a heavy smoke covers hills and valleys, and high temperatures become oppressive. However, everyone continues to work in the fields, clearing unburned debris and carrying additional firewood to the edges of the fields where it is stacked and later carried to the village (Plate 31). All now anxiously await planting time within a few weeks and then the arrival of the rains in May.

Each tribe has certain traditions and must follow carefully defined procedures during the planting period to maintain good relationships with the spirits and other forces that might affect their crops. The Lisu do not want to harm the "grandfather of the mountain" and the Akha are careful not to upset "the lords of land and water." The Akha also sing a song while planting their crops that is directed to both the ancestral god (*a poe mi yeh*) and to the birds and rodents. The ancestral god is asked "to look down" and protect the crop, and the birds and rodents are told that this is "people's food," so they should not eat it (Paul Lewis, personal communication). Usually both maize and rice seeds are planted immediately after the fields are burned and just before the arrival of the rains, though some prefer to begin planting rice after the rains have actually begun and the soil is moist. All cereals are planted with a dibble stick; this simple tool consists of a long bamboo pole upon which a curved iron blade about 5 centimeters (2 inches) wide has been attached (Plate 35). The planting team consists of a man and woman; the man slowly walks across the cleared, ash-covered field in a zigzag fashion beginning at the bottom, rhythmically moving the dibble or digging stick so that it strikes the ground about every 45 centimeters (18 inches), making a hole 2 to 3 centimeters (0.75 to 1.2 inches) deep (Plate 62). The long bamboo pole allows him to create a swinging pattern, thus digging a very consistently spaced series of holes of the same depth. He will vary the distance between holes, depending on the fertility of the soil. The woman, carrying a shoulder bag full of seeds, walks behind the man, dropping several seeds into each hole. No effort is made to cover the hole as the first rains and further walking in the field moves soil over the seeds. Taro, cassava, yams, and other tuber and root plants are often planted at the bottom of a field.

The rice or maize seeds sprout quickly once the rains begin, but so do many weeds. The months of June, July, and August are a period of hard work by all members of the family in combatting weeds, but the fact that the crops have been planted with a dibble stick rather than broadcast now means they can weed more easily between the clumps of cereal. Men, women, and children, wielding the small, L-shaped hand hoes (Plate 34), spend long hours each day slowly walking uphill between the growing rows of rice or maize digging out the undesired vegetation, but being careful not to disturb the soil more than 2 or 3 centimeters (0.75–1.2 inches) deep. As soon as one field is finished, the laborers move to another, for each field must be weeded two or three times during the season.

In September the maize crop and early varieties of rice are harvested. Children happily carry rat traps to the field and women erect scarecrows and other devices to frighten away birds. If the family grows opium poppies, the poppy seeds are broadcast between the rows of stalks as soon as the maize is harvested. Weeding in the maize fields is often done with regular hoes rather than the small hand hoes, thus tilling the soil to a greater depth and preparing it for the poppy seeds soon to follow. Husks covering the ears of maize are removed in the field; sometimes the kernels are also stripped from the cobs. The ears or kernels are then transported to the village in sacks or baskets and hung on the rafters or under the eaves of the house or spread out on split bamboo mats to dry (Plates 40–42). Later the grains are put in sacks for storage or sale.

October and November are dedicated to harvesting the later-maturing varieties of rice (see Chapter 4). Opium poppy fields are thinned and weeded for the first time during October and other crops harvested as they mature.

In December the long task of threshing the rice and carrying it from the fields is completed, often a formidable task as fields may be up to 5 or 6 kilometers (3 to 4 miles) from the village. Opium poppies are weeded for the second time, and other crops continue to be harvested as they ripen. The family's main activities now return to the village from the field; house repairs and construction begin, cereals are fermented for whiskey, fibers harvested and dyed, and weaving is begun.

CROP DIVERSITY

One of the most remarkable aspects of hill tribe farming is the amazing variety of plants that they cultivate (Plate 43). Whereas at first sight one might believe the swidden fields contain only rice or maize, I have determined that at least 88 different species of plants are grown in these fields. This number corresponds closely with Kunstadter's study (1978b) that tabulated 84 swidden species in a Lawa area in Mae Hong Son Province. Surprisingly, I have found more than 90 species of plants cultivated in gardens and within villages. In a 10-square-meter (12-square-yard) Lahu swidden field in Chiang Mai Province I recognized coriander, taro, field mustard, tobacco, castor bean, kidney bean, and maize. Some large fields may have up to 50 species growing over a period of several months.

Such crop diversity serves several purposes. First, these different plants provide many forms of foods for tribal diets, different fibers, and an important array of medicines for the treatment of many ailments. Second, the variety of plants reduces the effects of natural pests, which can disastrously harm a monoculture of the same genetic constitution. Third, diversity is also a form of harvest security, especially when considering basic food plants, such as rice, taro, potatoes, yams, and cassava within what, for many of the hill tribes, is still essentially a subsistence economy. If poor weather or a serious pest were to diminish or destroy one crop, such as rice, then the tuberous plants would provide sustenance and survival for the family.

Diversity is also seen in a single crop, such as rice. Rarely is only a single variety planted, because the farmers want varieties that mature at different times and have different tastes. Therefore, crop diversity is not only ecologically sound, but it also provides variety of diet, greater independence from lowland markets, and greater use of the now very limited swidden and irrigated land.

CASH CROPS

Hill tribe farming is changing slowly from subsistence agriculture in which virtually all the season's produce is used by the family to one that is part of the larger cash economy of northern Thailand. The highlanders have long produced some crops, such as opium, for cash or barter, but in recent years the influence of the lowland Thai economy and the ever-greater need for purchased products has forced many of the tribal people into growing mostly cash crops rather than simply rice for their own subsistence. This is not always bad, however, for in some cases it means villages may be able to rise out of severe poverty and the continual threat of starvation. Some Karen who live north of Mae Sariang, for example, have long cultivated land that is so poor they could never produce enough rice each year to feed everyone. They are now growing two crops of soybeans each year on this same land, and, with the help of pesticides, are producing enough to sell so they can buy all the rice they need (Rupert Nelson, personal communication).

At least 80 species introduced by various development programs and adopted from the lowland Thai are now grown by the tribal people as cash crops. Of course, no single village grows all of them (see Appendix 1), but, interestingly, only 16 species are native. The Akha village of Paya Phai Kao, for example, grows tea, coffee, peaches, soybeans, and maize to sell for cash. The Mien village of Pha Deua grows soybeans, maize, and ginger. Although the latter is a successful and valuable cash crop, it is very hard on the soil and the village recognizes that soon the soil will need to be fertilized and rotated or abandoned.

One of the most serious ecological problems to arise with the expansion of cash crop farming within the mountains of northern Thailand is the extensive and often uncontrolled use of chemicals, both as fertilizers and as pesticides. Often tribal farmers are poorly educated with regard to the use of chemicals (as are many lowland Thai), with the consequence that excessive amounts are being applied, thus allowing both the soil and water systems to become contaminated. An additional problem is that many of these new crops are being grown in a monoculture situation, thus enhancing the problems of pests and other diseases. Intercropping of some plants helps reduce this problem somewhat, but usually such a practice is only for short periods of time. For example, new coffee plantations are often intercropped with maize, wheat, ginger, taro, or some other short-lived plant, but as soon as the coffee trees reach sufficient height and breadth, intercropping is no longer possible (Plates 44–45).

Cash crop farming may also mean that pieces of land will not be allowed to go into fallow and that fertility levels will slowly drop, despite the addition of synthetic fertilizers, some of which wash into nearby streams. The land is also kept bare of competing vegetation, thus increasing the danger of erosion and requiring, at times, almost heroic measures to prevent serious gullying of hillsides. Aware of the dangers of erosion, tribal people resort to various procedures to reduce it, including establishing contour lines with logs or permanent plantings of leguminous perennials, disturbing the soil as little as possible, and avoiding the cutting of vegetation on the steepest slopes. More than half of the tribal farmers I queried claimed that erosion is a serious problem. The others said it was certainly a problem, but no worse than the problem of reduced fertility or insufficient land.

FIELD ROTATION

The ideal rotation plan for established or secondary swidden agriculture is to use the land for one, possibly two, years and then let it lie fallow for six to ten years. In very few villages did I find this being done because of the serious shortage of available farmland.

Most farmers in the Mien village of Pha Deua can now let a field lie fallow for only three years, whereas previously they could let it rest up to ten years. Weeds have become a much more serious problem, requiring considerably more labor to control; cogon grass (*Imperata cylindrica*) also invades and can be controlled only to a limited extent by burning. Thus, each family, in effect, has only three fields in the rotation scheme: two lying fallow and one being cultivated.

Swidden farmers in the Lahu village of Goshen can allow only about 20 percent of their fields to actually go into a fallow phase; the remainder are used every year, but with rotation of types of crops. Ginger, for example, can be grown for only one year; a different crop is then rotated onto that field or it is put into fallow for a year or two. The farmers would prefer to use the land for three years and then allow it to lie fallow for three, but they cannot afford to tie up the land for so long or people will go hungry. Farmers of a nearby Akha village typically use a field for three years, possibly four or five, if the soil is good and the crops rotated, with an intervening fallow period of five years.

The Hmong village of Cheng Meng near the Laos border has a one-year period of cultivation followed by five years of fallow.

In most cases a fallow period of five years or less is an insufficient time for adequate vegetation to regrow and soil fertility to recover from the period of cultivation, especially if the land is tilled for more than one year, which is almost always the case now (National Research Council 1982). Thus, slowly many of these mountainous fields will lose fertility, weeds will increase, and erosion will be more of a problem.

CROPPING SYSTEMS

As a rule, the hill tribes do not plant a single crop species in a field; rather, they tend to extensively mix a number of different plants, either growing them at the same time in a field or sequentially.

Mixed Cropping System

Mixed cropping is the most common method and rice is the primary crop, but other plants are grown in various areas within and around the field. For example, millet or sorghum may be planted along the margins of the field or along paths; chili, mustard greens, lemon grass, and eggplant are grown along the path so that they may be easily gathered when the workers return home from the fields. Squashes and melons are frequently cultivated near fallen logs or field huts where they will have support upon which to climb. This system of mixed cropping results in food being available in a given field for at least six months and if long-lived plants, such as cassava and taro, are also planted, then the field produces food all year. Though this system has been most commonly used by pioneer swiddeners, now almost all slash-and-burn fields, as well as some irrigated ones, have such a mixture of plants. Because so much labor must be expended to weed the now limited available land, a variety of plants must be grown concurrently, not only to produce enough food, but also to make full use of a limited labor supply.

Sequential Cropping System

A second method is the sequential growing of crops. The most common example of this system is a maize crop, followed immediately by opium poppies or rice in the same field. Another example of this system in an irrigated field is a rice crop followed by soybeans.

Single Crop System

The third method, the single crop system, has become much more common with the introduction of cash crops. Fields are devoted strictly to cabbages or tomatoes, and some villages may also grow two or more crops of rice a year in irrigated fields.

MAJOR CROP PLANTS

As nearly 90 different species are grown in swidden fields, and even more in village gardens and backyard plots, not all can be considered in detail in this volume. The following is a brief consideration of the top thirty crop plants grown by the hill tribes, excluding two of the most significant ones, opium and rice, which are dealt with in detail in separate chapters.

Allium cepa (onion). Highlanders grow onions for local consumption and as a cash crop. This species' place of origin is uncertain.

Allium sativum (garlic). One of several species of *Allium* grown in both gardens and fields for local consumption and as a cash crop, garlic is widely used as a condiment; the top of the plant may also be used as a vegetable. This common cultivar probably is derived from a native Asian species.

Ananas comosus (pineapple) (Plates 46, 62). This plant from the New World has been grown in village gardens for many years, but recently extensive plantations have been established at lower elevations and the fruits sold for cash.

Arachis hypogaea (peanut). A favorite leguminous addition to the diet and containing up to 31 percent protein, peanuts, native to South America, may be boiled or fried and served with rice, curry, and various greens. They are also an important cash crop, being grown mostly in swidden fields.

Brassica juncea (Chinese mustard; mustard green). This species, which probably originated in Africa but was introduced long ago into Asia, is extensively cultivated for local consumption, and large quantities are also sold for cash. Greens such as these may be grown both in village gardens or planted within swidden fields.

Brassica oleracea var. *capitata* (cabbage) (Plate 47). Introduced as an alternative crop for opium, the Hmong and Lisu have become particularly avid growers of this plant as a cash crop. It can be grown on hill sides but must be irrigated and treated extensively with pesticides and fertilizers. Most tribal groups eat cabbage as a green with their rice.

Camellia sinensis **var.** *assamica* **(tea)**. This native shrub or small tree is widely grown by the hill tribes and may be found in both fields and village gardens. Consumed by nearly everyone and grown as a cash crop, it is one of the most important cultivated long-lived plants.

Capsicum annuum **(chili)**. Almost no meal in the hills of northern Thailand is without chili, a native of tropical America (Plate 48). This highly pungent spice is prepared in a variety of ways and greatly enhances an otherwise bland, monotonous diet. Often found in household gardens, chili is also an important cash crop, being grown in swidden fields.

Carica papaya **(papaya)** (Plate 62). Though a native of the New World, papaya is now so widely grown in Asia that many think it is native. Virtually no tribal village is without these distinctive, somewhat succulent trees; they are found near houses, bordering streets, in gardens, and even along the edges of fields far from the village. This fruit is popular among the hill tribes; extra produce is sold for cash.

Coffea arabica, C. canephora, and *C. liberica* **(coffee)** (Plates 44–45). These plants, native to Africa, were introduced in opium replacement programs and have been adopted by several tribal groups as profitable cash crops. Virtually all the coffee beans produced by the hill tribes are purchased and used within Thailand, yielding a good quality coffee. However, farmers need considerable training in the cultivation of these long-lived trees whose propagation is so different from any of the traditional crops. Coffee plantations have helped stabilize many villages and provide good incomes for numerous families. Fields may also be intercropped during the first few years before the coffee plants come into production, thus enabling the highlanders to grow food or other cash crops, such as ginger and maize, before the coffee plants begin to produce. Coffee plants are now a common sight in most tribal villages.

Colocasia esculenta **(taro)** (Plate 49). This tropical Asian aroid produces large, readily digestible, starchy corms, and can be grown in either swidden or irrigated fields; it is also grown in village gardens. It is eaten by the highlanders and sold as a cash crop. The corms must be thoroughly cooked to destroy the acrid calcium oxalate crystals. The leaves are commonly fed to pigs.

Cucumis sativus and *C. melo* **(cucumber and melon)**. Like the lowland Thai, the tribal people often cook cucumbers and eat them with curry and rice; they also enjoy eating them raw. A variety of melons are grown for local use and as cash crops. These species are native to south Asia and Africa, respectively.

Cymbopogon citratus **(lemon grass)**. This highly aromatic grass from south Asia is widely cultivated throughout Thailand; highlanders use the leaves and stems for flavoring food and sell their excess to the lowland Thai.

Glycine max **(soybean)**. This important legume, native to Southeast Asia, is now extensively grown as a second crop after rice in irrigated fields. Some is consumed locally and development teams have shown tribal people how to make soy milk as a dietary supplement, but most soybeans are sold for cash. The very high protein content (up to 45 percent) of soybeans makes them a highly desirable supplement to the tribal diet. A few plants are occasionally found in village gardens.

Gossypium arboreum and *G. barbadense* **(cotton)** (Plate 161). Both of these species of cotton are widely grown, with the latter often planted in swidden fields. Tree cotton (*G. arboreum*) is planted in both gardens and along streets within villages; though probably a native of India, it appears to have escaped cultivation and is sometimes found growing in the forests near villages. Most of the tribal groups still harvest the fibers to make thread for weaving, and some cotton is also sold for cash.

Ipomoea batatas (**sweet potato**). Most of the hill tribes have adopted this good source of starch and sugar found within the large tuberous roots. The leaves are good as greens. Probably a native of tropical America, the sweet potato is now cultivated throughout the world, and the hill tribes have willingly adopted it as a supplement to their rice. Usually a family will plant a few hills of it in their garden, and some also grow it in their swidden fields as a cash crop.

Lagenaria siceraria (**bottle gourd**). Of great importance as containers, components of musical instruments, and sometimes as food, these Old World vines are grown in both gardens and along the edges of swidden fields. Some gourds are also sold for cash.

Lycopersicon esculentum (**tomato**). Another native of the New World, this plant has now become a popular cash crop for those tribal people with extensive irrigated fields. Some tomatoes are also eaten locally.

Mangifera indica (**mango**). This native of Southeast Asia is well-known and widely cultivated in villages and in gardens; a few households have established small mango plantations on swidden fields. Some fruits are eaten locally and others are sold for cash.

Manihot esculenta (**cassava**). Not preferred as a food by most tribal people, this introduced New World native is a short-lived perennial plant that is sometimes grown at lower elevations on poorer soils in fields and gardens. It is a source of starch and an important hedge against hunger should the limited fields not yield enough rice. Like taro, the underground roots of the bitter varieties must be treated because they contain cyanogenic glycosides, some of which are quite poisonous. Cassava is rarely grown as a cash crop by the hill tribes.

Musa acuminata and *M.* × *paradisiaca* (**banana**). These large monocotyledonous herbs are native to this part of the world, and apparently the many cultivars are of two different origins. The diploid and triploid varieties, such as dwarf or giant Cavendish, are probably derived from *M. acuminata*, which is found wild throughout northern Thailand (Plate 87). The hybrid varieties are referred to botanically as *M.* × *paradisiaca*. Bananas do well at lower elevations in the mountains, but even villagers in higher areas often attempt to grow them. The fruits are eaten locally and some are sold for cash.

Nicotiana tabacum (**tobacco**) (Plate 189). A source of both pleasure and income, tobacco is grown in village gardens and often as a second crop in irrigated fields. It was introduced from the New World in the sixteenth century.

Phaseolus vulgaris (**bean**) (Plate 50). Beans have been cultivated in small quantities by the hill tribes since their introduction into Asia from the New World. Recently, several new varieties, such as pinto and kidney beans, have met with considerable success as a cash crop after being introduced in opium replacement projects. More than one sack (120–130 kilograms; 265–287 pounds) of kidney beans can be produced on a rai of land (0.16 hectare or 0.4 acre), thus bringing in a cash income of at least U.S. $50 per rai. However, these profits are offset by a high price for good seeds (up to $0.75 per kilogram or $0.34 per pound), and it is necessary to use expensive pesticides. Other species of *Phaseolus,* such as *P. lunatus* (lima bean), are grown in small amounts. Most beans are grown in swidden fields.

Prunus persica (**peach**). A native of China, the peach is one of several fruit trees now grown in orchards on stabilized hill fields and in villages. Most of the produce is sold for cash.

Saccharum officinarum (**sugar cane**). Native to Asia, this large grass is found in nearly every tribal garden and is occasionally planted as a crop in fields. Most of the sweet stems are consumed locally, though some sugar extraction is carried out on a small commercial scale by lowland tribal people. Children often chew pieces of the stem as "sweets" or snacks.

Sesamum indicum (**sesame**) (Plate 51). The hill tribes grow large quantities of sesame in swidden areas and sell the harvested seeds as a source of oil. Some seeds are consumed locally, as, for example, in ceremonial rice cakes. The plant has an Asian origin.

Solanum tuberosum (**potato**) (Plate 52). This New World native is a fairly recent addition to slash-and-burn tribal agriculture, and many villagers still do not eat it, preferring to grow it only as a cash crop. Fungal diseases make potatoes a difficult crop to grow in the mountainous climate of northern Thailand. Potatoes bring a good price when sold in the lowland markets.

Sorghum bicolor (**sorghum**) (Plate 53). This native of Africa is cultivated in swidden fields for local use, as well as for cash. Humans rarely eat the grains of this cereal, but feed it to livestock and ferment it with other cereals to make whiskey.

Zea mays (**maize or corn**) (Plates 40–42, 45). This New World plant is cultivated by nearly all tribal people, but, surprisingly, it is not one of their primary foods. It matures before the rice crop, so it is occasionally eaten when the previous season's rice supply has run out and the family is without food. The Lahu, for example, pound the kernels, cook them in water for an hour, pour off the water, cook the kernels for another two hours, and then eat the mass which is something like sticky rice. Some Hmong enjoy roasting fresh ears and eating them as a snack. However, the primary use of maize is as livestock food, particularly for pigs. In Hmong households, perhaps as much as half the crop is sold for cash; the husked ears of the other half are hung in the rafters of the house, and then cooked for the animals when needed. Unfortunately, rats may eat as much as half the harvest that is stored in this way (Cooper 1984). Another important reason for growing maize in swidden fields is to prepare the land and then protect young opium poppy seedlings from the impact of heavy rainfall and from weeds. Well-prepared and highly fertile land may produce up to ten sacks (about 100 kilograms or 220 pounds per sack) of maize per rai, but normally prepared fields average only about two sacks per rai. Typically, they can earn about U.S. $8–10 per sack, selling it to the lowland Thai. It may be grown in either fields or gardens.

Zingiber officinale (**ginger**) (Plate 54). Widely eaten by the highlanders and now also commonly grown as a cash crop, this native of Southeast Asia may earn up to $0.75 per kilogram ($0.34 per pound). Unfortunately, ginger rapidly depletes the soil, so usually it is grown on a field for only one year. Ginger is also found in most village gardens as it is used for both food and medicine.

In addition to the crops described above, the following crop plants are important. A complete list of crop plants grown both for local consumption and to sell for cash is in Appendix 1.

Scientific Name	Common Name
Abelmoschus esculentus	Okra
Allium ampeloprasum	Leek
Anacardium occidentale	Cashew
Artocarpus heterophyllus	Jackfruit
Brassica oleracea var. *botrytis*	Broccoli; cauliflower
Brassica rapa	Turnip; Chinese Cabbage; *pak-choi*
Cajanus cajan	Pigeon pea
Cannabis sativa	Hemp; marijuana
Capparis spinosa	Caper
Capsicum frutescens	Bird pepper, bush red pepper
Ceiba pentandra	White silk cotton tree
Citrullus lanatus	Watermelon
Citrus aurantiifolia	Lime
Citrus maxima	Pomelo
Citrus reticulata	Mandarin; tangerine
Citrus sinensis	Sweet orange
Cocos nucifera	Coconut
Coix lachryma-jobi	Job's tears
Cucurbita maxima	Pumpkin
Curcurbita moschata	Pumpkin; gooseneck squash
Curcurbita pepo	Field pumpkin; gourd; marrow
Datura metel	Thorn apple
Daucus carota	Carrot
Dimocarpus longan	Longan
Dioscorea alata	Winged yam
Diospyros kaki	Japanese persimmon
Diospyros virginiana	Persimmon
Eleocharis dulcis	Water chestnut
Ficus carica	Fig
Foeniculum vulgare	Fennel
Fragaria sp.	Strawberry
Gladiolus × *hortulanus*	Gladiolus
Hemerocallis lilioasphodelus	Day lily
Hibiscus sabdariffa	Roselle
Hordeum vulgare	Barley
Lactuca sativa	Lettuce
Litchi chinensis	Litchi
Mentha arvensis	Mint
Mentha × *villosa*	Spearmint
Momordica charantia	Bitter gourd
Pachyrhizus erosus	Yam bean
Passiflora edulis	Passion fruit

Persea americana	Avocado
Petroselinum crispum	Parsley
Pisum sativum	Green pea
Prunus persica	Peach
Psidium guajava	Guava
Pyrus pyrifolia	Pear; Chinese pear; sand pear
Raphanus sativus	Radish
Ricinus communis	Castor bean
Sesbania sesban	Sesban
Setaria italica	Foxtail millet
Solanum melongena	Eggplant
Spinacia oleracea	Spinach
Tagetes erecta	Marigold
Tagetes patula	Marigold
Tamarindus indicus	Tamarind
Triticum aestivum	Wheat
Vicia faba	Broad bean
Vigna radiata	Mung bean; black gram
Vigna umbellata	Rice bean
Vigna unguiculata	Cow pea; yardlong bean
Xanthosoma violaceum	India kale

DIET AND NUTRITION OF THE HILL TRIBES

Food preparation and meal content vary considerably from tribe to tribe, but the following descriptions illustrate a typical tribal repast. Rice is the mainstay of all meals and huge quantities are served, especially from the perspective of a Western diet.

A Lahu main meal served to guests or on special occasions consists of some meat—usually pork or chicken, unless some wild game has been shot or trapped—that is boiled, often with cabbage or other vegetables. Chili is ground up with a mortar and pestle and added to the contents of the pot. After the mixture has cooked for a period of time, rice noodles and spices are added. Another bowl is prepared, containing lime, onions, tomatoes, pork rind, and chili; all are mixed together to be served with the meat and cabbage dish and rice. The final dish for the meal is an omelet made of eggs and onions fried in oil.

Sometimes the meal is more simple. Rupert Nelson (personal communication) described one Lahu meal in which cooked rice was piled on a low, woven bamboo table. A depression was made on top of the rice and curry put in it; people simply ate with their fingers, dipping the rice in the curry. After everyone had finished eating, the dogs were then allowed to clean up the table.

An Akha meal is typically served on a low, woven table, with the men and guests eating first, the women serving them. The meal consists of huge quantities of rice, usually placed in bamboo baskets on the floor, and several dishes of cooked food on the table: fried peanuts, hot sauce, chili and mustard greens, pickled greens, bamboo shoots, a curry containing some pork or dog, hot spiced soy cake, and a vegetable and chicken soup. The rice is eaten with fingers, the soup with a bamboo or metal spoon, and the curry and greens are eaten with split bamboo chop sticks. Tea or whiskey is usually served with the meal. Poor families cannot afford the luxury of meat with every meal, so malnutrition is a serious problem among many of them.

A Karen meal may consist only of rice, watery but hot curry with pumpkin, and tea.

The typical diet of any poorer tribal family would be a morning meal consisting of rice, salt, chili paste, and a green-leafed vegetable or pumpkin. The midday meal, if one is eaten, probably consists of cold rice and possibly some fruit that has been carried to the field. The evening meal is like the morning meal, possibly with some meat added. The animal protein might come from domesticated animals, frogs, snakes, fish, snails, rats, insect larvae, or some forest animal, such as a deer, bird, monkey, gibbon, wild boar, or jungle fowl.

Highlanders require 2160–2430 Kilocalories a day; most attain that caloric level but often do not consume sufficient protein and vitamins, with more than half of the tribal children suffering from at least some degree of protein deficiency (Robert 1987). Other nutritional problems include iron deficiency, which causes anemia, iodine deficiency resulting in goiter and cretinism; xerophthalmia, an eye disease causing blindness and other serious problems, due to lack of Vitamin A and the carotenes; and beriberi, caused by a Vitamin B_1 deficiency. There are also reports of a high incidence of angular stomatitis in children due to a Vitamin B_2 deficiency; this disease inhibits carbohydrate, protein, and fat metabolism. Pigeon chest, an ailment due to Vitamin D deficiency, is found in nearly 5 percent of the tribal children under age five (Robert 1987). One of the main causes of vitamin deficiencies is due to the tribal people changing from the traditional rice pounders to rice mills; the latter produces highly polished rice that has had most or all of the outer, vitamin-containing layer removed, whereas the rice pounder leaves it intact.

Malnutrition is increasing because many tribal people can no longer grow sufficient rice and other foods on the ever more-limited available land, wild game is almost gone, and less food can now be gathered from the reduced forests. Moreover, the change from a subsistence to cash economy has forced many tribal families to concentrate on growing cash crops, which often do not earn them enough money to buy the food necessary to make up for their short-fall in rice production. Government and church programs are actively working with the hill tribes in an attempt to improve their nutritional status.

The hill tribes have evolved an agricultural system that historically has had little adverse ecological impact within the mountains of Southeast Asia. Unfortunately, recent political, social, and economic events have so affected these people that many can no longer practice farming in their traditional ways. The results are serious environmental damage and insufficient food.

4

Rice: The Sustainer

It was mid-November and the rainy season had ended. In the distance I could see some dark figures working in a swidden field of mature rice. As I neared the figures, I recognized the typical Akha clothing—black short skirt, jacket, leggings, and ornate headpiece (Plate 55). Each woman rhythmically bent, grasped a bundle of nearly mature rice stems, and swung a small sickle or reaping knife with a curved blade, deftly cutting the plants about knee high from the ground. Laying the cut bundle on the stubble, the women took a step backward or sideways, and repeated the process. Occasionally a maize stalk towered above the meter-high rice plants.

The cut stems would be left for several days to dry, during which time the harvesters hoped it would not rain. Returning to their field, the Akha would gather the bundles of dried stems, and then hit them on a log with a woven bamboo mat underneath, to thresh the rice. The pile of grain would then be winnowed with large woven bamboo trays and placed in burlap sacks to be carried by ponies to the nearby village. This process of harvesting, threshing, and transporting rice is repeated annually during October and November in nearly every highland village.

Rice is the sustenance of the tribal people, and a good or poor harvest may very well be the difference between having enough food or going hungry. It is an integral part of their culture, the most important plant. For them, rice is life; it is— and long has been—their means of survival. An Akha saying emphasizes the importance of rice: "There are 100 kinds of plants, but there are 101 kinds of rice" (Paul Lewis, personal communication). The Karen word for food is the same word for rice (as it also is in Thai). Rice is security for the hill tribes (Phothiart 1989). Paul Lewis (1982) describes rice as one of the three basic themes of Akha culture; it is "The Quest for the Staff of Life."

The Akha legend of how the tribe first received rice involves a poor widow and her young daughter, who would dig tubers daily to have enough to eat. One day the daughter mysteriously disappeared, only to return later as the wife of the

Lord Dragon, who lived in the river. She invited her mother to join them in his river kingdom, which the mother did. Soon, however, the mother tired of life in the river and asked to return home. The Lord Dragon granted her request and gave her some magical rice seed wrapped in a leaf and a hollow reed, telling her that if she planted it she would always have enough to eat and drink. This she did, and that season the plants produced such a huge crop that she could not carry it all home. She went back to the Lord Dragon and asked him what to do. He told her to stand in the field and whistle three times and clap her hands three times. She did this and the amount of rice magically diminished in quantity so that she could carry home just enough for all her needs in just one day. The Akha use this legend to explain why they always carry their ceremonial rice seed in a leaf and hollow reed and why they never whistle or clap their hands in a rice field (Lewis and Lewis 1984).

This, and other stories, explain why the Akha, perhaps more than any other tribe, believe that rice has a soul; therefore, they have numerous rituals relating to the planting, growing, and harvesting of rice. It is also an integral part of many of their other ceremonies, which will be described in more detail in a later section.

THE BOTANY OF RICE

One of the most important cereal plants of the Poaceae (Gramineae), rice, *Oryza sativa*, whose wild ancestor was *O. perennis* Moench, was first domesticated in southern and southeastern Asia more than 4000 years ago. The genus *Oryza* is cosmopolitan in its distribution, and, of the 12–15 species currently accepted by botanists, only two are cultivated: *O. sativa*, which accounts for most domesticated rice cultivars, and *O. glaberrima* Steudner of West Africa, which originated from *O. barthii* A. Chev., a native of that continent (Zeven and Zhukovsky 1975). Best described as an annual swamp plant, *O. sativa* has been cultivated for so long that now there are literally thousands of cultivars or varieties.

Many rice varieties produce numerous hollow tillers or stems, usually up to 150 centimeters (60 inches) high, from the same fibrous root system. However, some deep water rice cultivars grown in central Thailand and southern Asia may produce stems up to 5 meters (16 feet) long! Each stem bears 10–20 leaves, and, in contrast to wheat and barley, produces a somewhat loose, branched panicle or inflorescence that does not stand erect when it matures (Plate 56).

The life cycle of a rice plant takes 100–210 days, depending on the climate (solar radiation, temperature, and moisture) and cultivar or variety. There are three phases of growth: the vegetative phase, lasting 40–90 days and ending when the inflorescence first appears; the reproductive phase of 30–55 days, during which time the flowers develop and accomplish pollination (rice commonly self-pollinates); and the ripening phase in which the grain or caryopsis matures, usually a period of 30–65 days after flowering. At this time the grains are tightly enclosed by the distinctive grass flower parts called the *palea* and *lemma*; collectively, these are commonly known as the *husk* (hull).

Grains with the husk are called *rough* or *paddy* (*padi*) rice. Pounding the dry fruit removes this husk, producing what is often called *brown rice*; milling removes both the husk and the bran (this includes the embryo, nucellus, and other important outer parts of the grain containing protein and vitamins, especially Vitamin B_1) (Cobley 1976) and is called *polished* or *milled* rice. Pounded rice contains approximately 8.5 grams (0.28 ounces) of protein per 100 grams (3.5 ounces), whereas milled rice has only 7.4 grams (0.26 ounces) (Maneeprasert 1989).

The main differences in rices are due to the proportions of the two types of starch present in the grain; in amylose the glucose molecules are arranged in a linear fashion, and in amylopectin they are branched. The more amylose present in rice, the drier and fluffier the grains are when cooked and the less they tend to disintegrate upon cooking. If the amylopectin portion is high, the grains become sticky and tend to disintegrate upon cooking. Sticky or *glutinous* rice, as it is also called, consists essentially of starch that is pure amylopectin.

Rice is divided into three main groups; two have been formally proposed as subspecies, the other is an ecotype.

1. *Oryza sativa* subsp. *indica* Kato. Known commonly as the *indica* type, these rice cultivars are characterized as tall and with droopy, light green leaves. They are cold-sensitive, so are found only in the tropics, particularly the monsoon areas of Southeast Asia. Because of their height, they tend to lodge fairly easily. They are also characterized as being short-day photosensitive and with relatively low yields of grain. However, they are more tolerant to drought than other varieties, and they also have greater resistance to disease and insect attacks. The grains are medium-long to long and with a fairly high amylose content: this means that when the grain is cooked it tends to be dry and fluffy. It is the type of rice grown by the hill tribes and in southern China.

2. *Oryza sativa* subsp. *japonica* Kato. This subspecies is most commonly grown in temperate climates and probably originated in China. The plant is more green and erect than the *indica* type, with greater resistance to lodging but greater susceptibility to disease and insect pests. However, it also responds better to fertilizers. The grain is shorter and the amylose content less; thus, when cooked this grain becomes stickier and disintegrates if cooked too long.

3. *Oryza sativa* ecotype *bulu*. Referred to as the *javanica* or *bulu* type of rice, this ecotype (Zeven and Zhukovsky 1975; Barker et al. 1985) is photoperiod insensitive and has a very long vegetative phase. It is adapted to the more equatorial tropical climates, so therefore has a restricted distribution, being found primarily in Indonesia, the Philippines, and Madagascar.

Although virtually all rice grown in Southeast Asia is of the *indica* type, there are numerous variations in flavor, maturing times, cooking qualities, and storage life that lead one group of people to prefer a particular type of rice over another. There are distinct regional preferences within Thailand; the central Thai prefer nonglutinous rice with a high amylose content, whereas the northern Thai (and Laotians) prefer the glutinous or sticky varieties with a high amylopectin portion.

The hill tribes grow both types but prefer to eat the nonglutinous varieties.

Other rice characters vary widely and are the basis of the hundreds of local cultivars found throughout Southeast Asia. Some of these features are as follows:

1. Length of growing season. Cultivars are described as early, middle, or late season forms. This length varies with the location in which the cultivar is grown and may result in quite different dates of maturation.

2. Color. Most rice is white, but grains may also be yellow, red, and purple.

3. Ability to resist shattering when harvested.

4. Disease or pest resistance.

THE HISTORICAL SIGNIFICANCE OF GLUTINOUS RICE

Glutinous or sticky rice is considered peasant food in central Thailand (Bangkok), whereas it is the preferred rice of many groups in northern Thailand, Laos, and southern China. Both the glutinous and nonglutinous forms of rice are of the *indica* type, but one of the most important features of the glutinous type is that it tends to require a shorter growing season than does the nonglutinous type. For historical and other reasons, glutinous rice tends to be grown in the more mountainous areas of Southeast Asia, which have somewhat shorter growing seasons, poorer soils, and more erratic rainfall. Though glutinous rice tends to disintegrate when boiled, those that regularly eat it, including the hill tribes, prefer to steam it; this reduces the stickiness and also eliminates some of the rice's sweetness (glutinous rice is referred to in Japan as sweet rice). Moreover, it tends to keep better after cooking. One major advantage of the higher sugar content in sticky rice is that it ferments better in the making of whiskey.

Apparently the early Tai-speaking groups adopted glutinous rice when living in China (Golomb 1976) and carried it with them as they migrated southward during the next several centuries. For many years the Tai (now represented as the northern or Lanna Thai, the Shan of Burma, and the Laotians) have lived among and interacted with various tribal groups, just as they do in northern Thailand today.

Glutinous rice is one of the ecologically sound plants that the tribal people adopted. It has proven to be a good crop, not only in the valleys, but also in the mountains of northern Thailand. It has great ceremonial significance to them.

TYPES OF RICE CULTIVATION

There are three basic types of conditions under which *Oryza sativa* can be grown: wetland, deepwater, and upland cultivation. Each method has many cultivars that are best adapted to that particular system.

Wetland Cultivation

The most common (and familiar) method is wetland cultivation, which is frequently called *paddy* or *padi* rice cultivation. However, the term *paddy* actually refers to unhusked grains of any kind of rice. Wetland rice may be rainfed or irrigated, but in both cases the fields are flat and surrounded by levees or dikes (bunds) enabling the water to stand in the plot (Plate 57). The land is cultivated after the rains have begun and water is standing in the field, first with a plow, followed by a harrow to break up the clods of wet dirt. Water buffalo traditionally have been the beasts of burden that pull the plow and harrow; however, many lowland Thai now use two-wheeled diesel-powered tractors or "iron buffaloes."

A nursery is planted from seeds that have been broadcast in a carefully prepared small field (Plate 58), and the seedlings are allowed to grow for 3–6 weeks to a height of about 30 centimeters (12 inches) while the main fields are prepared. These young plants are then transplanted by hand into the now completely flooded field, and the water level is retained at a depth of 15–30 centimeters (6–12 inches) through irrigation or, hopefully, sufficient rainfall. Fertilizer and/or manure are frequently added to the field, though blue-green bacteria (Cyanobacteria) in the standing water produce nitrogen. Some experimentation has also been carried out with *Azolla pinnata*, the small water fern that contains symbiotic blue-green nitrogen-fixing bacteria, by introducing it into the standing water of the wet fields.

The presence of water in the field for the entire life of the plant keeps down the weeds, and, interestingly, the water in the fields becomes fairly rich with small fish, providing men and women with an enjoyable pastime (and more protein in the diet) while they wait for the rice to mature.

The rice is harvested after the rains have ended and the fields are dry. The wetland method of rice cultivation is employed by more and more tribal groups. Swidden fields are thus reduced in number, the people move to lower elevations, and villages become more stable.

The Karen have long carried out wet rice cultivation, often on elaborately made terraces irrigated by the water from small mountain streams (Plate 59). I have also seen similar terraces near Akha and Lahu villages (Plate 60). The creation of such fields requires a considerable expenditure of both labor and capital, the latter sometimes being provided by the government, United Nations (UN), and mission projects. Families possessing such fields must also invest in animals and larger tools, which they do not need with swidden cultivation. However, the rice yield is about twice as great per rai and there is the possibility of growing a second, cash crop in the field.

Deepwater Cultivation

The second method of rice cultivation, found in central but not northern Thailand, is deepwater or floating cultivation in which specially adapted rice plants grow in fields with water up to 5 meters (16 feet) deep. The land is plowed and harrowed just as the rains begin, and the seeds are sown or broadcast directly onto the now-moist and harrowed soil. The seeds sprout and the plants produce remarkably long internodes that enable them to keep the main portion of the plant above the water that slowly rises during the rainy season. The plants flower above the water level, and then mature as the rains cease and the level of water drops within the field. The mature grains are harvested when the field has become completely dry. The hill tribes do not practice this form of rice cultivation.

Upland Cultivation

The third method, the most common form of rice cultivation among the hill tribes, is upland, hill, or dry rice cultivation. Water never stands on these upland fields because no bunds or levees are made to hold the water on the hill sides, and different cultivars of *Oryza sativa* are grown. The fields are prepared by the slash-and-burn technique described in the previous chapter. Just prior to or immediately following the arrival of the rains, seeds are hand-planted in holes in rows; no fertilizer is added, nor are the seeds covered in a deliberate action. This method is similar to that used in the cultivation of wheat, barley, and most other cereals. The propagation of hill rice will be described in detail in the next section.

THE CULTIVATION PROCESS

The growing of rice is carefully coordinated with the expected weather patterns of northern Thailand, although some minor variations occur from tribe to tribe because of certain ceremonial procedures that must be followed. The Karen, for example, have a rice goddess known as Phi Bi Yaw. This goddess is a rather benevolent female spirit that watches over the rice fields if she is propitiated and not insulted by loud noises in the fields (Rupert Nelson, personal communication). The fields are cut and burned during the dry season and field huts are constructed or repaired shortly thereafter. These small structures, built entirely of bamboo and with a thatched roof, are about 3–4 meters (10–13 feet) across (Plate 61). They provide a place for families to rest during the day, protection from the heavy rains, and shelter during the night.

By the end of April and first of May the planting process begins. Most families begin this step together, so often everyone in the village, except for the very young and old, walks to the fields each day or lives in the temporary field huts to avoid the long, time-consuming walk each way. Usually the male head of the household walks laterally across the field digging holes with his dibble stick, approximately 45 centimeters (18 inches) apart. This distance, also called a *cubit*, is measured from the elbow to the tip of the middle finger as an occasional check to see that the holes are correctly spaced (Sutthi 1989b). A woman from the same family, either the wife or an older daughter, follows the man, dropping 3–10 rice grains into each hole, making no attempt to cover the seeds (Plate 62). Children also go to the field at this time to protect the seeds from birds and rodents with their sling shots and small crossbows. For many it is a pleasant change from village life.

By the middle of May the rains begin and soon the rice sprouts as do large numbers of weeds. All the females and sometimes the men within the family must now dedicate themselves to the weeding process, which lasts from the end of May until October, working every day, unless there is a special ceremony in the village or it is raining very hard. The workers are careful to merely scrape the weeds out of the soil with their L-shaped hand hoes, thus disturbing the soil as little as possible to avoid erosion during the three times that most fields must be weeded. If a family has some wet rice fields, these several months are critical for both fields; the weeding must be done on the hill sides but the wet fields must also be prepared and the rice seedlings transplanted. Often farmers will hire other tribal people to assist them with this labor.

As the rice plants begin to mature, the children again flock to the fields to assist their parents in protecting the ripening grains from animal predators, and the family may settle in its temporary field hut to watch over its valuable crop. Usually these periods in the field are enjoyable, but occasionally robbers or even murderers may disturb the peace. Many devices, such as scarecrows, are used to frighten away birds, and streamers and bamboo clappers are attached by strings to the field hut; these are then shaken throughout the day by young girls. The boys set traps for both birds and rodents, and the men use their long-barreled, smooth-bore rifles to shoot larger animals.

Rice harvest begins as early as the end of September, especially if the rains come early and fast-maturing cultivars were planted. A family's fields may have from one to five different varieties, and usually small areas of glutinous rice are the first to mature and to be harvested. These early maturing varieties are important to the poorer tribal people who may have already eaten the last season's harvest and now have little to eat except tubers and forest products. However, it is difficult to harvest rice if the rains persist, and these early varieties also tend not to store well.

Two quite different methods of rice harvest are used by the hill tribes. The most common system is similar to that of the lowland Thai, in which a group of people slowly works its way through the field, from the top to the bottom in a zigzag fashion, cutting the still somewhat-green stems of rice with a sickle or reaping knife. Most gather several handfuls together into a simple sheath, which they lay on the stubble that is about 15 centimeters (6 inches) high; in this way the heads of grain do not touch the soil and air can circulate around them to dry them and prevent mold from forming if it should rain. Several days later the sheaths are tied with bamboo strips or rice straw and carried to the threshing site.

The Mien harvest rice in a manner quite different from other hill tribes or even the lowland Thai, although some groups in Yunnan and other parts of Southeast Asia practice it. Just the heads of the rice plants containing the mature grains are cut and put on a storage rack. A group of four harvesters slowly works through a field. Each uses a small, special knife (*gyip*) made of bamboo and steel (Plate 63). This knife is held in the hand in such a manner that, by grasping a few stems of rice and closing the first and second fingers around the rice tightly, the upper few centimeters of stem with the heads of rice are brought to the triangular-shaped blade and cut off, but still grasped by the fingers (Plate 64). Four handfuls are tied together to make a small bundle, then four small bundles are tied together, and this larger bundle is placed on the drying rack, usually constructed of bamboo

(Plate 65). The bundles of rice are placed so that only the stems are exposed, thus protecting the grains from any late rains. In the past the Mien would then take from the rack only what was needed each day and thresh it, an activity that would be followed for several months. This is no longer practical because the racks are in the fields often a considerable distance from the village and subject to pilferage and predation. Nowadays the same harvest method is used, but when the crop has been completely harvested all the grains are removed from the rack, threshed, and carried to the village for storage. The Mien claim they have used their method "since time immemorial."

The threshing process varies from tribe to tribe, especially the method by which loose grains are collected. Some Karen have adopted the northern Thai system, which utilizes a large woven bamboo basket, often up to 2 meters (9 feet) in diameter. The sheaths of mature rice are simply hit against the inside of the basket, thus knocking loose the individual grains, which collect at the bottom (Plate 66). A more common method among the hill tribes is to place a woven bamboo mat directly on a flat area of ground. A small sawhorselike structure made of bamboo, or a small log, a piece of bamboo, or a board is then placed on the mat; the sheaths of rice are hit against the sawhorse or log to break loose the grains, which pile up on the mat (Plate 67). Other tribes simply beat the stems on a woven mat using a heavy stick. The Mien also thresh on a woven bamboo mat, but they do so by removing several bundles of grains from the rack, laying them on the mat, and hitting them with a special mallet or a heavy stick.

Loose grains are winnowed prior to being sacked and carried from the fields. The usual winnowing device is a round, woven bamboo tray about 75 centimeters (30 inches) in diameter and with a rim 5 centimeters (2 inches) high (Plates 68, 71, 72). The paddy is expertly thrown from the tray into the air so that the grains fall back onto the bamboo structure and the breeze carries away the chaff and debris. Men or women will use large woven bamboo fans about half a meter (20 inches) in diameter to create a breeze on a still day. Some groups use a sievelike winnowing tray before they use the above method. This tray has holes large enough for the grains to fall through but small enough to catch the larger debris. A man may also simply kick the rice into the air with his foot while briskly sweeping the fan; the grains fall back onto the mat and the chaff blows away (Grunfeld 1982).

Once the paddy is winnowed, it is then put in burlap sacks or woven bamboo baskets (Plate 68) and carried to the village where it is often spread on woven bamboo mats on porches to completely dry prior to storage in the granary. The rice stubble is left standing in the field where it decays and thus provides some nourishment to the soil. The Mien spread the rice straw among the tall stubble in the field, then burn it in April or May, and allow the field to return to fallow.

VARIETIES OF HILL RICE AND THEIR SOURCES

Hundreds of cultivars of rice are grown by the highlanders of northern Thailand. In many conversations with villagers I learned about some of these rice varieties and tribal preferences.

Lao Si Kuan, a 70-year-old Mien living in Chiang Rai Province, said that his village relies on rice cultivars that originally came from China. All of it is dry or hill rice; very little is glutinous. Of the 12 different varieties of sticky rice known to the tribe, only a few of the "more popular" or better tasting ones are grown in any quantity. Lao described five nonglutinous varieties, which differ in time of maturation, taste, and ease of threshing. The *biew ziao* variety is fast-growing and produces a crop in a mere 90 days. A second variety, *biew zai*, takes 120–150 days to mature but tastes much better than does the fast-growing variety. The most tasty variety is *biew kia*, but the most popular variety is *biew tien*, which is slow-growing but also with excellent taste. Another popular variety, *biew biang*, takes 120 days to mature and has good taste.

Li Htu, an Akha headman in Chiang Rai Province, also claimed his village grows rice that came from China, all of it hill rice. Among the seven main varieties, some do better in the higher mountains, others at lower elevations. The varieties mature at different rates and include some glutinous forms. The Akha use sticky rice primarily in ceremonies.

A Law, headman of the Akha village of A Hai, commented that his rice stock came from his parents, who brought it from Burma. He usually plants three types of hill rice, one "soft" (early) and the other two "hard" (late). He has a few wet fields in which he plants four varieties of sticky rice. However, he does not mix the different varieties in the same field. His village has only enough land to allow people to grow crops in a field for three years and to then let it rest for five, but this short fallow presently enables them to grow all the rice they need.

The Karen commonly grow both wet and dry rice, but tend to have fewer varieties than some of the other tribes. However, they do have at least two varieties of traditional hill rice, which mature at different times, thus spreading out the harvest time. The Karen have adopted some wet rice varieties preferred by the Thai because they can easily sell part of it for cash. Two such varieties are *hom mali* and *muei nong*; however, the Karen themselves do not particularly like to eat these sticky varieties (Phothiart 1989).

Ca Nu, the 70-year-old headman of a Lahu Sheh Leh village in Chiang Mai Province, believes that all the cultivars of rice his village knows came from Yunnan Province in China. He grows only two varieties which he got from his parents, one glutinous (*she naw nyi*) and the other nonsticky (*ca hpuma*) (Plate 69). The nonglutinous form is preferred, with the glutinous variety being reserved for New Year's celebration and offerings to the spirits. Though this village has willingly used squash, pea, mustard, chili, and tomato seeds given to it by the government, it has not received any rice nor is it particularly anxious to try different types.

Aaron Pornsakulpaisarn, headman of the Lahu village of Goshen, said that his village uses varieties of rice from Burma. The people tried several varieties recommended by the Thai government, but did not like them. The variety they plant in wet fields during the rainy season tastes very good, but has serious caterpillar problems. They are therefore searching for a more resistant variety or will be forced to use a pesticide. Their second, or dry season, rice crop is also grown in the wet fields but has few pests. The headman claims that seed stock becomes impure after about three seasons, forcing the village to switch to another of the several varieties it has in storage. About 80 percent of the rice grown by this village is hill rice.

RICE YIELDS AND CONSUMPTION

Productivity varies widely, depending both on cultivation method and on cultivar and soil fertility, but generally irrigated fields produce about twice as much rice as do swidden or hill fields (Kunstadter 1978b). The tin is now the standard rice measure, based upon the square metal kerosene containers that are so common; it holds 20 liters (21.14 quarts) of paddy and weighs about 15 kilograms (33 pounds). A tin is more or less equivalent to the former basket (*tang*) of measure, which was only slightly smaller. Hill rice yields an average of 20–40 tins (*biip*) per rai (0.2 hectare; 0.4 acre), but as many as 50 tins may be produced per rai of irrigated wet rice fields (Maneeprasert 1989). This means that a rai of swidden field would produce 150–450 kilograms (330–992 pounds) of rice. Unfortunately, yields are steadily dropping as the land is overused. Typical yields of rice on swidden fields throughout Southeast Asia vary from about 10 to slightly above 30 tins per rai (Miles 1969).

A general rule of thumb is that one tin contains sufficient rice to feed an adult for 8–10 days; thus, an adult eats about one liter (one quart) (Judd 1969) or 700 grams (1.5 pounds) (Kunstadter 1978b) of uncooked, husked rice each day. A Lahu Sheh Leh headman said his household harvests about 300 tins each season, enough to feed eleven people for a year. He claimed that his annual yield was 70 tins of paddy per rai of glutinous rice and more than 100 tins per rai of nonglutinous. Unfortunately, his small village of 12 houses grew only enough rice to support five families for the year. Therefore, the villagers had to buy rice (Baht 40 or U.S. $1.60 per tin). A man's wages for one day buy one tin of rice.

The tribal people speak of yield in relation to the amount of seed planted rather than yield per area of land. A ten-fold yield means a farmer harvested ten times as much rice as he planted. One Akha claimed his typical yield was 70. A Mien elder expressed concern that his yields were lower each year. He remembers that 40 years ago the nearly virgin land yielded 80, but now the best areas produce yields of only 35–40, with less fertile fields or those at higher elevations yielding only 20. The headman of the Akha village of A Hai reported a typical yield of 50, but sometimes good fields produce a yield of 100.

Labor investment in growing rice is high, with nearly 165 days of a farmer's year devoted almost solely to his rice crop (Durrenberger 1979). Add to that the equal contribution of his wife and older children, who are involved in virtually every aspect of the operation, and one sees the great time demands placed on a hill tribe family.

Tribal people eat rice at every meal. This requires a daily trip to the village rice mill or, more typically, the women rise early to spend about a half an hour at the family's rice pounder. This important item is kept inside houses constructed on the ground, but beneath the kitchen area or outside if the house is built on posts. A typical rice pounder consists of a partially hollowed-out log set on end in the ground. A long, heavy pole is attached to an axle or pivot point slightly nearer one end, and a piece of wood is firmly anchored to the longer end at a right angle. The longer end of the pole is raised when one or two people step or stand on the short end; they then quickly step off, releasing the pole so that the end of the attached piece of wood falls heavily into the hollowed-out log containing the unhusked grains (Plate 70). Every few minutes some of the hulled grains are removed and put on a winnowing tray. When sufficient rice for the meal (or day) has been pounded (the husks removed) (Plate 71), grains are winnowed to remove small particles and the husks (Plate 72). The husks or bran are fed to the livestock. Some ingeniously designed rice pounders are constructed near streams so that a diverted flow of water fills a portion of a hollowed out pole, causing it to raise the long end, while the other end pivots upward over the container full of paddy; when the pole lowers to a certain point the water quickly flows out, allowing it to rise rapidly at that end while the other with the pounder attached at right angles smashes down to unhusk the grains.

Tribal people typically steam rather than boil their rice. Most groups soak the hulled rice overnight in a large container made of bamboo or metal, and then transfer the soaked grains from the container directly to the steamer without cooking it. However, the Akha sometimes boil it in the container for about ten minutes, squeeze the excess water from it, and then place it in a rice steamer to cook until it is soft. Often the steamer is made of woven bamboo, but it may be made from pieces of wood 30–40 centimeters (12–16 inches) in diameter. The Lahu like to use the wood of *i sho* (*Albizia chinensis*) for their steamers, but most tribal groups use any hard wood that is the right diameter and can be hollowed out successfully. A partition woven of bamboo or rattan is then tightly wedged into the hollow center near the lower end; this holds the rice above the water. I observed a Lisu woman fill her wooden steamer with soaked rice by simply scooping several handfuls from one container to the other (Plate 73). She then carried the steamer to the fireplace where she had already placed a large wok partially filled with water over the flames. She put the steamer in the wok, covered it, and cooked the contents for 20 to 30 minutes (Plate 74).

The amount of time the rice is steamed varies, depending on the variety and how a particular tribe likes their rice. The Akha, for example, may steam rice twice, the first time for an hour and the second time for as long as forty minutes (Maneeprasert 1989). Typically the Karen boil their rice in considerable water, then pour most of the water off, covering the container and again placing it over a low fire to cook.

Glutinous rice is often powdered and made into cakes, which are used in various ceremonies. The hulled rice grains may be pulverized in rice pounders, but more often smaller hollowed out pieces of wood are used. A woman then pounds the grains with the rounded end of a heavy piece of wood, which she lifts up with her arms and then lets fall into the hollow where the rice sits (Plate 75). The powder is then moistened and cooked as small cakes. Sometimes previously cooked sticky rice is pounded with water and then formed into the cakes.

CEREMONIAL USE OF RICE

Rice is a critical and intimate part of most tribal ceremonies, whether it be a ritual meal at a wedding, an offering to a spirit in the forest, or an item buried with a body in a funeral ceremony. The critical ceremonial roles played by this plant are illustrated by examples from three tribes.

Mien Ceremonies

The Mien, like other tribal people, use glutinous rice almost strictly for festive occasions and religious ceremonies. They make rice cakes by pounding rice and poppy or sesame seeds together; the oil from the seeds holds the cake together. The cakes are roasted over coals and used the next day or two in different ceremonies, particularly as offerings for spirit appeasement or in ancestor worship. These ceremonies occur at Chinese New Year (the date of which varies each year), during the third month when ancestors are worshiped, and in the seventh month or their "Little New Year." The latter ceremony may include both offerings to the spirits of ancestors who have departed and offerings by the young people to pay respect to the village elders.

The Mien believe that the soul of an unborn child lives in various locations around the village prior to entering the fetus at the end of a year; twice during this 12-month (not just 9!) period the soul lives in the family's rice pounder (Lewis and Lewis 1984). A newly born Mien baby goes through a blessing ceremony in which offerings of whiskey, water, some silver, a piece of white cloth, a chicken, and about half a kilogram of rice wrapped in white cloth are made (Chaturabhand 1988). A new mother is said to be "raw" following delivery, so she is given a special diet which includes fermented rice (Lewis and Lewis 1984).

Mien medicine men use rice grains as their "spirit soldiers"; they take the grains in their hands, which they slap together, thus spraying rice all over the room. Each of these "soldiers" then searches for the sick person's soul, which must be returned to the body to make the person well (Lewis and Lewis 1984).

Lahu Ceremonies

The Lahu use rice extensively in their ceremonies, too. The New Year festival is one of the most important times of the year for them, and one of the most significant symbolic activities is the communal making of rice cakes. Every family in the village contributes glutinous rice, which is steamed, pounded with sesame seed, and then made into small cakes about 3 centimeters (1 inch) thick and up to 25 centimeters (10 inches) in diameter. The donation of rice by each family demonstrates the interdependence and unity of the people.

Lahu Nyi holy days are observed in a similar communal manner; each family contributes a cup of uncooked rice to a basket at the home of the village priest. Some of this rice is thrown into the air as an offering to God, asking him to produce an abundant yield of rice. The priest's wife, who also puts rice into the basket, cooks the contents and gives some of it as an offering at the temple. Two of the most important holy days of the Lahu Nyi are celebrated at the times of planting and harvesting rice (Lewis and Lewis 1984).

Akha Ceremonies

Perhaps the Akha, more than any other tribe, consider rice a critical part of their culture. Within their well-ordered universe, they believe that all humans,

Akha ancestors, domesticated animals, and rice belong to the village realm, while the spirits and wild animals belong to the forest realm (Kammerer 1989).

For their ceremonies, the Akha must use "holy rice." To be ceremonially pure or "holy," the rice must be taken from the storage area by a clean or nontaboo person, steamed in water that has come from a purified source, and be the first handful that is taken out of the rice steamer (Lewis 1968–1970).

Rice is used as a blessing in Akha weddings, much in the same way it is used in Western weddings. Cooked rice is thrown over the bridal couple, asking the spirits to bless them and give them long life, good health, many children, many animals, and good rice harvests (Lewis and Lewis 1984).

An Akha woman is considered to be impure or unholy for 13 days following the birth of a child. People usually do not visit her house during this time, fearing that some of her unholiness might return to their home with them and adversely affect their rice supply. A mother takes her new child to a field on its 13th day along with some cooked rice wrapped in a banana leaf. She rubs a bit of the rice on the child's lips as its first rice meal (Lewis 1982).

After a water buffalo is killed in an Akha funeral ceremony, unhusked rice is poured over its head and neck of the animal to completely cover the upper part of its body and horns (Plate 76). Three shallow little lines are then made with a stick in the wet ground directly out from where the snout lies beneath the pile of rice. Some bracken fern (*Pteridium aquilinum*) is placed on each line, which represents a path that the soul can travel, and on top of it nine cowrie shells. The people then chant for the spirit not to take the upper or lower paths, as they will lead to a "bad" death for the deceased.

Rice plays a critical role when there is "terrible birth" in an Akha village. The father must kill the deformed child or twins by suffocating them with rice husks mixed with ashes.

The Akha believe so strongly in the ceremonies connected with the planting and growing of rice that if a family does not observe them, they may be expelled from the village (Lewis 1982). Great care is taken to choose the rice fields, and the family constantly watches for bad omens. Once the fields are chosen, the family then makes one of its nine annual ancestor offerings just before the rice is planted. The first grains to be planted are washed in a "purified water source" (Lewis 1982). The pure rice is carried to the field, where a small shrine or "field spirit hut" (Lewis 1968–1970) is built to serve as the rice soul's residence while the rice grows. The shrine is constructed almost at ground level, with two forked sticks upon which there is a small thatch roof; the opening faces down slope. The shrine may also be built on a pole at shoulder height (Plate 77). Tea leaves and fermented rice in a bamboo container are placed in the shrine. The first rice seeds are then planted in three groups of three holes (a total of nine, which is an important number to the Akha) just above this structure, using either a dibble stick or a small hand hoe. Throughout the season farmers are careful not to offend the soul of the rice, for if such a thing were to occur, then a ceremony would have to be performed to appease the rice soul who could ruin that season's crop or even a future crop.

In one instance a farmer found an odd-looking rice plant in his field and feared that the soul of the rice had been offended for some reason. A ceremony to appease the soul was held in that field, where a group of men gathered to slaughter a small female pig. Some time within the next three years a male pig would also have to be sacrificed, but in the village rather than in the field. A pan with the pig's blood, two bamboo containers with rice and rice whiskey, and water in a small dish with a gourd ladle in it were placed on the ground near the abnormal plant. Part of the pig's liver was stuck in several stems of the plant which had been tied into a clump (Plate 78). Some of the men built a fire using dry bamboo kindling; as soon as it was burning well the carcass of the pig was thrown into the flames to burn off the hair. The skin was then thoroughly scraped with a knife before the body was cut up on some banana leaves. For the next hour or two, with intermittent rain falling, the men prepared and cooked a meal in two kettles they had brought from the village. The head, feet, and other large pieces of the pig were placed with rice in the larger kettle, which was full of water, and brought to a boil. Other pieces of pork were chopped up and put, along with water, salt, spices, chopped leeks, and some of the partly coagulated blood, into the other, smaller kettle. Each man contributed some of the chili, salt, spices, or greens from a small bamboo basket or his shoulder bag. A bit later the large pieces of meat were removed from the kettles, chopped up, and returned to cook some more. One of the men set a table of woven bamboo which had been carried from the village, with metal spoons, chop sticks, Chinese-style tea cups, and plates.

Gifts were presented to the plant spirit before anyone could eat. The men carried several cups to the strange plant and offered tea, whiskey, water, and pieces of pork from different parts of the pig. The meat was put on the ground at the base of the plant and each liquid carefully poured from its container at the same place (Plate 79). Beginning with generous servings of whiskey, all the men now ate the sacrificial dinner, which consisted of chili mixed with greens, boiled rice, a basket of cold sticky rice (brownish in color because it had been damaged by a fire in the village earlier in the year), a dish of chopped up pork and spices, and a plate full of large pieces of pork. The pig brains were served and eaten as a great delicacy; in fact, all parts of the pig, including legs, jaw, and ears were at least chewed upon. Everyone was in good spirits and thoroughly enjoyed the excellent meal, finishing it with cups of hot tea.

Of the food that was left, a small part of each dish was then placed on the dirt at the base of the plant. The owner of the field took the bamboo container with the fermented rice, added a bit of water with the small gourd ladle, and blew through a small hollow reed into the bamboo container to mix the contents which were then poured onto the ground near the plant. The farmer gently bent the stems of the plant toward the ground, so that the abnormal heads were facing downward, and tied them with a bamboo strip. Next he placed the loosely woven bamboo container, in which the piglet had been carried to the field, over the rice plant, firmly staking it to the ground (Plate 80). It was mid-afternoon and the ceremony was finished, so the dishes were washed and all the items packed. The group made its way back to the village, a tedious climb of one and a half hours, with some men stopping to bathe in a stream.

When a rice crop is mature and ready to harvest, the priest in an Akha village, followed by all the families, makes a "new rice offering" (Lewis 1968–1970). To prepare for this offering, a day of abstinence is observed by everyone within the village. The priest then goes to the field and plucks three ripe heads of rice in the area just above the field spirit hut where the first seeds were planted in the nine holes. He carefully wraps them in cloth, for they will be placed in his ancestral altar when he returns to the village. The priest then cuts three more stems with mature heads of rice and places them in the field spirit hut. Finally, he walks throughout the field pulling off ripe grains and putting them in his shoulder bag. He returns home and gives this bag to his wife, who removes the loose rice, dries it over the fire, pounds, and cooks it. An ancestral offering is made of holy rice, new rice (both made from the rice just brought from the field), chicken meat, whiskey, and tea.

The men of the village then slaughter a water buffalo and a pig, which they cook and make into a curry. All the village gathers at the priest's house to eat the sacrificed animals, curry, and what is left of the food he prepared and offered to his ancestors. Each person ceremonially eats the special food of this "new rice offering," and often considerable amounts of whiskey are consumed. The families may then begin their rice harvest in the next day or two.

Additional ceremonies are performed before rice may be put in the storage house to keep the soul of the rice from running away. Finally, the family constructs a small spirit house by the granary and makes a special offering to the granary spirit when all the harvest is in the storage building (Plate 81).

Another example of the importance of rice in the spiritual lives of the Akha concerns their ancestral altar. Consisting of a section of bamboo stem or a shelf made of woven bamboo, the altar, which is hung on the women's side of the house's central partition (Plate 82), contains the first three heads of rice that are harvested. Four items hang below this altar: a small woven bamboo table upon which rice offerings are made to the ancestors, a stool, a bronze gong, and a shoulder bag containing holy rice and rice cake in small woven containers. On the floor is the ancestral altar basket; made of woven rattan, it contains some pieces of the grass *Thysanolaena latifolia*, from which their brooms are made, a piece of cloth, and three leaves of the wild banana, *Musa acuminata*. This altar is very sacred to the Akha and offerings are made to it nine times a year, using the special shoulder bag, table, and stool. Rice is always part of these offerings. The women cook "holy" glutinous rice, some of which is made into cakes. A chicken is killed and cooked, and rice whiskey, a special drink for the ancestors, is made by adding water to fermented rice. The offering is made, consisting of the cooked sticky rice in a small woven bamboo container, chicken meat, a rice cake, some rice whiskey in a small cup, and tea in another cup. Small amounts of the rice whiskey are sucked up three times (three is also a special number to the Akha) with a small piece of hollow bamboo and transferred to a section of bamboo which is placed at the altar. The head of the house feeds the bronze gong by rubbing it with a bit of each of the food items from the altar; he then strikes the gong three times. The family now eats the ancestral offering with great reverence and respect.

Rice, giver of life and important from the moment of birth until death, is as much a part of the life of the hill tribes as is bamboo or the forest. It is their sustenance and an essential part of their relationship with the spirits and ancestors.

5

The Forest: The Provider

Two Karen medicine men and I were on our way to the forest to search for healing plants when three women suddenly appeared around a bend in the trail. Two of them were married, for they wore brightly decorated black tunics, with Job's tears (*Coix lacryma-jobi*) sewn into intricate patterns, and red skirts with many horizontal stripes. The third was a girl of about ten years of age, and she wore a simple, long white shift decorated only with a band of red above the waist and some clusters of red fringe. All were heavily burdened and leaned forward to ease the weight of their baskets, which were borne mainly by a woven fiber strap around the forehead. Two of the baskets were filled with firewood, while the other was filled with a variety of green plants. Our guides knew the women, so we stopped briefly to talk, and this gave me an opportunity to look more closely at the basket of greens. I immediately recognized some banana stems, which would be used for pig food, but some of the other plants were not so easily identified. The woman explained through an interpreter that some of the plants were animal food and others were to be cooked and eaten with the evening meal. Holding up two herbs, she said that one was a medicinal plant and the other she planned to smoke. I thanked her for explaining what she was carrying and remarked to my companions how extensively the hill tribes utilize the forests.

The forest is a remarkable resource to the hill tribes and many essential items for their survival still come from it, as they have for generations. Of course, fields are necessary for growing rice and other crops, which provide both sustenance and cash—at least when things go well. Yet at times a family is not able to grow enough rice to feed everyone for the year because rats and birds eat part of the harvest or bad weather reduces the yield. If they do not have enough cash to buy rice from others more fortunate than they or from the lowland Thai, their survival may depend on foods from the forest. Knowledge of the forest and the plants it contains, a form of life insurance, cannot be forgotten, even in such technological and changing times as these.

The highlanders of northern Thailand are more alarmed by the disappearance of the forests than the lowlanders, for wooded areas are a source of many necessities for the hill tribes, providing foods that may make the difference between starvation and survival. These same forests also enable them to eke out a somewhat more-comfortable living in what is, at times, often a truly inhospitable environment, for they give the highlanders what could well be considered a wealth of items, such as lumber, dyes, fibers, medicines, bamboos, glues, and even poisons. The forest is also a place to bury the dead, a place for solace, a place of excitement during a hunt, and a place where the "lords of land and water" dwell. The forest is an essential part of their lives—as it always has been. It is a place with which they have developed an intimate and marvelous relationship, as well as a knowledge, that few people have today.

Hundreds of different plants grow in the forests of northern Thailand, and most are used in one way or another by the tribal people. Many of them are medicinal plants (see Chapter 8 and Appendix 2). The bamboos, an important component of most Thai forests, are of indispensable value to the hill tribes (Chapter 6). Some tribal people still gather natural fiber and dye plants from the forest (Chapter 10). Many of the trees are an essential source of timber for construction (Chapter 9). In addition to these numerous plants, there are many wild sources of food and other necessary plants that yield important household products.

The following material illustrates the wonderful forest lore of the hill tribes, particularly that of the Lahu Na. The mountain on which some of these people live, and where extensive but diminishing forests occur, lies at nearly 1000 meters (3300 feet) elevation just east of the Burma border in Chiang Rai Province. Much of this once beautiful area has now been heavily impacted by logging, road construction, slash-and-burn farming, and the incursion of lowland Thai from below.

Phrynium capitatum, a member of Marantaceae, a monocot family, produces broad leaves which the Lahu use as plates during ceremonies and celebrations when they do not have enough regular dishes. The same leaves are also pounded and used as a poultice on swellings caused by hypodermic injections.

Another monocot, *Molineria capitulata* of the Hypoxidaceae, also produces broad leaves which the Lahu use to wrap things in the same way they use banana leaves. The Akha, however, never use the leaves of this plant for an everyday function; instead, they only use the leaves of this plant to wrap the bodies of the babies that have just been killed because a "terrible birth" occurred.

Another fairly common shrub growing near the village is *Clerodendrum serratum*. The Lahu often boil a few leaves and drink the liquid as an antiacid or a soda mint for upset stomach.

The Lahu like the several species of *Acacia* that occur in their forest. *Acacia megaladena*, a shrub about 2 meters (6 feet) high, is used to make a poison, which is put in the water to kill fish. Three other species, *A. farnesiana*, *A. pennata*, and *A. concinna*, are used as food. In fact, the Lahu often make a delicious meal by mixing the young buds of any of these species with fish that have been killed by the poison of *A. megaladena*.

The favorite yam of the Lahu, *Dioscorea alata*, called *meu*, has a reddish-purple stem. The tuber of this species grows more than 2 meters (6 feet) into the ground, and the deeper it grows, the better it tastes! This yam is a particularly valuable food source for families that run out of rice before the new crop is harvested. Another yam, *D. bulbifera*, which the Lahu call *la hk'a*, also has an enlarged root. It is not as well known as *D. alata*. However, *D. bulbifera* is now used a great deal in China when the tribal people there do not have enough rice. The root of this species is dug up, cut into very thin slices, and put in water to soak overnight. The next morning the slices are winnowed on a winnowing tray, then mixed with rice, cooked, and eaten. The Lahu claim this yam actually tastes quite good, and one older man said they taught British soldiers how to eat it during World War II as one of many survival techniques in Asian jungles. The fruits of *D. bulbifera* are large and hairy and can be prepared in much the same way as are the roots, but they do not taste nearly as good. This plant cannot be dug up and eaten like a wild potato; it must be soaked overnight to take out some of the bitterness.

The hill tribes cook the very distinctive aroid *Amorphophallus campanulatus* and eat its large corms, often with curry. They use the leaves as pig food.

The Lahu call a small tree, *Baccaurea ramiflora*, the "fire fruit" (*ma fai*) because the little fruits hang down and look like exploding fireworks. The fruit is very sweet, one the Lahu like to eat very much. The flowers of *Vaccinium sprengelii* are a great delicacy among the Lahu, who eat them raw with great relish. They also like to eat the fruits, but not as well as the sweet-tasting flowers.

The Lahu claim a small vine, *Piper boehmeriaefolium*, is delicious when chopped up and cooked with chicken. One guide said the Akha particularly like to eat the leaves of this plant with fish.

Another common vine in these forests is *Smilax*. The Karen like to eat the young shoots of *S. ovalifolia*, which are also of considerable medical importance (see Appendix 2), but the Lahu use the dry leaves of *S. corbularia* to make stabilizers for their crossbow arrows.

The "bird lime tree" (*va pa a naw*) is *Mussaenda parva*, and it produces lovely inflorescences. The Lahu peel off the bark from the roots to expose a very sticky substance, which they rub on the branch of a tree. Birds landing on the branch get stuck and thus are captured. The Akha extract the glue in a somewhat different manner: by pounding the roots in water so that the gum goes into suspension. They then swirl a stick around in the water to make the wood sticky; this material can then be used as a glue for all types of repairs.

Macaranga gigantea, a fairly common tree with large, heart-shaped leaf blades up to 45 centimeters (18 inches) long, produces a sticky, mucilaginous material when pieces of the branches or trunk are cut and boiled. This sticky gum is used as a glue and as a red dye or ink; it may even be used for paint. Another tree, *Cinnamomum iners*, also produces a sticky sap or mucilage that is often used to repair baskets. The Lahu think of it simply as patching material and use it for all sorts of tasks around the village.

Litsea monopetala is a tree which the Lahu prefer for coffins and house posts.

Though the Akha do not seem to have any particular tree for making coffins, two other forest plants are essential for their funeral ceremonies and are included in their epic poem of creation (Paul Lewis, personal communication):

He [the Creator] then planted seeds of trees and vines.
  When the tree seeds had been planted,

the *si sa* tree [*Schima wallichii*] came up.
Do not say the *si sa* tree has no value.
For in the future when elders die there will be
dozens of buffalo tied to a *si sa* post, and sacrificed for those elders.
When the vine seeds had been planted,
the *la byoe* vine [*Bauhinia bracteata*] came up.
Now do not say that the *la byoe* vine has no value.
For in the future when elders die there will be
dozens of buffalo tied by means of the *la byoe* vine to the ceremonial posts (Plate 101).

The Lahu sing about *Buddleja asiatica*, the "old field tree," to symbolize all of life. When this tree blooms, the flowers first open only at the very bottom of the tree, gradually opening up the tree, with the flowers at the bottom beginning to wilt and then drop. This process is very unlike other trees in the forest whose parts usually flower at once. The Lahu say this tree is like people and life: at first they are pretty, strong, and attractive—but as they grow older they lose all their freshness (Paul Lewis, personal communication). The Lahu also preach about *Trichosanthes rubriflos*, saying that people should not be like the fruit: pretty on the outside and ugly on the inside.

The wood of a fig, *Ficus auriculata*, is popular with the Lahu for furniture making. However, this same tree also produces numerous pendulous sprouts, which are good tasting when young. Wild animals like to eat its fruits, so often hunters hide nearby to shoot the barking deer, sambur, and other animals that come to feed on them. One Lahu guide laughed and said that eating too much of this tree makes a person act as if he were drunk. Another fig, *F. fistulosa*, also attracts wild animals, so Lahu hunters construct blinds in which to hide and wait for them to come to eat the fruits, which are up to 5 centimeters (2 inches) in diameter. The Lahu slice the fruits and eat them with rice, but only if they are very hungry. Another fig, *F. hispida*, is used both for food and medicine (for the treatment of colds and to improve a mother's lactation). The Lahu boil the roots and use them as a spice with chicken, whereas the Karen simply eat the fruits. Likewise, *F. semicordata*, a common species in the north, produces leaves that are eaten by the Lahu, though the Karen prefer to eat the fruits. This species is even cultivated in villages.

The Lahu refer to the tall grass, *Thysanolaena latifolia*, which grows in open areas, as *mi shi ceh* (broom grass). From December to March most of the tribal people collect and bundle the young fruiting stems, which they sell for up to five Baht (U.S. $0.20) a kilogram (Plates 83–84). However, some families can earn more than $200 of extra income by gathering and preparing bundles of this grass. The developing seeds must be rubbed off prior to making the brooms, which are used throughout Thailand. Another plant, *Sida rhombifolia*, is also used to make brooms, but it mostly occurs in old village sites and fallow fields rather than at the forest edge or along the roadside. Sometimes the tribal people will even grow this plant near their houses so they can easily gather a bundle of the stems to use as a broom.

The Lisu gather branches of *Artemisia atrovirens* (*lu khwa*) for the special ceremonial purpose of sweeping off the shelf upon which the offering to the ancestors is placed (Plate 15).

A leguminous vine, *Archidendron clypearia,* is used to treat toothache. The leaves are boiled and then directly applied to the tooth to reduce the pain.

The "chicken lice plant" (*g'a she tzuh*) is a shrub, *Clausena excavata,* which has an unpleasant odor. The leaves are boiled and the liquid then used to bathe infected chickens. People may also bathe with the liquid if they have an infection or sore.

Sterculia pexa is a forest shrub that may provide some income for the tribal people. The bark is pulled off in strips, tied in 100-kilogram (220-pound) bundles, and sold to make cordage and bindings. It is still used a great deal in Burma, but plastic rope has now largely replaced it in Thailand. No part of this plant is eaten, but the Akha eat the leaves of a related species, *S. foetida,* even though they smell bad. Another member of the Sterculiaceae, *Helicteres plebeja,* has wood that the Akha like to use to make the shoulder yokes or boards for their baskets.

Saurauia roxburghii grows in the forest only near water. It is a large shrub that produces a good-tasting fruit. The bark can also be pounded and used as a poultice to make boils come to a head and to get poison out. Some claim it can also be used to treat broken bones.

The very rough leaves of a plant the Lahu call *ma naw hpa (Broussonetia papyrifera)* is used for sandpaper, and they say the bark is also a source of fiber. The Lahu describe this plant as having two types of fruit: the "mother" fruit is soft and very nice and can be eaten; the "father" fruit, however, always has two types of bugs, so it cannot be eaten.

The woody leguminous vine, *Derris elliptica,* which the Lahu call "tiger tail vine" (*la meh te*), contains a very powerful fish poison. The leaves are pounded and thrown into the water, making the fish "die like crazy." Occasionally humans are poisoned by this plant when it is harvested accidentally in the rice fields and pieces of it do not get winnowed out of the padi. The shrub, *Buddleja asiatica,* which they call *heh sha taw k'u,* also produces a powerful fish poison when the leaves are pounded and thrown into the water. However, the flowers of *Buddleja* are boiled in water to make a yellow liquid that is sometimes put on rice to color it. The Akha use the leaves of *Sesbania javanica* as a fish poison, but the Lahu are not familiar with it. The fruits of *Croton oblongifolius* are the source of a very powerful poison used by the Hmong, but it is unfamiliar to both the Akha and Lahu.

Solanum torvum is a wild member of Solanaceae, the nightshade family. It produces a fairly large green fruit that the Lahu like to use in curries, especially with beef, fish, and ginger. The bark of *Phyllanthus emblica,* a member of the Euphorbiaceae, is sometimes mixed in the same curry dish. The round fruits of *P. emblica* have many seeds that are taken out, beaten, and eaten. Sometimes the people gather and sell these fruits in the market.

Another species of *Solanum, S. erianthum,* gets to be a large shrub. Its wood has an important function for many tribal men (Plate 85) for it is made into charcoal, powdered, and added to sulfur and saltpeter to make gun powder for their home-made hunting rifles. The necessary sulfur and saltpeter are purchased or sometimes found locally.

Aralia chinensis occurs fairly widely throughout northern Thailand and is another source of food. The Lahu boil the fruits, mixing them with chili before eating, whereas the Akha prefer to eat the somewhat bitter-tasting leaves.

Oroxylum indicum of the Bignoniaceae produces an edible fruit when young. The winged seeds come out and float all over the place when the fruit matures and dehisces. The men pointed out two other plants that provide food. The first, *Rhus chinensis,* tends to grow in higher elevations and produces a purple- or lavender-colored fruit, which is fuzzy and very sour. However, the leaves are mixed with bamboo shoots and taste delicious. A common introduced plant, *Blumea membranacea,* is called "white man's weed" (*ka la mvuh*) by the Lahu because it came into northern Thailand about the time Caucasians arrived. The tribal people eat the plant and also feed it to their pigs. They call another plant, *Eupatorium odoratum,* "Communist weed" because they did not notice it until the Communists came into China. It is now quite common throughout the mountains and is widely used medicinally by the tribal people as a blood coagulant and to reduce swellings.

One very familiar plant is a wild raspberry, *Rubus blepharoneurus,* which the Lahu describe as "the raspberry that looks like the pubic hair of a woman" (*a law cha mvuh bui*). The edible fruit is yellowish red to orange in color, and the leaves are sometimes used as a substitute for the pepper leaf (*Piper betle*) when chewing betel nut. The tribal people also enjoy eating the fruits of two other *Rubus* species: *R. dielsianus* and *R. ellipticus.*

A fern, *Dicranopteris linearis,* is used for the treatment of broken bones and to help children sleep when they have a bad skin rash. The common bracken fern, *Pteridium aquilinum,* is not used as a medicine, but has two other uses; the young fronds are quite good to eat, whereas the old, dry fronds are excellent for making chicken nests.

Sometimes my informants were not sure of all the functions of a given plant. *Melastoma imbricatum,* for example, is said to be used for medicine, but none of my informants could tell me for what. When eaten, the fruit stains the lips red and they stay that way for quite a while. One of the Lahu names of this plant, *na kaw leh,* is the same as that of a bird in the owl group, which calls out at only special times. Another Lahu name for *M. imbricatum* is "broken back fruit" because when it becomes ripe it breaks open in such a way to suggest a broken back.

The Lahu fry and eat the young leaves of *Engelhardia serrata,* a beautiful tree in the Juglandaceae, the walnut family. They also mix the bark with meat as a spice.

The Lahu talk about a woody vine called *na ngo* that is said to be deadly poisonous. I collected material of this lovely plant in full flower, which was later identified as *Gelsemium elegans* of the Loganiaceae (Plate 86). I also learned that it produces fruits that turn yellow at maturity and reach a diameter of about 2 centimeters (0.8 inch). The Lahu say this plant is so poisonous that bees pollinating the flowers are killed. The roots are chopped up very finely into a pulp, which is used as the source of poison. Years ago, the pulp was used by tribal people to kill tigers when they bothered villagers. An animal was killed, the skin cut, and the pulp put in. The animal carcass was then placed out in the forest where the tiger would come to eat it. Lahu tradition says that sometimes frustrated lovers would make a pact to kill themselves; each then ate some of the pulp of this plant or drank the liquid in which the roots had been boiled. The poison is also used on arrows, which usually are notched so that they will break off in the prey, thus leaving the poisoned portion embedded in the flesh. The plant is too poisonous to be used in streams as a fish poison.

There is another tree called *kong* or *nong* from which a poisonous latex is collected. *Antiaris toxicaria* of the family Moraceae is not nearly as poisonous as *Gelsemium elegans,* so sometimes some *na ngo* is added to *kong* to make a particularly potent potion. This very tall, emergent tree commonly occurs to the west in Burma; however, I found a specimen in Mae Hong Son Province just east of the border. The latex of the bark is the source of the powerful poison, but, interestingly, that same bark is peeled off and made into mats, such as those used on elephants. I was also told that crossbow arrows were treated with poison from this tree and used against the Japanese in World War II.

A small herb of the Asteraceae, *Anaphalis margaritacea,* tends to grow only in cooler places and at higher elevations. It was a very valuable plant before the tribal people had ready access to matches, for it was made into tinder to catch the sparks made by striking metal against rock when starting a fire. The woolly plant parts were put above the fireplace on a rack, thus keeping them both dry and available when needed. *Eupatorium odoratum,* a very common introduced plant, has also been used occasionally for this purpose. The Karen use the hairy scrapings from the fruits of *Markhamia stipulata* as tinder for starting fires. They also eat the flowers but not the fruits.

The pith of the leaf petioles of the somewhat strange-looking plant, *Trevisia palmata,* is used like cork, sometimes for plugging bottles, but also in ceremonies. The pith absorbs much water and swells up after being dried, so for various kinds of sacrifices and ceremonies some of the pith is put in water, allowed to expand, and then flowers are put in it to keep them from wilting. The plant also produces edible fruits, and even the young leaves can be eaten. *Quisqualis indica,* a forest vine called Rangoon creeper, serves a similar function in that the leaves are commonly used by the Akha as stoppers for gourd containers.

The Akha live on the same mountain as do the Lahu Na and, like them, supplement their diet with foods from the forest. They will dig up the enlarged roots of the wild yam, especially when their rice supply is low or exhausted. They will also dig and eat the roots of *Oenanthe javanica, Houttuynia cordata* (the leaves are also edible), *Chlorophytum orchidastrum* (as well as the leaves), and *Monochoria vaginalis.* Other food plants include the ferns, *Lygodium flexuosum* and *Diplazium esculentum,* species of *Solanum, Fagopyrum dibotrys* (wild buckwheat), and the somewhat bitter-tasting wild lettuce (*Lactuca* sp.). The Akha are happy when they locate plants of *Garuga pinnata,* for the bark provides a wonderful flavoring for meat dishes.

The wild banana, *Musa acuminata,* is important medicinally but also provides a source of food (Plate 87). Although the fruits are full of hard seeds, the Akha will still boil and eat them. However, the Lahu believe the fruits are poisonous and will not eat them, preferring instead to eat the large bud bracts. The stems are fed to pigs.

The leaves of several trees and vines are commonly gathered as greens; these include *Anomianthus dulcis* (the inner part of the fruit is also eaten), *Crateva magna* (usually cooked or pickled before eating), *Dracaena fragrans, Dregea volubilis, Eryngium foetidum,* some *Ficus* species, *Mucuna macrocarpa, Zygostelma benthamii,* and *Polygonum chinense.* The Karen will also eat the young fronds of the beautiful cycad, *Cycas siamensis.*

The Akha seek a variety of plants from within the forest to make teas. They collect the trifoliate leaves of *Dumasia leiocarpa* to boil and drink as a forest tea. The leaves of *Rubus dielsianus* are harvested, dried, and boiled with water to make a tasty drink which they call "thorn tea" (*a gah law baw*). *Polyalthia simiarum*, a large tree in the Annonaceae, produces not only good-tasting fruits, but the bark is chopped up and boiled to make a sweet-tasting drink. The Akha also make a tea from the vine *Clitoria macrophylla*.

There are a number of forest plants that provide edible fruits and nuts for the tribal people. Most nuts of the Fagaceae, such as those of *Castanopsis diversifolia* and other species of that genus, are narvested and eaten, as well as the fruits of several palms, such as *Arenga pinnata* and *Corypha umbraculifera*. The Hmong roast and eat the prickly fruits of *Spondias pinnata*; they like to use the wood for construction, as it is not eaten by termites. The sweet, red berry of a species of *Antidesma* is quite popular among the Lisu, and *Millingtonia hortensis* fruits are popular with the Karen. *Canarium subulatum* has edible fruits, and the Karen also eat the shoots of it; two species of *Dillenia* produce edible fruits which the Lahu and Karen eat: *D. indica* and *D. pentagyna*.

Clearly, there is a wealth of foods from both virgin and secondary forests, which the tribal people know and utilize during the appropriate times of year; the above plants are strong evidence that the forest provides a significant supplement to their diet, as well as food to keep them from starving at certain critical times.

Another group of plants, the rattan palms (the two most common species are *Calamus kerrianus* and *C. rudentum*), have been of great importance to the hill tribes, for not only have they used the fibers for weaving baskets and other articles where they wished to have very durable and long-lived products (Plate 88), but they have also eaten the young shoots. Unfortunately, the rattans are now very rare because they have been overexploited and in many areas have completely disappeared (see Chapter 10).

Forest trees have been widely used for lumber (see Chapter 9). However, government regulations now prohibit the felling of trees in many areas. This prohibition greatly affects the tribal people in certain necessary activities. An old Akha man bemoaned the fact that if he can not cut down trees he could no longer live—or die—in Thailand: he must cut down trees to make swidden fields to grow rice to have enough to eat, and when he dies his family must be able to cut down a tree to make his coffin (Paul Lewis, personal communication). Almost any dead branch or fallen log will also serve as firewood, and considerable time is spent salvaging branches and other wood during the clearing of swidden fields before they are burned (Plate 89). As mentioned earlier, collecting firewood is usually part of each daily trip to the forest. The tribal people also utilize the resinous wood of *Pinus merkusii* by building a fire at the base of the tree and later stripping off the damaged, pitch-filled wood. This material makes an excellent fire starter and is also a good source of cash income when sold in bundles to the lowland Thai, who use it to start charcoal fires.

The disappearance of the forests of northern Thailand is indeed distressing, not only to conservationists and those concerned about the loss of timber and watershed, but it is even more disastrous to the tribal people who are so dependent upon the forests for their survival. Though the hill tribes remove forests to make swidden fields when desperate for areas on which to grow rice, they do so only with great reluctance. To them the forests are not only a provider of food and many other natural products essential for their health and survival, but they are also a source of enjoyment and an important link to the past through the worship of the ancestors and spirits who dwell in the remaining timberland near their villages.

6

Bamboo: From Cradle to Grave

Long before I could see the village, I knew I was near it, for I had just come to several taboo signs warning evil spirits to stay out of the village. Made of bamboo, these "stars" were attached to stakes driven into the ground by the side of the trail (Plate 90).

As I approached the village I could clearly see that the houses, rice storage buildings, and livestock pens were constructed primarily of bamboo; the fences were bamboo, and so was the water system, consisting of half sections of hollowed-out bamboo attached end to end, at the outskirts of town (Plate 91).

I went to the house of Pima Ana Cheu Meu, the medicine man with whom I was going to stay, and was soon invited into the men's side of the house for refreshments. Walking by some woven bamboo chicken cages (Plate 92), I climbed up a bamboo ladder to the uncovered porch with a split bamboo floor. To one side lay a bamboo mat covered with freshly harvested peanuts that were drying, and overhead was a long piece of bamboo that served as a clothesline. Ducking as I passed through the bamboo-framed door into the dark interior of the house, I sat on the split bamboo floor near the fireplace, where water was being heated in a piece of green bamboo using bamboo as firewood. My host offered me boiled peanuts and hot tea in bamboo cups. Except for the main posts, cross beams, and thatch, virtually the entire house was made of bamboo: rafters, walls, flooring, doors, and shelves. The cogon grass (*Imperata cylindrica*) thatch had been tied with pieces of split bamboo onto other larger pieces of bamboo to make "shingles," which, in turn, were attached with bamboo ties to the rafters. The most sacred item in the house, the ancestral altar, was a piece of bamboo hung on the women's side of the bamboo dividing partition (Plate 82).

The kitchen of an Akha house is normally on the women's side of the central partition, but the headman permitted me to sit in the open area dividing the house where I could watch dinner being cooked. Bamboo and gourd water containers leaned against the wall near the fireplace along with bamboo winnowing trays and fan. A kettle sat on an iron tripod over the fire, and lying next to it in the hot coals was a section of green bamboo in which water was heating. I watched as a woman took lard from a bamboo storage container and dropped it into the kettle (Plate 93); she then took her bamboo mortar and pestle and proceeded to grind chilies

93

and other herbs to put into the kettle. She deftly cut bamboo shoots and added them to the meal, using a knife with a 35 centimeter (14 inch) long blade and bamboo handle. She also added a few pinches of salt from a bamboo container lying next to the hearth. Scooping up some soaked rice from a bamboo container with a bamboo cup, she put it in a bamboo steamer, which she placed in a wok partially filled with water that sat on the fire in the men's side of the house. Some of her ladles, spoons, and skewers were also made from bamboo.

Soon one of the medicine man's daughters brought a low round table and several tiny stools, all made of woven bamboo. While she set our places with bamboo dishes, including split bamboo chop sticks, her father poured some rice whiskey into the bamboo cups; the rice had been allowed to ferment in a bamboo container before being distilled. Giving his guests small bamboo straws, we drank this powerful home brew to his—and the village's—health. The meal was soon brought to us on woven bamboo trays and in bamboo containers: piles of rice, curry, boiled peanuts, some meat, lots of greens, and bamboo shoots. Some of the bamboo shoots had been pickled in sections of bamboo that contained either salt-water brine or the water in which rice had been soaked; the outer, tough parts of other bamboo shoots had been removed and the center part boiled in water until tender. One small dish contained several small caterpillars that had been removed from some of the young bamboo sprouts; these are considered a great delicacy by the Akha and other tribal people.

Following the meal several neighbors arrived. One began to smoke a water pipe made from two sizes of bamboo, while yet another rolled a cigarette using a very thin bamboo sheath as cigarette paper. Later in the evening a man played a mouth harp made from bamboo. Other musical instruments appeared, including a flute made from bamboo and another that combined a gourd and several small pieces of bamboo.

The following day a few men went to the forest to check their animal traps, many of which were made of bamboo. Some traps employed sharpened bamboo spears, which were hurled by bamboo "springs;" others were snares made of finely woven bamboo cord; and a few involved deadfall logs carefully held up by bamboo under tension and ready to be sprung by the animal (Plate 94). Many years ago the tribe caught tigers in bamboo cages, but no one had seen a tiger in the area for many years. Two men carried their homemade black powder rifles with 2-meter-long barrels, carefully loading charges with split bamboo ramrods. Another had a crossbow, in which the bow part was made of bamboo; even the arrows (and "feathers") were made of split bamboo (Plate 95). One of the men also had some bamboo bird-calling whistles.

While the men checked their animal traps, some children happily ran off to check bamboo bird and rat traps they had set near the village. Other younger boys began to play with their bamboo "coasters," riding wildly down one of the steep, wider trails within the village. A few older boys headed for the stream, carrying split bamboo fish traps and creels at their waists to hold their catch.

Meanwhile, several older women began weaving with bamboo looms. The cotton thread had been spun by the women as they walked along the trail, carefully pulling fibers from bamboo thread-spinning containers attached to their waists and rapidly rotating the bamboo thread spinners on their thighs. An old man sat on his porch smoking a bamboo pipe and soon began to weave a low table from bamboo strips (Plate 96). It and some newly woven baskets would be stored above a fireplace to cure in the smoke, thus protecting the bamboo from insects.

One group of men and women left for the fields to finish harvesting, threshing, and carrying the season's crop of hill rice. The intricate, woven bamboo head-pieces the women wore were covered with coins, dyed feathers and fur, beads, silver, and buttons (Plates 13, 41, 97). Each woman also carried a bamboo basket borne on a wooden shoulder yoke and with a plaited bamboo cord and head-piece. Before returning home that evening, they would fill these baskets with greens, bamboo shoots, firewood, or banana stems for pig food. Two men wore hats of woven bamboo and each had a bamboo-handled knife in a scabbard made of bamboo or rattan. Most of the group wore rubber thongs, but one older man wore bamboo clogs. One woman also carried a rain cape made of the leaves of *Pandanus furcatus* (screw pine) and bamboo rolled up in her basket.

Accompanying this group, I passed a bamboo altar to the rice spirit in one field. Each of the small horses with us, which had been fed bamboo leaves the night before (Plate 98), was led by a plaited bamboo rope, and carried a pack frame whose cruppers around the tail were made of bamboo. Later that day each animal would carry several sacks of threshed rice back to the village.

Arriving at the field, the rice stalks were cut and their stems tied with bamboo strips; later the rice would be threshed by beating the stalks bearing the heads of grain on several pieces of bamboo or wood propped up on a bamboo mat in the field. Winnowing would be done with split bamboo trays—and bamboo fans if there was not enough wind.

During my visit to this village and others like it, I learned of an Akha creation myth that emphasizes the significance of bamboo. First, three white rocks appeared, and from these came water. From under the rocks four different plants arose, two of which were bamboos (Lewis 1968–70).

Special events within an Akha village, as well as daily activities, also involve bamboo. A "marriage house" is sometimes constructed in the center of a village (Plate 99). This structure, made completely of bamboo, is a single small room about 5 meters (16 feet) off the ground built on tall bamboo posts. A newly married couple spends its first night in this public, yet private, place in the middle of the village.

Moments after the birth of a new child, the umbilicus of a newborn baby is cut with a special knife made from dry, well-cured bamboo, which the expectant mother has made and carried with her (along with string to tie off the umbilical cord) for several days prior to delivery (Plate 100). The bamboo is sharpened on its edge by shaving it with a knife. As soon as the child is born, the midwife places the still-attached umbilicus on a piece of charcoal (which the Akha believe is sterile because it has been burned), and quickly cuts it with the bamboo knife. Then she ties it with the bit of fine string that the mother has been carrying. The baby may then be placed in a woven bamboo cradle. The father takes the placenta and the special bamboo knife, puts them in the piece of bamboo used to remove ashes from the house's fireplaces, and buries them under the house, directly beneath the ancestral altar. The mother must stay near the fire on a special bed that the father has made of bamboo.

Should the baby die before it is named, which usually is done immediately after it is born, the body is placed in a bamboo water carrier that has been split into two halves. The binding cloth is wrapped around it and it is then buried in the forest. The bodies of children born in "terrible births" are also buried in bamboo containers.

Suicide is thought to be a terrible form of death to several groups of tribal people, which fear the spirit or ghost of the dead person may cause bad fortune for others. Therefore, long bamboo stakes are driven into the heels of the dead person to "hobble" the spirit or ghost (Campbell et al. 1981).

Bamboo is an essential element in the funeral of an adult. An animal to be sacrificed during the funeral ceremony, such as a water buffalo, is tied to a post made from the trunk of *Schima wallichii*, and killed with a spear having a bamboo handle; to calm the animal just prior to death a "blinder" made of a large bamboo culm sheath may be held by a stick to keep the animal from seeing the elder approach with the spear (Plate 101). The length of the coffin, carved from a single large piece of wood, is measured by a special piece of the bamboo *Dendrocalamus strictus*. Three bamboo containers with whiskey, water, and tea (mixed with ginger) are carried by the men when they go to the forest to make the coffin. Two pieces of bamboo protrude from the casket (Plate 102). One, a long hollow piece that passes through the roof of the house in which the casket lies during the funeral ceremony, allows gases or "odors" to escape from the interior of the coffin, for the body is not embalmed and gases quickly build up in the sealed wooden casket. The other hollow piece of bamboo comes out of the bottom of the coffin and goes into the ground beneath, allowing liquids to drain from the body as the coffin lies in the house of the deceased for several days prior to burial. Upon completion of the funeral ceremonies within the village, several men carry the coffin to the burial ground on a bamboo pole. Thus bamboo pervades the lives of the hill tribes, from the moment of birth until a body is laid to rest in the forest soil.

Without doubt, the various bamboos are some of the most important plants for the survival and livelihood of the hill tribes, and they permeate every facet of tribal life. In a sense, the highlanders of Thailand are no different from other groups of people living in Southeast Asia, for all employ these amazing grasses in a variety of ways. Yet, somehow the hill tribes seem closer to the forest and more dependent on these giant grasses than are the lowland people, who must often go some distance to the mountains to obtain bamboo or replace it with steel and plastic.

CLASSIFICATION OF BAMBOOS USED BY THE HILL TRIBES

It is a challenge to describe accurately the variety of bamboos in northern Thailand for several reasons, but the main one is that bamboos are poorly understood by botanists. Because of the difficulty of collecting specimens and the wide range of characteristics which bamboos exhibit, few plant scientists study this group. Moreover, most botanical classifications are based on characteristics of reproductive structures, namely the flowers and fruits. However, for bamboos this is a difficult challenge at best because many do not flower for decades. Thus, the plant systematist must often be satisfied with only vegetative structures. Despite these challenges, we now have a fairly good idea of which bamboos occur in northern Thailand.

Another serious challenge to understanding the bamboos used by the hill tribes is the confusion of vernacular or common names. The tribal people sometimes do not distinguish between closely related species within the same genus, giving them the same name. This is particularly the case when two plants have similar growth patterns and uses. All bamboos have one or more Thai names and many have a name in the northern Thai language. In many instances, the tribal people have their own names, but in other cases they borrow—and corrupt—names from either the Thai, the Burmese, or the Shan. A Thai or tribal name may therefore be the name which is widely understood in the lowlands or a modification of that name through mispronunciation or misunderstanding. In several instances my informants did not know the names of certain species, perhaps because the bamboos did not occur in the area or were not used by them. With the exception of a few Hmong and Mien names that are spelled in the orthographic form used by linguists who have worked with those languages, spellings are phonetic and tend to follow the commonly used system of the Lahu. Thai names are from Lin (1968) and Smitinand (1980).

Bamboos are large perennial grasses of the Poaceae (Gramineae), subfamily Bambusoideae (Plate 103). There are about 50 genera and at least 1000 species of bamboo distributed around the world between 46° north latitude and 47° south latitude. This subfamily, which is believed to be the most primitive of the grasses, is often referred to as "tree grasses," and, indeed, some, like *Dendrocalamus giganteus*, do attain great height and develop culms or stems up to 30 centimeters (12 inches) in diameter (Plate 104).

Bamboos are characterized as having woody, long-lived, hollow stems (culms); complex branching systems; active rhizomes or underground stem systems; and amazingly infrequent flowering. Some species, such as *Dendrocalamus strictus*, flower every 20–44 years; *D. giganteus*, on the other hand flowers about every 75 years. The commonly cultivated bamboo, *Bambusa vulgaris*, is believed to flower at intervals of more than 150 years (Janzen 1976). Amazingly, flowering and subsequent seed production is synchronized so that all individuals of a given species reproduce at the same time wherever they are located in the world.

The rhizomes may be long and slender, more or less cylindrical in shape, and usually smaller in diameter than the culms themselves; this **leptomorph** type of bamboo usually forms groves of more or less evenly spaced culms and tends to be most common in cooler, temperate areas. The bamboos of warm, tropical regions have the **pachymorph** type of rhizome; it is short, thick, somewhat fusiform or top shaped, and much larger in diameter than the culms which arise from it, usually in dense clumps (Soderstrom 1981). Only a few bamboos, such as the genus *Dinochloa*, have solid culms; the vast majority of stems are hollow in the internodal region but with a solid septum at each node. The remarkable strength of the bamboo culm is not the result of true wood as in dicotyledonous trees, but rather

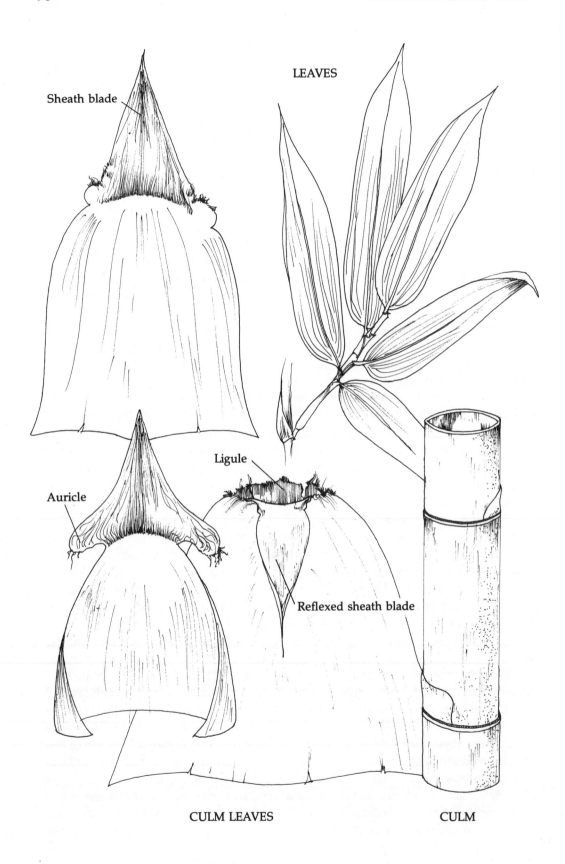

LEAVES

Sheath blade

Ligule

Auricle

Reflexed sheath blade

CULM LEAVES   CULM

due to the vascular bundles capped with sclerenchyma tissue and embedded in a lignified, thick-walled ground tissue. Thus, each culm has resistance to bending and twisting, but can be easily cut with a sharp knife. Bamboos usually grow in association with woody plants and occur in open, human-created grasslands in northern Thailand only where they have migrated from the surrounding forests.

A large number of characters are used in identifying bamboos, but unfortunately most are based on features of vegetative structures that are unique to the grasses, and thus poorly known to most people (see facing page). Important characters of the **culm** or stem are its total length, growth habit, diameter, length of the internode, nature of the **node** (prominent or insignificant), nodal branching patterns, wall thickness, and whether the culm is hollow or solid. The **culm sheath** or **culm leaf** is an important structure that is usually available throughout the year (Plate 105). Its size, shape, color, persistence, and the nature of the hairs covering it are helpful characters in identification. The **sheath blade,** which is borne at the tip of the culm sheath, is characterized by such features as orientation (reflexed or erect), shape, size, hairiness, and the nature of the edges. Typically, the sheath blade bears **sheath auricles,** which may be conspicuous or nearly absent, of varying shapes, and with differing margins (i.e., bristles, hairs, etc.). The culm sheaths also bear **ligules** at their upper edge, interior to the blades; these vary greatly in size, shape, and nature of their margins. **Leaves** vary with regard to persistence, color, shape, length and width, number of lateral veins, and surface features. Many features of the inflorescences are valuable in identifying bamboos, but infrequent flowering makes the use of any reproductive characters difficult, especially in the field.

The literature on Thai bamboos is quite scanty, and that which is available is based on outdated publications dealing with bamboos of other countries. No original taxonomic work has been done on Thai bamboos. Two summary mimeographed papers (Ruaysungneun and Purivirojkul 1980; Smitinand and Chaianant n.d.) and a small booklet exist in Thai (Boonkert 1985), and two papers published in English (Lin 1968; Smitinand and Ramyarangsi 1980). The paper by Lin is a report of several field trips made by a team of foresters and botanists in 1967 who identified 18 species of bamboo in northern Thailand. The other paper deals briefly with the status and management of Thai bamboo forests. I have located and collected 16 of the 18 species reported in Lin's paper, including *Dinochloa maclellandii* (Munro) Kurz, but, so far as I can tell, this species is not used by the hill tribes. Lin reports two species, *Bambusa burmanica* Gamble and *Dendrocalamus longispathus* Kurz, as occurring in northern Thailand. However, my tribal informants do not believe these species cross the border from Burma into Thailand. Soejatmi Dransfield at the Royal Botanic Gardens, Kew, identified three of my specimens as species which were not reported by Lin (1968) as occurring in the north; they are *Bambusa tuldoides, Gigantochloa wrayi,* and *Phyllostachys* sp. Two species, *Bambusa vulgaris* and *B. ventricosa,* are used by the hill tribes, but have been introduced, the former from an unknown origin and the latter probably from China.

I have determined there are 20 native species of bamboos in northern Thailand that are used by the hill tribes:

Bambusa arundinacea   *Dendrocalamus latiflorus*
Bambusa pallida   *Dendrocalamus membranaceus*
Bambusa polymorpha   *Dendrocalamus strictus*
Bambusa tulda   *Gigantochloa latifolia*
Bambusa tuldoides   *Gigantochloa wrayi*
Cephalostachyum pergracile   *Melocalamus compactiflorus*
Cephalostachyum virgatum   *Oxytenanthera albo-ciliata*
Dendrocalamus brandisii   *Phyllostachys* sp.
Dendrocalamus giganteus   *Thyrsostachys oliveri*
Dendrocalamus hamiltonii   *Thyrsostachys siamensis*

NAMES AND TRIBAL USES OF THE BAMBOOS

Bambusa arundinacea, Giant thorny bamboo

Thai: *phai, phai-pah, phai-nam, mai-phai, phai paa*; Shan: *mai-sang-nam*; Akha: *a cu, za ju*; Hmong: *tsong*; Karen: *waa-chuu*; Lahu: *yaw ceh*; Lisu: *mai hio*; Mien: *hau gim* (LAUX NQIMV) (*gim* = thorn).

These plants occur as large clumps, with bright green culms up to 30 meters (98 feet) high. The culms are up to 18 centimeters (7 inches) in diameter and have very thick walls and therefore a small cavity; the internodes are 15–45 centimeters (6–18 inches) long, and the nodes are prominent. The most distinctive feature of this bamboo is that the branches are armed with 2–3 recurved spines. The culm sheaths are large and nearly as broad as high, deciduous, rounded on top, and thickly ciliate when young. Sheath blades are triangular, sharply pointed, more or less flexed, and densely covered with hairs on the inside. Sheath auricles are poorly developed; the ligules are narrow and fringed with hairs. The leaves are linear to lanceolate in shape, 17–20 centimeters (7–8 inches) long, 0.5–2 centimeters (0.2–1 inch) broad, and have 4–6 lateral veins.

This species is widely distributed at lower elevations throughout Southeast Asia, and westward to Pakistan. It occurs in mixed deciduous forests in the eastern portion of northern Thailand, but not in Chiang Rai Province.

The Hmong use the culms for their gourd and pipe musical instruments.

The Lahu use the culms for crossbow arrows and for spears in animal traps. During the Vietnam War they used this bamboo to make spears for human booby traps. Lahu women prefer this bamboo for making the bobbins on which they wind their thread because it is very smooth. It is also used for the small part of the water pipe, which holds the tobacco (or cigarette).

The Mien use this species to make pig pens and chicken coops, and for some parts of house construction.

The shoots are edible but somewhat bitter.

Bambusa pallida, Graceful bamboo

Thai: *pai-song-kham, mai-phiu, phai-phiu, pai-song-dam*; Shan: *meh mo, mai mo*; Akha: *ma bu*; Hmong: *gi dong klu*; Lahu: use Shan name; Lisu: *pi-aw cu*; Mien: *hau kiam*.

These plants form clumps with olive green culms up to 18 meters (60 feet) high and 5–8 centimeters (2–3 inches) in diameter, with thin walls and large cavities. The internodes are 45–75 centimeters (18–30 inches) long and the nonprominent nodes have numerous branches. Culm sheaths are 18–30 centimeters (7–12 inches) long and about 25 centimeters (10 inches) broad, glabrous or with only minute hairs on the outside, broadly truncate at the end, and bear very long, triangular-shaped, acuminate sheath blades that are often longer than the sheath itself. Sheath auricles are small, rounded, and with a few bristles; the ligules are narrow. Leaves are linear-lanceolate in shape, 10–20 centimeters (4–8 inches) long, and 1.5–2 centimeters (0.6–0.8 inch) broad, and have 4–6 lateral veins.

This species occurs at higher elevations in northern Thailand and adjacent Burma.

The Akha make strips from this bamboo to tie various items together. No other tribe reports a use for it, although most have a name for it and know it as a distinctive type of bamboo.

Bambusa polymorpha, Grayish-green bamboo

Thai: *pai-hom;* Shan: *mai-sa-lawn;* Karen: *waa-kheh.*

This densely clumping bamboo tends to shed its leaves in the dry season; it has gray to grayish green culms 15–25 meters (50–82 feet) high and 6–15 centimeters (2.3–6 inches) in diameter. The culms have thin walls, internodes that are 50–90 centimeters (20–35 inches) long, and thickened nodes from which numerous branches arise above. The culm sheaths are thick, 15–18 centimeters (6–7 inches) long, but 30–35 centimeters (12–14 inches) broad; they are densely covered on the back with white hairs and rounded truncate at the end. The sheath blades are reniform in shape and much broader than long. The sheath auricles are small and bristly, and the ligules narrow and entire. The leaves are small and linear in shape, 7–18 centimeters (3–7 inches) long, and 1–2 centimeters (0.3–0.5 inch) wide; lateral veins number 4–6.

This species occurs in lower mixed forests with teak, and at moderate to higher elevations in northern and northeastern Thailand; it is also in Burma.

The Karen like to use this bamboo for very heavy walls, bridges, houses, and fences, though the culms are relatively small in diameter and have walls that are not notably thick.

Bambusa tulda

Thai: *phai-bong, mai-bong;* Shan: *bong, mai-wawng, mai-wang;* Akha: *ma bu;* Hmong: *mai bong;* Karen: *waa-su, waa khu, waa-sho-wa;* Lahu: *meh baw, mai baw;* Lisu: *pi-aw cu;* Mien: *hau lai* (LAUX MBONG).

These plants are evergreen or deciduous, forming tall clumps up to 22 meters (72 feet) in height. The gray-green culms are up to 10 centimeters (4 inches) in diameter, have internodes 30–60 centimeters (12–24 inches) long and walls up to 1.5 centimeters (0.6 inch) thick; the nodes, which are not swollen, tend to produce fibrous roots below and numerous branches above. The culm sheaths are bluish green, as broad as long (15–25 centimeters or 6–10 inches), densely hairy, either rounded or triangularly truncate at the end, and bear triangular-shaped, somewhat reflexed sheath blades. The sheath auricles are large, rounded, wavy-wrinkled, and with bristles; the ligules are narrow and entire. The leaves are linear-oblong to linear-lanceolate in shape, 15–25 centimeters (6–10 inches) long, 1.5–4 centimeters (0.6–1.6 inches) broad, and with 6–10 lateral veins.

This species occurs at moderate elevations in northern and western Thailand, and extends westward into Burma, where it is quite common, and thence further westward into India.

Bambusa tulda is readily confused with *B. tuldoides* and it is uncertain whether the tribal people distinguish between them so far as their names and uses are concerned.

Bambusa tuldoides, Verdant bamboo

This plant forms clumps, with culms 6–8 meters (20–26 feet) high, 3–4.5 centimeters (1.2–1.8 inches) in diameter, and with somewhat nodding tips. The culms tend to be whitish at first, but then become distinctly green in color and thick-walled; the internodes are 30–36 centimeters (12–14 inches) long, and the nodes appear fairly prominent. Branches emerge distinctly upward from the basal first node, with one thicker and longer than the others. The culm sheath is glabrous, prominently ribbed when dry, and arched-convex at the tip. The sheath auricles are slightly unequal, with the outer one slightly larger; the ligules are minutely fringed and up to 3.5 millimeters (0.14 inch) high. The leaves are lanceolate to narrowly lanceolate in shape, 10–18 centimeters (4–7 inches) long, and 1–1.7 centimeters (0.4–0.7 inch) broad, and with 3–6 lateral veins.

This species occurs at moderate and somewhat higher elevations in mixed deciduous and evergreen forests in northern Thailand, northward into southern China, and westward into Burma.

The Karen use the dried culms for the handles of agricultural tools because they are very durable. They also use the culms for ties in making bundles of rice seedlings when preparing to transplant them into wet fields.

The Lahu use this species to make tie-downs for house construction and wherever else strong ties are needed. They use the culms to make fences around pigpens and in their yards, as weevils do not seem to attack this bamboo as readily as some of the others. The Lahu claim this bamboo will last for years. There is a report that the Lahu Sheh Leh use this bamboo for making crossbow arrows.

Tribal people also employ this bamboo for house rafters, shoulder poles to carry loads, animal enclosures, fish traps, and carriers or creels which they wear around the waist and in which they hold fish.

Bambusa ventricosa, Buddha's belly bamboo, Buddha bamboo

This is a highly unusual bamboo sometimes consisting of dwarf culms, up to only 1 meter (3.3 feet) in height, and rarely exceeding 4 centimeters (1.6 inches) in diameter. However, at other times the culms reach up to 5 meters (16 feet) in height and up to 8 centimeters (3 inches) in diameter. The internodes are short and distinctly swollen, 10–20 centimeters (4–8 inches) long, and inconspicuous branches occur at each node. Culm sheaths are inconspicuous, bearing 2 unequal auricles, and a small ligule. Leaves are ovate-lanceolate to oblong-lanceolate in shape, 12–21 centimeters (4.7–8 inches) long, 1.5–3.3 centimeters (0.6–1.3 inches) broad, and with 5–9 lateral veins.

This bamboo is a native of China but some plants have been imported into Thailand as horticultural oddities.

The Akha make tobacco pipes from the culms.

Bambusa vulgaris

Thai: *phai-luang, phai-luong, phai lueang*; Northern Thai: *chan-kham*; Akha: *za shaw, ya shu*; Karen: *waa baw*; Lahu: *va shi* (= yellow bamboo); Lisu: *za khao*; Mien: *hau giem, hau kwai, hau bin* (the spectacular striped one) (LAUX DORNKH NIE).

The plants are moderate in size, producing rather widely spaced culms 6–15 meters (20–50 feet) high, and arching at the tops. The culms are 5–10 centimeters (2–4 inches) in diameter, with fairly thin walls, and internodes 25–45 centimeters (10–18 inches) long; they produce nodes that are not prominent, and from which arise numerous branches. The culm sheaths are streaked green and yellow when young, nearly as broad as long (15–25 centimeters or 6–10 inches), rounded at the top, with dark brown hairs on the outside, and with entire margins. The sheath blades are triangular in shape, acuminate apically, and hairy and narrow basally (Plate 105). Sheath auricles are very conspicuous, ear-shaped, and fringed with bristles; the ligules are narrow and dentate. The leaves are linear to lanceolate in shape, pale green in color, 15–25 centimeters (6–10 inches) long, 1.8–4.5 centimeters (0.7–1.8 inches) broad, and with 6–8 lateral veins.

The origin of this species is unknown, though some believe it is a native of Madagascar; it has now become so widely cultivated throughout the world that there is doubt that its actual origin will ever be known. It is almost certainly not a native of northern Thailand.

This bamboo is used for a variety of purposes, as are the moderately large native bamboos.

Bambusa vulgaris var. *striata* (Lodd.) Gamble is widely cultivated as an ornamental bamboo. It has pale yellow culms with narrow longitudinal green stripes.

Cephalostachyum pergracile

Thai: *phai-kaolarm, pai-kaolam, mai-pang, phai khaao laam*; Shan: *mai-khao-lam*; Akha: *ma ka la*; Hmong: *tsio maw jan*; Karen: *waa-blo, waa-phlong, waa blaw*; Lahu: *ma hka la*; Lisu: *mai kao lam*; Mien: *hau tze*.

This is a deciduous, clumping bamboo of moderate size that produces glaucous, green culms 9–12 meters (30–40 feet) high, 4–8 centimeters (1.6–3 inches) in diameter, and with very thin walls. The internodes are typically 30–50 centimeters (12–18 inches) long and the nodes are scarcely thickened. The culm sheaths tend to be deciduous, 15–20 centimeters (6–8 inches) broad and 10–15 centimeters (4–6 inches) long, and densely covered with black hairs on the outside. The sheath blades are cordate to ovate in shape, densely hairy on the inside, and with a wavy fringe at the top. Sheath auricles are rounded and bear distinct white bristles along their margins; the ligules are very narrow and entire. The leaves are thin, linear to lanceolate in shape, 15–30 centimeters (6–14 inches) long, 2.5–6 centimeters (1–2.4 inches) broad, with 7–13 lateral veins, and distinctly pubescent beneath.

This species occurs throughout northern Thailand in mixed forests, often with teak, and extends westward into Burma.

This bamboo flowered throughout its range in 1987, causing severe problems among the tribal people because of the population explosion of rats, who fed on the abundant bamboo seeds. The following season the rats invaded the rice fields, causing severe loss of their crops.

For all the hill tribes, as well as for the Thai, the primary use of this bamboo is for cooking sticky rice. The green, thin-walled culms are filled with soaked rice and baked over coals. The walls are then peeled away and the rice eaten.

The Akha make rice steamers from this plant and use the culms for roof framing.

The Hmong make ties from it with which to attach roof parts.

The Karen will eat the shoots, though they are somewhat bitter tasting. They also make mats, hats, and fencing from it, and like to use it for tie-downs because it is flexible and soft.

The Lahu use it for roof poles or rafters and eat the shoots (which they claim taste good). They, too, make matting from it because it can be finely split and is soft and flexible.

The Mien also use it for tie-downs and will eat the shoots.

Cephalostachyum virgatum

Thai: *phai-hiar, phai-hiae, mai-hiar*; Akha: *ya ka, za ka*; Hmong: *tsio plau kai*; Lahu: *mai phu*; Lahu Sheh Leh: *mai ho*; Lisu: *ma he*; Mien: *hau bia* (*bia* = thin).

This bamboo, like *Cephalostachyum pergracile*, also produces slender culms up to 12 meters (40 feet) high and with thin walls. The culms are 4–6 centimeters (1.5–2.4 inches) in diameter, have internodes 40–60 centimeters (16–24 inches) long and prominent nodes. The culm sheaths are densely covered with golden hairs on the outside but otherwise are similar to those of *C. pergracile*. The sheath blade is recurved and with a cuspidate tip. The auricles are truncate and bristly, and the ligules narrow and with fimbriate edges. The leaves are linear-lanceolate to oblong in shape, 15–30 centimeters (6–12 inches) long, 2.5–5 centimeters (1–2 inches) broad, and with about 10 lateral veins.

Cephalostachyum virgatum also occurs in mixed forests, often with teak, in northern Thailand and adjacent Burma.

The Akha claim this bamboo has a poison which will kill an animal if it is stabbed with it. The Akha get sores from this type of bamboo if they are cut with it, perhaps because of this poison. The walls of the culms are so thin that they are almost razor-sharp when broken or cut; thus, it is easy to be cut by them.

Most of the tribes use this bamboo for house siding, roof framing, mats, ties, containers to boil water, knife handles, trays, baskets, animal cages, and even rafts for floating down rivers (Plate 106). Some use it to make water pipes for smoking. The shoots are not eaten.

The Akha use this bamboo to make the frame for the woman's headpiece (Plate 97).

The Akha, Lahu, and Lisu all use small stem pieces of this species to make their gourd-pipe musical instruments. Larger pieces can be constructed into their flutes.

The Mien feel *Cephalostachyum virgatum* is particularly good for the weaving of mats.

Dendrocalamus brandisii

Thai: *mai-bongyai, mai-sang-mon, phai bong yai*; Northern Thai: *mai-po*; Shan: *mai-puk*; Akha: *ma bu*; Karen: *waa-khlue, waa-khlue-pho, waa-khru*; Lahu: *meh va bvuh-eh, va bvuh-eh*; Lisu: *ma baw da ma*; Mien: *hau biang*.

This large bamboo occurs in clumps, with gray-green culms 18–37 meters (60–120 feet) high, 12–20 centimeters (5–8 inches) in diameter, and with thick walls. The internodes are 30–40 centimeters (12–16 inches) long, and the swollen nodes produce numerous branches. The culm sheaths are thick and leathery, up to 60 centimeters (24 inches) long and 56 centimeters (23 inches) broad, and are covered on the outside with flattened white hairs. Sheath blades are linear-lanceolate in shape, up to 46 centimeters (18 inches) long and 13 centimeters (5 inches) broad, spreading or reflexed, and hairy on the inside. The sheath auricles are small, but the ligules are 10–20 millimeters (0.4–0.8 inch) high, narrow, and denticulate. The leaves are oblong to lanceolate, 20–30 centimeters (8–12 inches) long, 2.5–5 centimeters (1–2 inches) broad, and with 10–12 lateral veins.

This species occurs only above 1000 meters (3300 feet) elevation in the far northern part of Thailand, probably only Chiang Rai Province. It also occurs in Burma, Laos, and Vietnam.

The Akha and Lahu do not like to use this bamboo, but will make tie-downs with it if no better bamboo is available.

The Mien, on the other hand, like it, and use it for general construction, such as siding and rafters.

The Karen use it only for making matting.

Dendrocalamus giganteus, Giant bamboo

Thai: *phai-pao, mai-po, pai-poh*; Akha: *za pi-euh, za jeh*; Hmong: *dong ngeh*; Karen: *waa-kwaa*; Lahu: *mai pao, mai pok*; Lisu: *ma khaw neh*; Mien: *hau yiang* (they claim an even larger variety, called *hau tzang*, occurs in China and northern Laos).

This is the largest of all the northern Thai bamboos, with long, gray-green, straight culms reaching more than 30 meters (100 feet) high. The culms are 20–30 (or more) centimeters (8–12 inches) in diameter, with fairly thin walls, internodes 35–40 centimeters (14–16 inches) long, and hairy nodes with numerous branches. Culm sheaths are very large, 50 centimeters (20 inches) long and broad, early deciduous, with golden brown hairs on the outside but usually appearing purple when young, rounded at the top, and hard. Sheath blades are up to 40 centimeters (16 inches) long, 9 centimeters (3.5 inches) broad, and reflexed. Sheath auricles are inconspicuous or lacking; ligules are very conspicuous, 6–15 millimeters (0.2–0.6 inch) high, distinctly toothed, black, and with fringed edges. The leaves are highly variable in size, oblong to lanceolate in shape, 15–50 centimeters (6–20 inches) long, 3–10 centimeters (1.2–4 inches) broad, and with 12–16 lateral veins.

This huge bamboo occurs in mixed forests at lower elevations in northwestern Thailand, and thence westward into Burma and India (Plate 104).

Dendrocalamus giganteus is very popular for house posts, especially among the Lahu, who say that it is best when it is green. A hole is cut in the culm wall so that water can be put in the lower part, thus keeping it from drying out. As long as the bamboo stays fresh and moist, it will not deteriorate. However, it is very important that the ground also be kept moist.

The Karen use this bamboo for the same purpose, claiming that it will last for 5–6 years if used green and then kept moist. They also like to use it for flooring, rice steamers, and water troughs.

Because of its large size, there are numerous other uses of this bamboo. It is split to make floors and walls, it is laid on top of roofs to hold down thatch on the outside, smaller pieces are used for rafters, and young stems can be made into strips for weaving into baskets and other containers. Larger pieces may be made into mortars and pestles, dishes, rice steamers, pig troughs, drums, water containers, and fermenting and pickling containers (Plate 107).

The plant also serves as a food because the shoots are good to eat; an individual shoot of this species may weigh several kilograms.

Dendrocalamus hamiltonii

Thai: *phai-nual-yai, pai-nuan-yai*; Northern Thai: *pai-hok, mai-hok, mai-po, mai-phieo*; Shan: *mai-hok, mai-hok-khu*; Akha: *ya baw, za baw*; Karen: *waa-klu, waa-gler*; Lahu: *va sha, va k'aw*; Lisu: *ma khaw neh*; Mien: *hau yiang pen, hau pang*, (LAUX LINX).

This is one of the larger, more widespread clumping bamboos, usually occurring in scattered stands, and consisting of tall, erect culms up to 25 meters (82 feet) high. The culms are grayish white when young but become green with maturity, 10–18 centimeters (4–7 inches) in diameter, with thick walls, and having internodes 30–50 centimeters (12–20 inches) long. The upper nodes are much-branched. The culm sheaths are up to 45 centimeters (18 inches) long and 20 centimeters (8 inches) broad, usually without hairs, and truncate at the top. Sheath blades are ovate to lanceolate in shape, hairless except at the bases, pointed at their tips, and with incurved margins. Sheath auricles are small; ligules are 5 millimeters (0.2 inch) high, smooth, and entire. Leaves are quite large, broadly lanceolate in shape, 20–38 centimeters (8–15 inches) long, 3–6 centimeters (1.2–2.4 inches) broad, and with 6–17 lateral veins.

Dendrocalamus hamiltonii occurs widely in the north in mixed forests at moderate elevations. It is also found in Burma, Laos, Vietnam, and extends westward into India.

The hill tribes use *Dendrocalamus hamiltonii* for food (though it is not preferred), water carriers, house construction, bridges, fences, baskets, water pipes (for smoking), matting, cigar papers, scabbards, trays, drinking straws, deadfall animal traps, drain pipes, chicken cages, clogs, salt containers, siding on houses, and for ties (good for anything around the house, especially for thatch) (Plates 108–109).

The Akha make the containers from this bamboo for the bodies of twins and deformed babies who are killed immediately following a "terrible birth." This bamboo is sometimes used for the ancestral altar (Plate 82), though more commonly *Dendrocalamus strictus* is used.

The Karen claim this is the strongest type of bamboo. However, it needs to be water-cured or bamboo weevils will eat it. If cured properly, they claim it is then very durable and strong.

The Lahu say that this bamboo can be made into very fine strips and is best for weaving baskets, mats, and dozens of other things which require a very fine quality of bamboo strip. They also feel it is the best bamboo for crossbow arrows. Interestingly, a hairless caterpillar, which the Lahu believe is a wonderful delicacy to eat, lives only on this bamboo.

The Mien will eat the shoots of this bamboo, but do not believe them to be as good tasting as *Dendrocalamus brandisii* and *D. membranaceus*.

Dendrocalamus latiflorus, Sweet bamboo

Lahu: *va meu* (a gray form, called *va meu hpu*, grows only in Burma); Shan: *mai pok*; Mien: *hau nia*.

This large, poorly known bamboo is quite distinct from the other larger types: it has very straight culms up to 25 meters (82 feet) high, which curve only at the tips. The culms are 8–20 centimeters (3–8 inches) in diameter, with thin walls and large cavities. One of the most distinctive features is that the internodes are very short, only about 15–25 centimeters (6–10 inches) long; the nodes are indistinct. The culm sheaths are about as broad as long (25–30 centimeters) (10–12 inches), and truncate at the top. Sheath auricles are linear, with low, toothed ligules. The leaves are very broadly oblong-lanceolate in shape, 17–25 centimeters (7–10 inches) long, 3–7.5 centimeters (1.2–3 inches) broad, and with 6–9 lateral veins.

This species is native to southern China and Burma; whether it has been introduced into Thailand or actually occurs just inside the country along the Burma border is uncertain. One informant was certain that the plant is native in Burma and is cultivated only in Thailand. However, I collected a specimen from a population near the Burma border in Mae Hong Son Province which was almost certainly native.

The Lahu cultivate *Dendrocalamus latiflorus* and claim it is the second largest and probably the tallest of all the bamboos they use. They use this species for house posts. They have many uses for this bamboo when it is split, such as for floors, walls, and mats. It also makes an excellent water container. The Lahu cook rice with this bamboo, using a piece consisting of two green culm joints. Water is poured into the cavity of the bottom joint through several small holes in the partition separating the two joints, and the top joint is lined with strips of bamboo. The "rice cooker" is placed upright on the fire so that the water in the bottom half is heated; it produces steam which goes through the holes in the partition into the upper joint containing the rice, thus cooking it. When the rice is cooked, it can then be pulled out with the strips that lined the interior of the upper part.

This bamboo is also used to make containers for measuring rice, to make dinnerware (bowls, plates, spoons, ladles, and even mortars and pestles) and pickling barrels, and is also woven to make low tables.

Dendrocalamus membranaceus

Thai: *pai-nual, mai-nuan, pai-naan, pai-sang, mai-lai-lo*; Northern Thai: *saang doi*; Shan: *mai-sang*; Akha: *za cui*; Hmong: *xyong* (XYOOB DAG) (DAG = split); Karen: *waa-mee, wa-mu, wa-mi*; Lahu: *naw van*; Lisu: *ma khaw neh*; Mien: *hau biang*.

This is another fairly large bamboo of the north, with loosely clustered, erect, green culms attaining a height of about 20 meters (65 feet). The culms are 3–10 centimeters (1.2–4 inches) in diameter, with walls up to 1 centimeter (0.4 inch) thick, and internodes that are 20–40 centimeters (8–16 inches) long; there are prominent nodes, each producing numerous branches. The culm sheaths are longer (30–60 centimeters or 12–20 inches) than the internodes, but narrow (13–20 centimeters or 5–8 inches); they are covered with dark brown hairs on the outside and have cilia along their margins. Sheath blades are broadly lanceolate in shape, up to 40 centimeters (16 inches) long, but only 2.5 centimeters (1 inch) broad, and distinctly reflexed. Sheath auricles are small or lacking, but sheath ligules are very conspicuous, denticulate, and fringed with bristles. Leaves are lanceolate in shape, hairless above, 10–25 centimeters (4–10 inches) long, 1.2–2 centimeters (0.5–0.8 inch) broad, and with 4–7 lateral veins.

This bamboo occurs in mixed forests below 1000 meters (3300 feet) elevation throughout Thailand, and westward into Burma and India.

The Karen seem to be the tribe that uses this bamboo to the greatest extent. They eat the shoots, make water carriers and mats, and use it for building houses (splitting the culms for walls and flooring) (Plate 110). They make fencing with it too. Some Karen also make water pipes with this bamboo for smoking tobacco.

The Akha eat the shoots, but feel this bamboo is too small for most other things.

Dendrocalamus strictus, Male bamboo

Thai: *pai-zang, mai-sang, mai-nuan, pai-sang, phai sang nam*; Shan: *mai-sang*; Akha: *ma sa*; Hmong: *xyong*; Karen: *wa-me-pre, wa-mi-loe*; Lahu: *meh shan, mai san*; Lisu: *ma khaw neh*; Mien: *hau jong*.

This is a moderately large, deciduous, dull green or yellowish-green bamboo that reaches a height of 15 meters (50 feet). The culms are 3–8 centimeters (1.2–3 inches) in diameter, with very thick walls and a very small or no cavity; the short internodes are 30–45 centimeters (12–18 inches) long, and with prominent nodes that produce numerous slender, leafy branches. The culm sheaths are variable, but rounded on the top, up to 30 centimeters (12 inches) long, yellow-green when young, and with dark, golden-brown hairs on the outside. Sheath blades are erect, hairy on both sides, and narrowly triangular in shape. Sheath auricles are small and inconspicuous; the ligules are narrow and toothed. The leaves are linear-lanceolate in shape, 3–30 centimeters (1.2–12 inches) long, 1–3 centimeters (0.4–1.2 inches) broad, densely woolly beneath, and with 3–6 lateral veins.

This species occurs in the dry, mixed forests of western and northern Thailand, and extends westward into Burma and India.

Dendrocalamus strictus, a very popular bamboo, is widely used for a variety of purposes because the culms have considerable strength and thick walls. They are used to make mats, flooring, house siding, roof framing, building pillars or posts, slippers, clogs, roofing ties, water pipes for bringing water into a village, water pipes for smoking, scabbards, containers for fermenting rice, drain pipes, chicken cages, containers for boiling water, dibble sticks, crossbow parts, shoulder baskets of various kinds, rice containers, rice winnowers, and various types of animal cages. The fine fibers are woven or twisted into rope.

Kitchen utensils made from this bamboo include trays, drinking straws, tea cups, rice steamers, spoons and chopsticks, mortars and pestles, containers for water and various other liquids, cups, salt and pepper containers, ladles, and sieves (Plate 93). Containers made when the bamboo is still green will not burn when placed in the fire, so it is used to cook rice and curries, and to boil tea and medical potions.

The Akha commonly use this bamboo for making ancestral altars, mouth harps, and drums; the women use it to make the cotton containers which they wear at their waists, as well as the spindles for spinning the cotton and the framework for the loom on which to weave the cotton thread.

The Lahu use this bamboo to hold down thatch roofing. They particularly like to split the stems into very fine strips which they weave into mats and baskets. They make a type of catapult from this species, as well as a long bow. Historically, they made sharpened, fire-hardened, "pungi sticks," which were used in war. They also use this bamboo to make spring traps, fish trap containers, and creels to carry fish at their waists when they go fishing.

The Mien feel this bamboo is one of the best for making the spring of a crossbow. They select culms that are at least two years old and carefully dry them after splitting them to the proper dimensions.

The shoots are very sweet-tasting and greatly preferred by all the hill tribes for eating.

Gigantochloa latifolia

Thai: *naw van*; Akha: *ya coe*; Lahu: *naw van*.

This moderately large bamboo reaches a height of 15 to 20 meters (49 to 65 feet). The culms are up to 8 centimeters (3 inches) in diameter, with walls up to 14 millimeters (0.5 inch) thick; the internodes are 30–40 centimeters (12–16 inches) long and few branches arise from each node. Culm sheaths are 15–20 centimeters (6–8 inches) long and rather thin, with a few dark brown hairs at first. Sheath blades are erect, 9–15 centimeters (3.5–6 inches) long, and 4–8.5 centimeters (1.5–3.3 inches) wide. Sheath auricles curve downward along their edges and sometimes have bristles; ligules are up to 1.5 centimeters (0.6 inch) tall, sometimes incised towards the base. The leaves are oblong to lanceolate in shape, 23 centimeters (9 inches) long, 5 centimeters (2 inches) wide, and with 10–12 lateral veins.

I collected this poorly known species in Chiang Rai Province near the Burma border. It is reported to extend from southern Thailand northward.

The Lahu claim this is a "sweet bamboo" and like to eat it. They also use the culms in a variety of ways, including construction.

Gigantochloa wrayi

Akha: *ya pyeu*.

This bamboo forms clumps and is of moderate size, attaining heights of nearly 20 meters (65 feet). The culms are 2–7 centimeters (0.75–3 inches) in diameter, with fairly thick walls and internodes of average length. The culm sheaths are 27 centimeters (10.5 inches) or more long. Sheath blades are reflexed, green, relatively narrow, 12–22 centimeters (4.75–8.75 inches) long, and 3.5–4 centimeters (1.4–1.6 inches) wide. Sheath auricles are generally low but rising near their outer ends and bearing a few bristles; the ligules are 6–10 millimeters (0.25–0.4 inch) high and deeply divided into lobes, with their tips becoming bristles. The leaves are oblong to lanceolate in shape, 9–40 centimeters (3.5–16 inches) long, 1.2–6 centimeters (0.5–2.4 inches) broad, finely hairy beneath, and with 6–10 lateral veins.

Another poorly known species, this bamboo was first reported from Malaya but clearly extends northward, probably on both sides of the Thai-Burma border to at least the northernmost limits of Thailand. My single collection of this species was made in the mountains west of Chiang Rai.

The Akha use this bamboo for a variety of purposes, including construction, ties, fencing, and food.

Melocalamus compactiflorus

Thai: *phai-hang-chang, mai-hang-chang*; Akha: *ha gui*; Hmong: *shong man lai*; Karen: *waa-boh, waa-baw*; Lahu: *meh eh*; Lisu: *peua cu*; Mien: *hau la-o* (*la-o* = vine).

This bamboo had a spreading and arching, often climbing, habit of growth, and the culms tend to be zigzag in their orientation. The culms are gray-green, up to 30 meters (100 feet) long and 1.5–3 centimeters (0.6–1.2 inches) in diameter; the internodes are 30–60 centimeters (12–24 inches) long and the prominent nodes bear large branches. Culm sheaths are about 15 centimeters (6 inches) long, 6–8 centimeters (2.4–3 inches) broad, hard, brittle, and tending to be glabrous. Sheath blades are long, reflexed, and hairy towards the base. Sheath auricles are crescent-shaped, reflexed, and bearing bristles on the edge; ligules are narrow and entire. The leaves are large, oblong to lanceolate in shape, 15–30 centimeters (6–12 inches) long, 2.5–5 centimeters (1–2 inches) broad, with 8–12 lateral veins.

This climbing bamboo occurs mainly in the dry forests at 1000–1500 meters (3300–4900 feet) elevation in northwestern Thailand and into adjacent Burma.

The culm sheaths, like those of several other bamboos, make the skin itch, so tribal people do not like to work with this plant in the forest. It is generally too small for most purposes, but several tribes use it for basketwork. The Lahu believe it is the best bamboo from which to weave the carrier straps for their baskets.

The Hmong use the growing tips for medicine. They boil the tips in water and use the liquid to bathe body parts as an antiallergy treatment, such as for hives. They also eat the shoots.

The Karen use the long culms as cross pieces in their fences.

Oxytenanthera albo-ciliata, Field bamboo

Thai: *phai-rai, phai-khai, mai-lai, mai-khai*; Shan: *mai-lai*; Akha: *ma la*; Hmong: *xyong chai*; Karen: *wae-kli, waa gleh*; Lahu: *mai lai, ma leh*; Lisu: *peua cu*; Mien: *hau lin*.

A bamboo of confused relationship, it has been included in both *Gigantochloa* and *Oxytenanthera*. Soejatmi Dransfield (personal communication) believes it probably should be placed in *Dendrocalamus* but no formal combination has yet been proposed.

This fairly small, gray-green, dense bamboo makes extensive stands, often at the edges of cleared fields, and with somewhat curved culms up to 10 meters (33 feet) high. The culms are 1.5–3 centimeters (0.6–1.2 inches) in diameter, with fairly thick walls; the internodes are 15–40 centimeters (6–16 inches) long, and the prominent nodes bear numerous slender branches. The culm sheaths are 10–20 centimeters (4–8 inches) long, 15 centimeters (6 inches) broad, very hairy at first, becoming deciduous. Sheath blades are lanceolate in shape, longer than the sheath, and acuminate. Sheath auricles are small and bent; the ligules are 1.2–2.5 centimeters (0.5–1 inch) long, toothed, and truncate. The leaves are linear to lanceolate in shape, fairly small, 15–20 centimeters (6–8 inches) long, 2–2.5 centimeters (0.8–1 inch) broad, and with 6–8 lateral veins.

Oxytenanthera albo-ciliata occurs in mixed forests in both the lowlands and mountains of northern and western Thailand, especially along the Burma border and into that country.

The plants produce extremely hairy young bracts that can cause considerable itchiness when the stems or bracts are handled. However, the shoots are an excellent food, so most tribal people are willing to put up with some discomfort to harvest the young shoots.

This bamboo is used for making bird traps, ties for all sorts of things, knife and other implement handles, dibble sticks, snare springs, ram rods for homemade black powder hunting rifles, cruppers on the harnesses of horses, and props for fruit trees. It is also split and woven into baskets, mats, and similar items.

The Lahu believe this is the most delicious of the bamboo shoots.

The Karen construct small weirs with this bamboo for irrigating their wet rice fields; they simply pound several stems into the ground, creating an obstacle to divert the water from one channel to another.

The Mien like to use it for fencing.

Phyllostachys sp.

Thai: *paang-puk*; Shan: *mai-pang-puk*; Karen: *mep-we*.

Only a single collection of this plant was made near Boh Kaew in Chiang Mai Province. It has been identified by Soejatmi Dransfield as belonging within *Phyllostachys*, a genus not reported by Lin (1968) and others as occurring in Thailand. It is possibly *P. mannii* Gamble, but a positive identification is impossible without reproductive structures.

This bamboo is shrublike and with yellowish culms only 4–6 meters (13–20 feet) long and 2.5–3 centimeters (1–1.2 inches) in diameter. The internodes are 15–20 centimeters (6–8 inches) long and slightly flattened on one side. The culm sheaths are papery, 20–23 centimeters (8–9 inches) long, 2–4 centimeters (0.8–1.6 inches) wide, and rounded at the top. The sheath blades are 2–4 centimeters (0.75–1.5 inches) long, recurved, and narrow; the ligule is broad. The leaves are 10–13 centimeters (4–5 inches) long, 1.2–2 centimeters (0.5–0.8 inches) wide, and with 4–6 lateral veins.

This species apparently occurs at moderate elevations in mixed forests in the mountains of northern Thailand and westward into Burma.

The Karen eat the young shoots and sometimes use culm strips as a substitute for rattan (*Calamus* spp.) when weaving baskets.

Thyrsostachys oliveri

Thai: *pai ruak, phai ruak dam, mai huak*; Northern Thai: *phai raakdam*; Akha: *ka to*; Hmong: *kang tang*; Karen: *waa-bo-suu, waa-suu-der*; Lahu: *mai hi*; Lisu: *ma ca*; Mien: *hau yim*.

This is the tallest of the two species of this genus, with its straight culms reaching a height of 25 meters (82 feet); the culms are bright green in color, turning dull greenish yellow with age, 5–8 centimeters (2–3 inches) in diameter, and with rather thin walls. The internodes are 30–60 centimeters (16–24 inches) long, and the prominent nodes bear several slender branches. The culm sheaths are thin, persistent, green when young but becoming brown with age, densely hairy on the outside, and have margins fringed with cilia. Sheath blades are long and awl-shaped. Sheath auricles are inconspicuous, and the ligules narrow and toothed. The leaves are linear to lanceolate in shape, light green, 17–20 centimeters (6.8–8 inches) long, 1.2–2 centimeters (0.5–0.8 inch) broad, and with 5–7 lateral veins.

This species of *Thyrsostachys* occurs at lower elevations in mixed forests and teak forests throughout northern Thailand and into adjacent Burma. It is not common but is sometimes planted in villages.

Apparently this bamboo is not used extensively by any of the hill tribes. The seeds are eaten in times of famine, if they are available. It requires considerable effort to harvest the seeds, which do not taste particularly good. The shoots are edible but taste slightly bitter. The Lahu claim that a good-tasting grub grows inside the culm.

The Lahu use the small stems to make broom handles, and the Mien use mature culms for rafters because they are so straight.

Sometimes the larger culms are used for house construction.

Thyrsostachys siamensis

Thai: *mai-ruak, phai-ruak*; Northern Thai: *mai-huak*; Shan: *mai-ti-yo*; Akha: *a ma*; Karen: *wae-pang, wae-baang, wa-bo-bo*; Lahu: *ma pfuh*; Lisu: *wa ma*; Mien: *hau tzong* (LAUX CAANGZ).

These deciduous plants have densely tufted, gray-green culms, up to 13 meters (43 feet) high, 3–7.5 centimeters (1.2–3 inches) in diameter, and with very thick walls; the internodes are 20–28 centimeters (8–11 inches) long, and the nodes are inconspicuous. The culm sheaths are up to 28 centimeters (11 inches) long, 20 centimeters (8 inches) broad, persistent, thin, soft, and with a few hairs on the outside. Sheath blades are erect, narrowly triangular in shape, and with recurved margins. Sheath auricles are small or inconspicuous, and the ligules finely toothed. The leaves are linear to lanceolate in shape, 7–15 centimeters (2.8–6 inches) long, 1–1.2 centimeters (0.4–0.5 inch) broad, and with 3–5 lateral veins.

This species often occurs in pure stands, sometimes associated with hardwood trees at moderate elevations, in western and northern Thailand and into Burma. It is commonly propagated in other parts of Thailand.

All the tribal people eat the young, small shoots.

The Lahu use the culms as spools for yarn, make combs for their hair, and fashion small pieces into reeds for one of their most popular musical instruments, which consists of a gourd and series of bamboo sections of different lengths, each with a separate reed.

The Mien often plant this species in their villages, using it for fencing and food. They say that the split culm also makes a good spring for the crossbow.

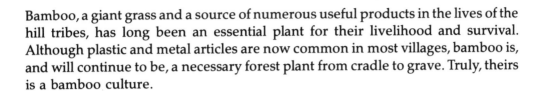

Bamboo, a giant grass and a source of numerous useful products in the lives of the hill tribes, has long been an essential plant for their livelihood and survival. Although plastic and metal articles are now common in most villages, bamboo is, and will continue to be, a necessary forest plant from cradle to grave. Truly, theirs is a bamboo culture.

7

Opium: Cash, Cure, and Curse

Addiction to opium is, unfortunately, a fairly common problem in some tribal villages, especially among the men. These people living within the Golden Triangle, an area of approximately 225,000 square kilometers (86,875 square miles) (Boucaud and Boucaud 1988), have long had the reputation of growing opium as an important cash crop.

The opium poppy, *Papaver somniferum* (Plate 111), was well known to the ancient Greeks, who at first considered it magical, even attributing to it considerable religious significance. Three of their gods, Hypnos, Nyx, and Thanatos, were portrayed with poppies (Duke 1973). In fact, most civilizations of the Near East were familiar with the plant, which slowly took on medicinal rather than religious properties. From the fourth century B.C. onward several famous Western medicinal practitioners, including Hippocrates, Dioscorides, and Galen, recognized opium as an important pain reliever (White 1985).

Opium was first transported to Asia from the eastern Mediterranean region by Arab traders, probably in the seventh century A.D., and early reports from China indicate the plant was used to cure dysentery and to relieve pain. By the seventeenth century opium was still limited primarily to medical use among the elite; the Emperor of China, surprisingly, was more concerned about the use of tobacco, a plant that was introduced to China from the New World only a century earlier, but was widely smoked throughout the country. When he issued a prohibition against the use of tobacco, those addicted to it began to mix opium with the rare and forbidden tobacco. As more and more opium was added, in time many smoked pure opium and became addicted to it. Thus, widespread opium smoking by the Chinese was the result of a prohibition to smoking tobacco (Lewis and Elvin-Lewis 1977).

By the beginning of the nineteenth century millions of Chinese had become addicted to opium, creating a huge market for the drug, which prior to this time, was supplied primarily by the Portuguese. The British, recognizing the huge profits to be made, supported the cultivation of vast tracts of opium poppies in

India and took over this monopoly. British merchants, wishing to obtain large quantities of Chinese tea and silk and now having huge amounts of Indian-grown opium, began to trade extensively, though illegally, with Chinese Hong merchants. Chinese lords clearly saw the terrible effects opium smoking had upon the people, and attempted at various times to halt the importation of this plant product; both the Hong merchants and the British ignored the edicts.

The matter came to a head in 1839 when large quantities of opium were confiscated in Canton and destroyed on the orders of a commissioner sent by the Chinese emperor, an amount estimated to be 21,000 chests of opium with a weight of 2.6 million pounds (1.2 million kilograms) (Wakeman 1975; Lewis and Elvin-Lewis 1977). British merchants complained bitterly to their government, which retaliated by invading the harbor of Canton and quickly defeating the Chinese; this short, uneven conflict has been called the First Opium War. The Chinese were forced by England to legalize opium in China, and even though a high tariff was imposed on the drug, it is estimated that by 1886 approximately 15 million or about a fourth of the Chinese people were addicted to the drug (Boucaud and Boucaud 1988).

For decades the Chinese government attempted to negotiate with the British an end to this terrible trade, but it was not until the first decade of the twentieth century that the British finally stopped shipping opium to China. However, this created another serious problem for the Chinese, one that had slowly been growing until then. With the gradual reduction and final termination of these shipments of opium from British India, millions of Chinese addicts no longer had a ready source of the drug. Seeing this steady rise in the demand for opium, the Chinese themselves resorted to growing the poppies. One of the best regions for its production was in the mountains of Sichuan and Yunnan provinces where the hill tribes lived. Probably the Miao (Hmong) became involved in this crop by the middle of the nineteenth century (Kesmanee 1989).

By as early as 1875 about one-third of the arable land in Yunnan Province was devoted to the growing of opium poppies (Boucaud and Boucaud 1988), and by the end of the century it constituted one of the main crops in southern China and southward into what are now Burma and Laos. This considerable quantity of opium was grown by both the Chinese and the tribal people or minorities. It is estimated that by 1883 the Chinese grew twice as much opium as they imported because there was by that time such a huge number of addicts to support (Tapp 1986).

The Akha claim to have begun smoking opium only about four generations (90 years) ago (Ajopho 1990). As the tribal people subsequently migrated southward into the areas that are now Laos, Burma, and Thailand, particularly following World War II, they brought the seeds and a knowledge of how to grow the poppy. China did not declare opium illegal until 1949.

Thailand did not prohibit opium until even after 1949. However, it must be emphasized that Thai rulers, from as early as 1360, disapproved of its use except for medicinal purposes. In the nineteenth century the government of Thailand set up a state monopoly on opium in an attempt to control the import and sale of the drug. In 1947 it restricted the growing of the poppy to northern Thailand, and in 1959 created the Opium Law, which made the growing, smoking, and trading of opium illegal. Unfortunately, this law helped lead to the greater use of heroin (Kesmanee 1989). Thus, in the first half of this century opium became extensively cultivated in Thailand's portion of the Golden Triangle, and for some hill tribes, particularly the Hmong, Mien, Lahu, and Lisu, it has been the major source of cash income since that time. As a trading economy has developed, many families have resorted to growing opium for cash to buy rice to eat, as well as other necessary items, such as pots and pans, gunpowder, salt, cloth, and meat. Unfortunately, the high price of opium, coupled with the ever-diminishing amount of land available for agriculture, has made opium-growing for cash more attractive than the traditional rice crop in a subsistence economy.

ORIGIN OF THE OPIUM POPPY

The origin of the opium poppy is shrouded in mystery. M. D. Merlin (1984), in an exhaustive study of the plant, believes *Papaver somniferum* actually arose in central Europe and then spread southward during the late Bronze Age into the eastern Mediterranean. However, there is still no consensus among paleobotanists as to the actual origin of this species. Clearly *P. somniferum* and *P. setigerum* DC. are closely related; some suggest they are separate species, while others believe the latter is a subspecies of the former. The relationship of *P. somniferum* with *P. rhoeas* L. is also a problem. Merlin (1984) suggests that *P. rhoeas* was first introduced into the eastern Mediterranean also for human use; subsequently, it was replaced in both its habitat and in human use by the introduction of *P. somniferum*. It is probable that *P. somniferum* entered all the cultures of the eastern Mediterranean and from there spread eastward into the Orient.

The Akha have a legend about the origin of opium. There was once a girl so beautiful that men came from all over the world to court her. Of these many men, only seven gained her affection and became her lovers. One day all seven arrived at her house at the same time. She decided to make love with all of them, even though she knew it would make her die, because it was better than choosing only one man, thus making the others bitter and causing conflict. The girl asked her people to care for her grave, promising to send up a beautiful flower (opium) that would grow from her heart. She also said that anyone who tasted the fruits of this flower would want to taste them again and again. Finally, she warned them to be very careful, for the fruits bore both good and evil (Ajopho 1990).

OPIUM ALKALOIDS

The opium poppy contains more than 25 alkaloids in its latex, of which 6 are important to humans. Alkaloids are basic, heterocyclic compounds, usually obtained from plants; many are physiologically active. Some of them depress, whereas others stimulate the central nervous system. Others are mind-altering in their effects. The six most important alkaloids are described below.

Morphine. The amount of morphine in the latex of the poppy capsule varies widely, from 3 to 23 percent. This alkaloid, which is a powerful analgesic, narcotic, and stimulant, was first isolated in 1803, and has been used as a significant pain-killer by physicians for more than 150 years. Unfortunately, it is strongly addicting. Because of its wide use in the U.S. Civil War, more than 45,000 soldiers became addicted to the drug, the condition being known as the "Soldier's Disease" (Duke 1973). Morphine acts on the central nervous system, producing a drug-induced analgesia followed by a diminished awareness of pain (Lewis and Elvin-Lewis 1977). Commercial morphine is obtained from both the latex and the straw of the poppy plant itself (Duke 1973).

Heroin. This alkaloid is synthesized from morphine by the addition of two acetyl groups. In response to the serious addiction problems of morphine at the end of the nineteenth century, chemists developed a semisynthetic compound, called heroin, which they believed had greater analgesic effects than morphine but without its addicting qualities. However, it soon became apparent that heroin was even more addicting than its parent drug. Although its effects are similar to those of morphine, it also causes disruption of blood flow, infections, and collapsed blood veins (Simpson and Conner-Ogorzaly 1986). Because of its use leading to terrible addiction and the resultant socio-economic effects, the manufacture or possession of heroin is illegal in the United States and most other countries. As a rule of thumb, 4–6 kilograms (8.8–14 pounds) (equals 2.5–4 *viss*, or *joi* in Shan and Thai) of raw opium, depending on the quality of the latex, will produce one *jing* or unit of top-grade heroin weighing 700 grams (25 ounces) (Vinai Pornsakulpaisarn, personal communication).

Heroin is generally manufactured from morphine and can be accomplished with almost-cookbook simplicity by anyone with some training in chemistry. There are five steps in the process: (1) heat equal amounts of morphine and acetic anhydride at 84°C for 6 hours; (2) treat the solution, now called diacetylmorphine, with water and chloroform to remove impurities; (3) drain the solution into another container and add sodium carbonate until crude heroin particles begin to solidify and precipitate to the bottom; (4) filter the heroin particles with suction and purify them in a solution of alcohol and activated charcoal, which is then heated to evaporate the alcohol and leave the relatively pure heroin at the bottom of the container; and (5) dissolve the powder in alcohol, adding ether and hydrochloric acid to make white flakes of heroin, which are then filtered under pressure and dried. This relatively simple process makes "number 4 heroin," which is 80–99 percent pure (McCoy 1972). As a general rule, 6 kilograms (13.2 pounds) of raw opium latex produce about one kilogram (2.2 pounds) of heroin (Davies and Wu 1990).

Papaverine. This alkaloid, first isolated in 1848, constitutes only 0.5–2.5 percent of the latex and has very little narcotic or analgesic action. Rather, it tends to relax the involuntary muscles of the body. The heart rate and blood vessels are so affected that there is actually an increase in blood flow.

Codeine. The most extensively used of all opium alkaloids because of its frequent employment in cough medicines and decongestants, codeine is less of a stimulant of the spinal cord and lower part of the brain than is morphine. Fortunately, it is also far less narcotic, euphoric, and constipating than is morphine.

Small amounts of codeine tend to be soporific or sleep-inducing; however, larger doses may cause considerable restlessness and an increase in reflex excitability. More than 85 percent of the world's legal opium crop is for the production of codeine analgesics (Duke 1973).

Narcotine. This little-used alkaloid is a mild narcotic and spinal stimulant, but it speeds up respiration. After morphine, it is the most abundant opium alkaloid (Duke 1973).

Thebaine. This alkaloid may cause tetanic spasms similar to those resulting from strychnine; it is best described as a convulsant rather than a narcotic. Its main use is as a narcotic antagonist in the treatment of heroin addiction (Duke 1973).

GROWING OPIUM

The opium poppy does well in the region of the Golden Triangle only when it is grown above 850 meters (2800 feet) elevation (Plates 111–113). In fact, the best opium is that which is grown above 950–1000 meters (3100–3300 feet). Many tribal villages are located below 850 meters (2790 feet) elevation, but most of those above that level still choose not to grow the opium poppy. It is presently estimated that fewer than 20 percent of the hill tribe villages in Thailand actually grow opium as a cash crop, producing fewer than 40 tons (36 metric tons) of the raw drug per year on about 3500 hectares (8649 acres) of land. In Burma, on the other hand, vast highland areas are dedicated to this crop, probably in excess of 60,000 hectares (148,260 acres). Their estimated yield is at least 600 metric tons (661.5 tons) (White 1985).

Typically, mountain slopes are cleared and burned during the late winter and spring when there is no rain. Maize is first planted in the future poppy field just as the rains begin in May. In September and October the maize is harvested and the poppy seeds are sown in the soil already prepared and cleared of most weeds for the maize crop. The maize stubble provides some protection for the young seedlings, but as the poppies grow taller the maize stems are sometimes gently removed. The poppy fields continue to be carefully weeded.

Today there are few large poppy fields in Thailand, only small ones of a few hundred square meters or less. Often growing with the distinctive bluish green, waist-high poppies and maize stubble are many other plants, such as peaches, Chinese pears, wild beans, taro, sorghum, green peas, and a variety of herbs and spices such as ginger, coriander, lemon grass, mustard, fennel, and mint. Marijuana is often grown as a border plant in Hmong poppy fields.

About the end of December the opium poppies begin to flower, making a lovely sight with colors varying from pure white to deep reddish purple. Some flowers have fringed petals; others are variegated red and white (Plates 111, 113). The tribal people carefully watch the flowering process because the season for harvesting the valuable latex occurs for only a critically short period of time. When the capsule turns from green to slightly gray-green, it is ready to cut. This period lasts from only a few days up to a week or so after the last petals drop from a flower, depending on weather conditions, so everyone must work rapidly to get as much latex as possible (Plate 115). Some hill tribes, such as the Mien, stagger the planting of poppy seeds so that the flowers in a field will mature over a longer period of time. Others, like the Hmong, try to plant an entire field at the same time, thus creating a tightly compacted harvest period. They claim that, although the work load is intense, it is easier to simply move through the field tapping all the plants rather than having to select only the mature ones in several sweeps of the field.

The actual tapping process is carried out with a special knife consisting of three to five razor-sharp blades bound together so that one stroke of the knife on the maturing capsule makes the same number of incisions, each about one millimeter apart, to release the milky-white latex (Plate 114). Both men and women tap poppies as harvest must be done at a critical period of time. In fact, labor can be critical and often members of other tribes, especially addicts desperate for money or opium to feed their habits, are hired to assist in the tapping and harvesting. The laborers carefully move backwards through the field making a single stroke of the knife on each young fruit with one hand while gently holding the bobbing capsule with the other (Plates 115–116). Harvesters must be careful not to cut too deeply, allowing the precious latex to ooze into the interior of the capsule, or to let their clothes come in contact with the fresh latex, which sticks readily to anything it contacts. Almost immediately a thick, white latex begins to ooze from the two or three slits; this slow bleeding of the plant's alkaloid-filled liquid continues through the night (Plate 117).

The following day laborers return to the field with special scrapers with which to remove the viscous latex that turned a dark yellowish brown overnight. A single expert stroke of the spatula removes the night's accumulation of latex, which is then scraped onto a piece of bamboo or banana leaf (Plates 118–120).

The cutting and harvesting process is repeated three or four times for each capsule in as many days. Each day's harvest is carried from the field to the village where it is carefully bundled in a packet weighing a *viss*, covered with paper made from mulberry bark (*Broussonetia papyrifera*), and tied (Plate 121). The value of such a bundle may be up to U.S. $175 and requires approximately 3000 poppy flowers to produce (White 1985). Tribal people calculate they can produce about one kilogram (2.2 pounds) of raw opium per rai (6.25 rai per hectare or 2.47 rai per acre) (Richard Mann, personal communication). The price of raw opium fluctuates widely from year to year and production levels are inconsistent due to climatic effects and the number of years a crop has been grown on a piece of cleared land. Production is high the first year, but drops considerably the second; usually it is not worth planting a crop on the land after the third year because of reduced fertility and a terrible weed problem.

The farmers allow the capsules containing huge numbers of tiny seeds to mature. Fruits from the most-robust plants are then harvested and the seed saved for planting the following year. The hill tribes do not use the seed for food or oil as is commonly done in the Near East (Duke 1974).

The cultivation of opium is now illegal in Thailand, and tribal people are subject to considerable pressure from the government to refrain from growing it. Despite this prohibition and the threat of having their poppies destroyed, about 40 tons of raw opium are still produced annually in northern Thailand. The government has been relatively lenient if a few people in a village grow a rai or two of poppies "for local consumption," that is, as a medicine; it also allows older, incurable addicts to have a small plot for their own use. Unfortunately, some younger tribal people prefer to smoke heroin rather than the cumbersome, messy raw opium. They simply take a plug of tobacco out of a cigarette and insert a small amount of heroin costing less than a dollar. Thus, they do not use needles as do Western addicts. In a sense, they have returned to the sixteenth-century custom of mixing tobacco and opium, but with a much more addicting substance.

Drug dealers pressure hill tribe families to grow opium poppies. The Haw Chinese and the Shan are among them. An Akha headman near the Burma border became the victim of a Chinese drug dealer who told the headman he wanted the village to grow opium for him. The Akha firmly refused, saying that his village was "clean" and that he wanted nothing to do with this illegal crop. The furious drug dealer drew a pistol and shot the headman before fleeing. Fortunately, the Akha did not die from the wounds, but knowing that he and his family were in mortal danger, they hurriedly moved to Burma to avoid further problems. These pressures from drug dealers continue to exist, especially in the more isolated villages, which can be easily visited by representatives of the opium war lords, and which are without any protection from the Thai border patrols.

OPIUM SUBSTITUTION PROGRAMS

In the late 1960s the Thai government made a survey of opium growing in their portion of the Golden Triangle. The findings shocked many, and officials began seeking ways to curtail poppy growing by the hill tribes. In 1973 a pilot program called the THAI/UNFDAC Crop Replacement and Community Development Project was begun under the auspices of the United Nations and with the blessing of the Thai government. Its goal was to induce the tribal people to grow substitute crops for opium, thus allowing them to continue to reap some cash benefits from their farming. In addition, the UN program included drug treatment, rehabilitation, and education. At first it involved 30 tribal villages in five key areas in the north. Roads were constructed and a variety of crops introduced; agricultural experts trained the tribal people on new farming techniques, and reforestation projects were initiated. One of the most important thrusts of this program was the creation of a marketing infrastructure to allow the new products to be transported and sold in lowland markets. In 1980 the program was changed from the UN Narcotics Division to the UN Development Program and was called the THAI/UN Highland Agricultural and Marketing Project (HAMP). In 1985 the United Nations withdrew from the program, but it was taken over in large part by Norwegian Church Aid using most of the same personnel. Portions of this program still continue (Mann 1990). Of the various substitute crops introduced, the most productive (and profitable) have been coffee, red kidney beans, taro, potatoes, and vegetables such as cabbages and tomatoes. The Thai-Norwegian Church Aid Highland Development Project, as it is now called, has been one of the most successful of the opium replacement programs; administrators of the program reported that within the project area, the total land on which opium was grown dropped from 5194 rai (831 hectares or 2000 acres) in 1984 to a mere 157 rai (25 hectares or 60 acres) in 1988 (Thai-Norway Highland Development Program 1988).

Some of the other opium replacement programs, financed from a variety of sources, have met with varying degrees of success. Often, major factors in unsuccessful programs have been lack of available roads and markets rather than a hesitancy on the part of tribal farmers to convert to other crops. The hill tribes are acutely aware of the difficulties in growing poppies and the dangers inherent in its illegal trade. As soon as they see that cabbages, coffee, potatoes, or beans can be grown for a profit, they are often willing to change to the new crops.

USE OF OPIUM BY THE HILL TRIBES

Opium has not been used traditionally as a mind-altering drug in the sense of Western drug cultures; rather, it has been used primarily as a medicine.

Smoking is the usual method by which the tribal people use opium. The pipe has a clay or wooden bowl at the end of a hollow bamboo stem or the hollowed-out woody stem of a *Capsicum* (chili) plant that is several years old. Each addict usually has his or her paraphernalia for smoking opium; it consists of a wooden or metal tray, a small container for the raw opium, a small lamp (or candle if it can provide a steady enough flame) to produce the necessary heat, a small metal skewer, and a small curved spatula. Animal or vegetable oil (from opium or sesame seeds usually, but also from seeds of *Ricinus communis* or *Jatropha curcas*) is also needed for the lamp, but poverty forces some addicts to use kerosene or candles, which produce toxic vapors that are inhaled (Ajopho 1990). This paraphernalia is highly cherished by the smokers because it permits them to gain the greatest benefit from the drug. As an addict slips farther and farther into poverty, it may be necessary to sell or trade some of these items for the drug. Thus, severely addicted individuals frequently use poor equipment or even borrow items from others. The addict may even resort to mixing aspirin or other inexpensive analgesics with the opium to distend the blood vessels, improve the absorptive capacity of the lungs, and increase circulation (Wongsprasert 1989).

Addicts prepare to smoke by first rolling the raw opium into small balls with their hands. Each ball weighs about 3 grams (0.1 ounce) and is just enough for a single smoke. He then lies on his side by the candle or lamp, places the pea-sized ball of latex on the tip of the skewer, and slowly heats it over the flame until it begins to bubble and smoke. Quickly and deftly he places the ball into the hole of the pipe bowl, puts the stem in his mouth, and moves the bowl to the flame to allow the opium to continue to heat (Plates 122–123). In a moment or two the material vaporizes and the addict quickly and deeply inhales the heavy white smoke. Experienced smokers are able to trap the vapors in their lungs by closing the glottis, thus ensuring the maximum absorption of the drug into the blood stream through pulmonary alveoli. The immediate effects of opium include making the smoker pleasantly drowsy, yet seemingly with an increased sense of awareness. The worries and distractions of daily life slip away; smokers feel intensely aware of their surroundings and believe they can actually sense things that are imperceptible. However, heavy doses result in lethargy, slurred speech, and poor muscle control. If more than the usual dose is consumed, a deep sleep results (Westermeyer 1982). Soon the effects of the drug wear off and the user is plunged back to reality. Another pipeful is smoked and the user once again ascends away from worldly problems, thus making an endless circle of greater and greater drug dependence.

Up to seven of the small balls can be consumed in only an hour and a half, and severe addicts may smoke up to 20 pipefuls during the day. Sometimes tobacco or even dry shavings of bamboo are mixed with the ball of latex to facilitate its burning and vaporizing. Occasionally a small spatula is used to clean the sticky, unburned deposits from the bowl of the pipe.

Raw opium may be eaten rather than smoked, though most addicts claim that smoking is more enjoyable. In fact, more of the opium's active compounds are taken in by the body when it is chewed, so if the latex is eaten, nearly all the morphine and codeine it contains is absorbed in the gastrointestinal tract (Westermeyer 1982). The small balls of opium are eaten directly, or dissolved in warm water or rice whiskey and drunk. Occasionally, the latex is dried and taken as snuff through the nose.

When children are given opium for medicinal purposes, it is usually dissolved in water or given orally in pellet form, or an adult blows opium smoke in the child's face to inhale. Many tribal people still consider opium to be the best analgesic available, and they use it much as Westerners take aspirin. Most tribal people believe that it is a very effective treatment for pain, diarrhea, cough, and sleeplessness.

Tragically, opium is quite deadly when taken in large quantities. On occasion Hmong girls have used opium to commit suicide when they were forced to marry someone they did not love (Boyes and Piraban 1989).

ADDICTION

For most tribal people, opium is neither a recreational nor a spiritual drug, and addiction is considered a serious social and personal problem. Unfortunately, people of all ages are susceptible to addiction, which can occur following a serious illness or injury in which the drug has been used for a period of time.

With addiction comes a literal physical craving for the drug. Withdrawal effects are horrible, and include vomiting, stomach cramps, and often excruciating pain in the head and other parts of the body. One addict described his opium deprivation as "freezing to death while heaped with blankets by a blazing fire in midsummer" (Fay 1975). It is no wonder that the addict will often resort to violence to secure more of the drug.

Tragically, opium addiction has become fairly widespread among the hill tribes, especially the men (a ratio of about ten men to one woman). Current estimates are that nearly 9000 people in the six main tribes, or about 1.7 percent, are opium addicts. Fewer than one percent of the Karen and Lisu are addicted, but well over 2 percent of the Hmong, Lahu, and Mien, and over 4 percent of the Akha, are addicted (Viriyapanyakul 1990).

Older tribal people remember the pleasures of opium smoking and how it helped them relax after a long walk or a hard day in the fields. Some claim that opium smoking made a person mild and gentle; women especially liked this aspect because their husbands treated them less harshly and did not seek a second wife. Addicts often became the dreamers and thinkers in the village. Some of the elders even believe that opium smokers contracted illnesses less frequently. They also thought that opium was an effective treatment when someone was bitten by a snake or stung by a scorpion or wild bees. Years ago many addicts were arrogant, believing that one could not be "fully a man until you are able to smoke opium" (Ajopho 1990).

However, the realities of the late twentieth century have made opium addiction a much more serious problem within the tribal societies than might be perceived in the previous description. An ever-increasing number of tribal people, especially men, have fallen victim to opium as a means of escaping the oppression and cultural upsets they are experiencing. It is no longer a means of showing one's manhood; most likely, it is the most troubled, least-adaptable of the highlanders who become addicted. Like the Native Americans who succumbed to alcohol in large numbers as a result of their loss of identity and forced acculturation, the hill tribes of Thailand also seek to escape the degrading and discriminatory practices they often experience; many think that alcohol or opium can make this possible.

Westermeyer (1982) reported that over 50 percent of the addicts treated in a Laotian treatment center claimed they had become addicted because of curiosity. This would not seem to be the case in Thailand among the hill tribes. Probably half of the addicts among the highlanders became addicted after an illness or injury in which the drug was used to treat ulcers, tuberculosis, intestinal problems, and pain. Psychological problems, including loss of land, poverty, marital problems, low life expectancy, and discrimination, have led others into addiction.

Peer pressure to experiment with opium does not seem to be a significant factor among tribal men and women, probably in fewer than 10 percent of addiction cases (Wongsprasert 1989). Rather, it seems clear that there are strong social pressures among most of the tribal people to avoid opium smoking, including even the expulsion from a village for doing so. Indeed, addicts often gather together for support, but this seems to occur only after they have fallen prey to the drug. These people can be easily recognized in a village because of their sallow complexion, extreme thinness, and usually ragged clothing.

In one Hmong village two addicts were attempting to run a small store, but villagers complained that often they were absent or there were few things to buy. Unfortunately, most addicts are unable to care for their own fields of rice or maize, so often their only source of income is to hire themselves as laborers or to borrow from relatives. The families of addicts slowly sink into ever-greater poverty as the victim sells everything of value for the drug. In some cases the spouse and children abandon the addict, hoping to survive without the constant financial drain of getting more opium. Thus, addiction can be a serious problem for the family or even the village in which there are addicts.

Opium is a curse for many, yet for others it may be the difference between starvation and survival. The sale of a few *joi* of the raw drug may provide just enough money to buy rice when a harvest has run out and there is nothing to eat. Or the money might even be used to purchase school uniforms and books for a child. There is no simple solution to the problem of growing opium by the hill tribes, yet the crop replacement programs and other economic incentives show at least limited success. Perhaps, in time the opium poppy's curse may be only a memory for most of the hill people in northern Thailand.

8

Plants That Cure

I was tired, dirty, and frustrated. The day had not gone well and I was stranded in an Akha village because of the heavy rains that had fallen throughout the day. I had tried to get to the forest to collect samples of the medicinal plants used by Law Jo, the headman, and his wives, but the trail was so muddy and slippery that I collected only about a dozen specimens before returning to Law Jo's home (Plate 124).

Suddenly Mark, my Akha interpreter, came around the corner of the partition dividing the interior of the house. He asked me if I would like to see Law Jo heal a woman with a sore stomach. I quickly followed him to the other side of the partition, where the headman was working by the fireplace and a woman lay on her side on the floor. Law Jo mixed a solution containing ginger root (*Zingiber officinale*), which they call *tsaw tsui*, in an enamel basin. A piece of sheet metal lay on the floor near the hearth and the metal part of a large Akha hoe was in the blazing fire. Dropping two silver coins in the basin, he removed the red-hot hoe blade from the fire with some metal tongs and placed it on the piece of sheet metal. After dipping his bare foot in the solution in the basin, the headman stepped lightly on the red-hot hoe blade and then placed his foot on the woman's stomach. It was almost unbelievable, for the sole of his foot was not burned when he placed it on the red-hot metal! Law Jo repeated this amazing procedure several times before stopping, declaring that the woman's stomach ache would soon go away.

This example is just one way the hill tribes use plants to cure illnesses or injuries. In other villages I have watched men and women boil tea or make poultices of plant material for curing a vast number of ailments. Such importance is placed on healing plants that when I asked tribal people to show me the plants they use, nearly everyone assumed I wanted to see medicinal plants only. I had to specifically ask them to point out or collect plants used for food, fibers, dyes, construction, and other things. Indeed, theirs is a huge pharmacopeia of plants that are used medicinally, and I have now collected and identified nearly 700 medicinal plants used by the hill tribes. Clearly, it is impractical to list or describe all or even

127

most of these plants in this chapter, though they are all listed in the appendices. Rather, I want to consider several common ailments which afflict the hill tribes and describe the procedures they follow to cure them with medicinal plants. I will also delve into the fascinating use of plants during pregnancy, childbirth, and those critical days immediately following it. However, first it is appropriate to consider briefly the concept of disease believed by most of the tribal people in northern Thailand.

THE HILL TRIBES CONCEPT OF DISEASE AND CURING

Disease and injury in northern Thailand—and the processes of curing them—involve parts of folk, traditional, and Eastern healing systems according to Western medical researchers (Pake 1986). Folk medicine is readily accessible, trusted, inexpensive, and successful. As it has long been an important component of the cultures of the hill tribes, folk medicine is therefore well-known and more readily understood (McKenzie and Chrisman 1977).

The health systems of the hill tribes are described as traditional as opposed to modern; that is, medical practitioners attain their skills informally, often through extended apprenticeships of many years, whereas modern medicine involves extensive formal training (and also a very formal apprenticeship). Traditional systems are also described as non-Western in scope (Jones 1979).

The hill tribes concept of disease and curing is Eastern in approach. It tends to be more holistic than that of Western medicine, which clearly separates the mind and body. Western medicinal practices also sanction subduing nature by human action, but Eastern medicine encourages people to live in harmony with nature; in fact, to many people in the East, good health is the result of a harmonious relationship with nature. Significantly, Eastern medical practitioners use whole natural plant and animal products as medicines whereas Western practitioners rely primarily on synthetic pure drugs (Pake 1986).

The hill tribes concept of disease and curing is much less sophisticated than the concept of the Northern Thai, who have developed an extremely elaborate system of traditional medicine (Brun and Schumacher 1987).

Generally, the hill tribes (and many other Southeast Asians as well) believe there are three causes of disease: natural, supernatural, and metaphysical (Tung 1980).

Natural Causes

Natural causes of disease and illness are obvious things such as knife or bullet wounds, broken bones, and food poisoning from eating bad food, like rotten meat. Every household has an arsenal of herbs with which to combat these types of ailments.

Supernatural Causes

Supernatural causes of disease and illness are the result of the actions of spirits. For example, the Akha believe that people will be healthy if they are "spiritually potent," but sickly and often attacked by the spirits if they are "thin-souled" (Lewis 1968–70). Spirits, in other words, seem to be one of the main causes of disease, and most tribal people believe that spirit propitiation is necessary once the spirit causing the problem has been identified. Most hill tribes therefore employ a variety of rituals to divine the spiritual origin of a disease. For example, the Karen use chicken bones, feathers, eggs, and rice grains, whereas the Lisu "read" the lines of a pig's liver or the alignment of holes in the leg bone of a chicken (Lewis and Lewis 1984). Tribal people do not feel that the spirits of nature are malevolent unless they are offended; much time and effort is thus spent keeping the spirits happy. Evil spirits, on the other hand, are expected to cause problems unless a person is protected by ancestral spirits. Thus, most highlanders continually offer gifts to their ancestors, honoring them and seeking their continued protection from the evil spirits.

Metaphysical Causes

The tribal peoples' belief in metaphysical causes are clearly the result of Chinese influence (Tung 1980). Arising from the Taoist concept of Yin and Yang, which includes the idea of a balance between hot and cold, some tribal medicines are said to be "hot" or "cold." Herbal medicines are cold whereas Western ones are hot (Tung 1980). Hot diseases must be treated with cold medicine; cold diseases, on the other hand, must be combatted with hot medicine. One Lisu medicine man, commenting on a "five-month medicine," which is made by boiling the roots of *Desmodium oblongum,* said it is taken to "cool the body down." This does not necessarily mean that the victim has a fever; rather something has made the body "hot" and to cure this hot disease the victim must take a cold medicine. The Lisu also make a tea, by boiling the roots of *Clerodendrum fragrans,* for the patient who is hot inside.

CURING SYSTEMS

The curing systems of the hill tribes involve two important parts. First, they contain the actual medicinal treatment of the ailment, usually employing herbs alone or a combination of plant and animal materials. Second, the healing system as a whole takes place within a well-understood cultural context, thus giving definite social and personal meaning to the experience; this is strong evidence that tribal practices are, indeed, folk medicine.

HEALTH CARE SYSTEMS

Most hill tribe health care systems have three phases or arenas in which illness is experienced and responded to: popular, folk, and professional (Kleinman 1978). These three phases are clearly evident in Akha medicine. The first or popular phase is strictly nonritual and takes place by and within the household; it may account for more than 70 percent of medical treatment within a village. Medicines of this phase are understood by many tribal people, both men and women, and include herbs, minerals, and other substances for the treatment of common ailments, such as colds, infections, cuts, bites, and sprains.

The folk phase involves extensive ritual practices performed by specific practitioners, such as priest-doctors, who use herbs and even manipulation as significant parts of the treatment process. The knowledge of these cures is restricted to only a few people.

The third or professional phase is what might be conceived of by some as magical, and in these instances the medical practitioner uses mysterious, unrecognizable cures, usually involving elaborate rituals and even trances. The Akha believe pills, injections, and modern treatments with Western medicine belong in this phase (von Geusau 1979).

The introduction of Western medical practices among the hill tribes has clearly modified the traditional ways in which they have dealt with illness and injury. Lewis and Lewis (1984) describe four levels of curing that are now found in northern Thailand among the hill tribes. The first is ritual curing, as described above; this level also includes finding the spirit responsible for the ailment. The second level of curing is traditional or nonritual medicine; this level extensively involves herbal cures. The third level is that of the "injection doctor" or "injectionist." In recent years Chinese and Thai quacks have begun to travel from village to village selling injections of one type or another. These "magical" drugs are almost revered by the tribal people, but often serious problems arise because the "doctors" have little knowledge of what they are doing and of the serious side effects that may be caused, especially considering what can now be transmitted by a dirty needle. The fourth level of curing is that of legitimate Western medicine, which has been steadily gaining in acceptance as it has become more available. Government health teams and church-trained paramedics now regularly visit many tribal villages, and, with the improvement of roads, more and more highlanders are going to lowland Thai hospitals for treatment of serious injuries and diseases. It needs to be emphasized that more than one of these levels may be utilized; a frantic tribal person seeking to save a loved one may resort to a combination of these cures, hoping that at least one will lead the victim to recovery.

Western medical practitioners are often skeptical about the curative powers of herbal remedies. Although modern scientific proof of the efficacy of these tribal herbs is often lacking, investigations are taking place in Thailand, China, India, Japan, and the United States. A few of the plants used by the hill tribes are also used by the Chinese and Indians, who have a long history of traditional medicine involving herbal cures. Indeed, many Chinese and Indian herbal drugs have been evaluated scientifically and found to contain clinically active compounds. A major computer data base, called NAPRALERT, has been created by the School of

Pharmacy at the University of Illinois, Chicago, to record what plants are used in folklore, whether these plants have been tested biologically, and what chemical compounds have been found within them. Several hundred plants listed in this data base are used medicinally in Thailand, but only a few are used by the hill tribes. However, the small number of scientific studies of traditional Southeast Asian medicinal plants that have been made and their long use by tribal people and others suggest that many medicinal plants contain pharmacologically active substances with real curative properties and that they should be further investigated.

TRIBAL MEDICINAL KNOWLEDGE AND ITS ACQUISITION

Like practitioners in other traditional medical systems, hill tribe practitioners are trained informally and have lengthy apprenticeships. Usually they attain the knowledge of how to use herbs to heal early in life and practice this skill until they die. Most go to the forest themselves to gather the materials, for they want to be sure they collect the correct things (Plates 125–127). However, when they become too old to gather the plants themselves, they often send a trusted son or another relative to collect the materials. Slowly this younger person learns which plants to use and how to properly mix and prepare them. Some herbal practitioners who wish to keep their knowledge secret from others swear the relative to secrecy, even teaching him or her to grind up the plant material so finely that no one can recognize it.

Nowadays it is sometimes difficult for elderly medicine men and women to find younger people who want to apprentice themselves. One Lahu Sheh Leh medicine man said that he had tried to teach his younger sister-in-law about traditional medicines but said that she "did not have the right temperament." He therefore was trying to teach a niece about medicinal plants and how to use them. This lack of interest or ability in finding new practitioners certainly does not bode well for the future.

AVAILABILITY OF MEDICINAL PLANTS

Not only are there fewer and fewer people who know the traditional uses of healing plants, but the medicinal plants themselves are harder and harder to find. A Lahu elder commented that in the past many medicines came from large trees, but now so many forested areas have disappeared that many of these trees are gone. He believes at least 19 of his 20 favorite medicinal trees no longer occur in the area around his village. However, he also pointed out that many of the simple, most commonly used medicinal plants are cultivated or grown around the village. He was also happy to report that the tribe carefully protected some forested areas where other "special" medicinal plants grew.

A Karen medicine man said sadly that he must go farther and farther from the village to find the plants he wants because they tend to occur in less disturbed areas. Some valuable plants are right around the village, but they are only useful for simple medical problems, such as headache or stomach ache; the better plants are far away in undisturbed fallow fields or forested areas.

A Mien elder remarked that medicinal plants used to be a very important part of tribal life, but that traditional medicine is followed less and less for two reasons. First, many of the plants are no longer available, and second, most people in his village are more interested in Western medicines, which are now available. The few people in the village who know how to use herbal medicines do not actively practice healing and must be sought out.

PREPARATION OF MEDICINAL PLANTS

There does not seem to be any distinct method by which a tribe prepares or uses plants for curing, although the Hmong think the best medicinal plants occur along mountain streams. The Mien, in contrast, prefer to gather medicinal plants along open mountain trails. Many Hmong medicinal plants are boiled in water and drunk as a tea, or soaked in whiskey, which is then drunk. Other tribes prepare their medicines in a wider variety of ways.

Tribal medicinal practitioners use all types of plant parts, including roots, stems, leaves, growing tips, flowers, fruits, and seeds. Sometimes entire herbs are utilized, including the roots.

The main types of preparations are as teas or infusions that are drunk; by boiling the herbs and using the liquids to wash the body part; as poultices and juices made by mashing up the whole plant or squeezing a selected part of it (Plates 128–129); by eating the plant or plant part, usually with eggs or some other food; and by boiling the plants for steam baths and inhalants. In some cases a single plant makes up the cure, but in other cases several plants are mixed. Occasionally, as many as eight or ten plants are mashed up as a poultice or boiled together to make a tea which is drunk or a liquid with which to bathe the body. However, some tribes have superstitions about mixing plants; the Akha, for example, never use an even number of plants in a mixture.

CONSISTENCY OF USAGE OF MEDICINAL PLANTS

Pake (1986) observed that Hmong herbalists do not share their knowledge of plants with one another and therefore often do not use a given plant for the same purpose or in the same way if treating the same ailment. I noted the same thing early in my study; in fact, plant use, both within a tribe and among the hill tribes, is surprisingly inconsistent. Two evident exceptions are the introduced plants *Eupatorium odoratum*, which is used by all tribal people as a blood coagulant, and *Aloe vera*, which is rubbed on burns (Plate 130).

I have also compared parts of my list of medicinal plants with other references dealing with medicinal plants of Asia and elsewhere (Perry 1980; Department of Medicinal Plants 1982; *A Barefoot Doctor's Manual* 1985; Duke 1985; and Hsu et al. 1986). The list was also compared to those plants listed in NAPRALERT for Thailand. Nearly 80 percent of the medicinal plants in my data base were not mentioned in the above sources as having the same medicinal uses, and more than 35 percent of the plants were not listed as having medicinal properties in any of the references. I have identified 40 plant species that are used to treat malaria, but 10 of them are not mentioned in any way in the above six references. Even more interesting is the fact that only one of the 40 plants (*Alstonia scholaris*) is listed by any of the references as used to treat malaria. Ten of the 13 plants that I have identified as used for nausea and vomiting are listed by the six references, but only two species, *Morinda tomentosa* and *Plumbago zeylanica*, are listed by any of the six as used for those specific ailments. Clearly, the medicinal plants used by the hill tribes are quite different from those used elsewhere, partly because of the distinctiveness of the flora in which the hill tribes live, but also because few researchers have extensively studied these people previously or their close relatives in southern China. Probably there are a number of plants worthy of extensive research and analysis.

THE TRIBAL TREATMENT OF AILMENTS

Few of the tribal people have any formal education in health and medicine; consequently, their approach to disease, injury, and healing is quite unsophisticated, meaning that they do not know the names and symptoms of most diseases, but rather think in terms of ailments or afflictions. Obvious exceptions are certain well-known diseases such as malaria, chicken pox, polio, venereal diseases, and the common cold.

In gathering information about how the tribes use plants for medicinal purposes, I therefore asked for what purpose a particular plant was used; in most cases, the answer was in terms of an ailment or symptom. My data base was constructed so that medicinal plants could be analyzed from a list of 75 ailments, based in part on a list created by the International Rescue Committee (Keith Feldon, personal communication). In fact, a few of the categories used in the total list were not strictly ailments, such as breast milk stimulant, massage, fish poison, post delivery, or animal medicine. Nearly 700 plants have now been identified as used by the hill tribes for medicinal purposes, so it is impractical to describe all their uses in this chapter. Instead, I have selected a few ailments or injuries that the highlanders treat with plants and will discuss them in the rest of this chapter. Then in Appendix 2 I have listed each medicinal plant, the tribe that uses it, the ailments treated with it, the part used, and what method is used to prepare medicine from the plant.

Malaria

One of the most debilitating and deadly diseases afflicting the hill tribes is malaria, and virtually everyone knows its symptoms. Although much of northern

Thailand is now free of this disease, it is still widespread in the Shan states of Burma where the Akha, Lisu, and Lahu still live; many pass back and forth between the two countries. When more than 2000 Akha were forcibly repatriated to Burma in 1987, almost everyone who went to Keng Tung State, Burma, contracted malaria (Paul Lewis, personal communication).

Forty species in 32 plant families are used by the hill tribes to treat malaria. My Mien informants did not show me any plants used for this disease, and the Akha have only two: *Clerodendrum serratum* and *Desmodium triquetrum*. With both species the Akha boil the roots and drink the bitter infusion. The Hmong and Lahu also believe that *Clerodendrum serratum* is an important malaria medicine.

The Hmong utilize numerous plants to treat malaria. *Blumea balsamifera* is not taken internally, but the entire plant is beaten into a pulp, which is wrapped around a wrist and the opposite ankle; the Hmong do the same thing with *Solanum erianthum* and a mixture of *Citrus maxima, Clausena excavata, Cuscuta reflexa, Polygonum barbatum*, and *Prunus persica*. The Hmong also prepare infusions from *Paederia wallichii, Psidium guajava* (the leaves of which are mixed with the leaves of *Punica granatum*), *Saurauia roxburghii, Scoparia dulcis*, and *Viscum articulatum*, which they drink. They make poultices from *Eryngium foetidum, Lonicera macrantha, Oroxylum indicum* (particularly good for treating an enlarged spleen due to malaria as is *Mussaenda pubescens*), and *Plumbago zeylanica*. They prepare another malaria medicine by mixing the pulverized leaves of *Euodia glomerata* or *Houttuynia cordata* with eggs, steaming the concoction, and eating it.

The Karen boil the bark of *Alstonia scholaris* and drink the liquid to combat malaria. They also chop up the stems and roots of *Cleidion spiciflorum*, boil, and drink the liquid; boil the bitter tasting of roots of an undetermined species of *Rauvolfia* and drink the liquid; and boil the whole plant of *Trichosanthes tricuspidata* and drink the bitter-tasting liquid.

The Lahu treat malaria by boiling together *Brassaiopsis ficifolia* and *Pteris biaurita*, and allowing the patient to breathe the vapors in a steam bath. The Lahu also drink an infusion from the bark of *Croton robustus*, but boil the roots of *Helicteres elongata* (bitter-tasting), *Melastoma normale, Millingtonia hortensis, Polyalthia cerasoides* (the liquid is extremely bitter-tasting), and *Trema cannabina* (which is also bitter-tasting) for other infusions.

The Lisu report boiling the roots of *Cassia occidentalis, Chloranthus elatior*, and *C. nervosus*, and drinking the liquid. They boil the leaves of *Eupatorium odoratum* or *Gnetum* and use the liquid to bathe a person feverish from malaria. *Sambucus javanica* plays a significant role in the Lisu treatment of malaria: its roots are boiled and the tea drunk to stop the fits and seizures caused by prolonged illness with malaria.

Many tribal people are familiar with quinine and other synthetic Western drugs used for the treatment of malaria, most of which are bitter-tasting. It is no surprise, therefore, that many of the plants that they use in their treatments are also bitter tasting. The bitterness probably indicates the presence of alkaloids. Some of these substances may have therapeutic effects and further investigations are certainly warranted.

Broken Bones

Few drugs are used for healing broken bones in Western medical practice, but people in various regions of the world who practice traditional medicine use herbal remedies. For example, the leaves of *Ehretia cymosa* are used to heal fractures in Ghana (Lewis and Avioli 1991).

A well-educated Lahu headman commented to me that he was particularly impressed by the large number of herbs which are used for the treatment of bone fractures by the hill tribes. Having worked as a paramedic on a church-sponsored health team, this man strongly believes that scientists should seriously look into this matter. He emphasized that even compound fractures can be healed with some of the plant-derived medicines used by the highlanders.

The hill tribes use over 50 species of plants to treat broken or "aching" bones.

Achyranthes aspera **(Amaranthaceae)**. The Mien crush the leaves and use them as a poultice for painful bones.

Amalocalyx microlobus **(Apocynaceae)**. The Mien boil the whole plant and drink the liquid to treat broken bones, separated tendons, or simply very tired muscles. They also bathe the sore parts with this liquid.

Artemisia atrovirens **(Asteraceae)**. The Akha make a poultice from the leaves, which they put on broken or aching bones.

Blumea membranacea **(Asteraceae)**. The Lahu boil the leaves with the stems and drink the liquid to alleviate the pain in bones. The plant is also used to treat bladder infections, fever, stomach pain, and vomiting from eating bad food.

Bryophyllum pinnatum **(Crassulaceae)**. The leaves of this plant are made into a poultice which is placed by both the Akha and Mien on a broken bone.

Buddleja asiatica **(Loganiaceae)**. The Akha crush the leaves and apply them as a poultice to broken bones.

Cassytha filiformis **(Lauraceae)**. The Hmong treat "arthritis" with this plant Pake (1986).

Celosia argentea **(Amaranthaceae)**. The Akha mix this with *Ipomoea batatas* to make a poultice that is put on broken bones.

Cissus hastata **(Vitaceae)**. The Lisu shave the root to make a poultice which is employed to set broken bones or treat joints that are severely out of place. The poultice is also used on wounds.

Cissus repanda **(Vitaceae)**. The Hmong scorch the leaves and tie them as a poultice directly over a broken bone.

Coelogyne trinervis **(Orchidaceae)**. The Hmong mix this orchid with another, *Dendrobium* sp., to make a poultice which they apply directly to a broken bone.

Crateva magna **(Capparidaceae)**. The Lahu mix this plant with other herbs to make a poultice for broken bones.

Crinum asiaticum **(Liliaceae)**. This plant is mixed with *Pedilanthus tithymaloides* and *Kalanchoe*, mashed up, heated over a fire, and used by the Hmong as a poultice on the area where there is a broken bone.

Croton oblongifolius **(Euphorbiaceae)**. The Lahu mix this plant with lemon grass leaves (*Cymbopogon citratus*) (Poaceae), pine needles, and salt, then pound it into a poultice, and apply it as a plaster to the area of a broken bone, securing it in place with a piece of cloth.

Cydista aequinoctialis (**Bignoniaceae**). The Akha make a poultice from this vine to treat dislocated joints and broken bones.

Desmos sootepense (**Annonaceae**). The Akha pound the leaves and apply the mass to the area of a broken bone (Katherine Bragg, herbarium label #59).

Dumasia leiocarpa (**Fabaceae**). The Akha first dry this herb, then boil it to make a tea that is drunk for broken bones. They claim this tea is also an excellent strength medicine.

Equisetum debile (**Equisetaceae**). The Hmong make the stems into a poultice which they apply directly to the area of a broken bone. The Akha mix this plant with *Selaginella roxburghii* (Selaginellaceae) to make a poultice for broken bones.

Erythrina subumbrans (**Fabaceae**). The Mien mix the leaves of this plant with those of *Turpinia pomifera* (Staphyleaceae); the mixture is heated and applied to the area of a broken bone.

Erythrina variegata (**Fabaceae**). The Lahu boil the bark and then apply the liquid to a broken bone to facilitate manipulating it back into place.

Eupatorium odoratum (**Asteraceae**). The Karen mix and boil the roots of this plant with the leaves of *Spatholobus parviflorus* (Fabaceae) (Plate 131); the liquid is both drunk and used to bathe the area above a broken bone.

Fagopyrum dibotrys (**Polygonaceae**). The Akha believe this is a very important medicinal plant that should be included in any concoction that is made up to treat broken bones.

Gynostemma pedatum (**Cucurbitaceae**). The Lahu combine fresh parts of this plant with parts that are heated over a fire. The mixture is applied as a poultice to a broken bone, sprain, or very sore muscle.

Iresine herbstii (**Amaranthaceae**). The Akha pound this plant with *Peucedanum* sp. (Apiaceae), *Solanum nodiflorum* (Solanaceae), and *Strobilanthes anfractuosa* (Acanthaceae), and make a poultice which they apply to broken bones.

Kalanchoe **spp.** (**Crassulaceae**). Both the Hmong and Mien crush the whole plant and use it as a poultice on broken bones. The Akha also use it, but in combination with *Ricinus communis* (Euphorbiaceae) and *Sambucus javanica* (Caprifoliaceae), as a poultice in setting a broken bone. The Mien use *S. javanica* alone as a poultice, heating it over the fire just prior to use.

Litsea glutinosa (**Lauraceae**). The Lahu make a poultice from it, which they put on the area of a broken bone.

Machilus parviflorus (**Lauraceae**). The Akha pound up the leaves and apply them to treat a broken bone (Katherine Bragg, herbarium label #77).

Mitragyna speciosa (**Rubiaceae**). The Hmong pulverize the leaves and stem, heat the plant material over a fire, and apply it as a poultice to a broken bone.

Mussaenda parva (**Rubiaceae**). An important medicinal plant of the Akha, one of its uses is to be pounded into a poultice for setting broken bones.

Nothaphoebe umbelliflora (**Lauraceae**). The Akha pound the bark into a poultice and use it on broken bones.

Oroxylum indicum (**Bignoniaceae**). The Akha make a poultice for the treatment of broken bones.

Phlogacanthus curviflorus (**Acanthaceae**). The Hmong pulverize the roots, mix in some alcohol, and then apply the plant material three times as a poultice on a broken bone after it has been set.

Phragmites karka (**Poaceae**). The Hmong take a shoot with mature leaves, pound it, heat it over a fire, and apply it as a poultice to a broken bone.

Plantago major (**Plantaginaceae**). Katherine Bragg reports (herbarium label #16) that the Akha put the pounded leaves on broken bones.

Sansevieria trifasciata (**Agavaceae**). The Hmong use it when giving a massage to relieve painful bones.

Saurauia roxburghii (**Actinidiaceae**). The Lahu make a poultice from the bark to treat broken bones.

Schefflera alongensis (**Araliaceae**). The Lahu make a poultice from the leaves, which they pound, then heat, and while the plant material is still hot, they wrap it tightly onto the area of a broken bone with a cloth.

Siegesbeckia orientalis (**Asteraceae**). The leaves and stems of this plant are applied as a poultice by the Akha.

Stephania glabra (**Menispermaceae**). The Hmong pulverize the vine, soak it briefly in alcohol, and use it as a poultice on fractures, severe sprains, or pulled ligaments, binding it tightly in place with a cloth.

Stereospermum colais (**Bignoniaceae**). The bark, leaves, and young stems of this plant are made into a poultice by the Lisu, who apply it to sprains, bruises, and broken bones.

Symphorema involucratum (**Verbenaceae**). The Hmong call this the "seven-day cure." The plant is pulverized, mixed with alcohol, and applied as a poultice to wounds and broken bones. The same poultice is also used to treat internal injuries.

Tacca **sp.** (**Taccaceae**). The Lahu pulverize the whole plant, heat it over a fire, and then apply it as a poultice on a broken bone or bad sprain. They even cut the skin slightly so that it bleeds a little and allows the medicine to go into the limb where it can help in the healing process. The Lahu use *Tinomiscium petiolare* (Menispermaceae) in the same way.

Thunbergia grandiflora (**Acanthaceae**). The Akha make a poultice from the leaves and put it on a cut or broken bone to reduce swelling and hasten healing.

Tinospora sinensis (**Menispermaceae**). The Akha pound up the stem and apply it to the area of a broken bone (Katherine Bragg, herbarium label #62).

Bites and Stings

There are numerous forest animals, both large and small, with which the highlander must contend. Some, such as the cobra, are quite dangerous. The highlanders are constantly aware of these threats, especially as they walk barefoot through heavy vegetation wearing rubber thongs. No matter how careful they are, people are still bitten and stung, so, not surprisingly, the tribal pharmacopoeia contains many plant-based medicines for treating bites and stings. In all cases, the plant material is made into a poultice which is applied directly to the wound.

Snake bites are treated by *Afzelia xylocarpa, Clausena excavata* (Plate 132) (which the Hmong specifically call "snake bite medicine"), *Curcuma aeruginosa, Dioscorea hispida, Homalomena* sp., *Lobelia angulata, Neptunia oleracea, Piper agyrophyllum, Pteridium aquilinum, Selaginella helferi,* and *Thunbergia laurifolia.* One Karen elder stated that eating the raw rhizome of *Curcuma* protects a person from severe snake bites.

Scorpion stings are serious injuries and may be treated with *Curcuma aeruginosa, Dioscorea hispida, Homalomena* sp., *Kaempferia* sp., and *Piper agyrophyllum*. If a person is stung by a scorpion some distance from the village, the latex of *Kaempferia* is quickly put on the wound until the victim can get home for better treatment.

Large centipedes roam the forests and often hide within villages; they give a painful bite and instill great respect from the tribal people. The bites are treated with *Ageratum conyzoides, Dioscorea hispida,* and *Syzygium cumini*.

Many types of caterpillars cause serious welts and itching if they are handled or crawl upon a person's skin. Leaves of *Aglaonema* sp., *Anthurium andraeanum, Canthium parvifolium, Cissus hastata, Congea tomentosa, Paederia wallichii, Sansevieria trifasciata,* or *Tetrastigma lanceolarum* are mashed, heated over a fire, and applied, as hot as the victim can stand it, upon the area where the caterpillar crawled.

Insect bites, including bee stings and even mosquito bites, are treated with *Alpinia galanga, Broussonetia papyrifera, Croton oblongifolius, Eryngium foetidum, Lygodium flexuosum, Machilus parviflorus, Microlepia herbacea, Pteris venusta,* and *Tetrastigma lanceolarum*.

Skin Parasites

Although the tribal people make great efforts to keep themselves clean and free from parasites, they nonetheless must constantly battle lice, scabies, and other arthropods that particularly attack the scalp. Fungi are also a serious problem, especially in the rainy season.

The Akha make an infusion by boiling the leaves of *Clerodendrum colebrookeanum* with salt; the liquid is then used to bathe the skin. *Elaeagnus conferta, Rhus succedanea,* and *Stemona burkillii* are used in a similar way but without salt. The Hmong boil the entire plant of *Dregea volubilis* and bathe the affected parts; they also make infusions from *Phlogacanthus curviflorus* and *Verbena officinalis,* which are used to bathe the body for lice and scabies. The Karen prefer to use *Crypsinus cruciformis;* they grind the rhizome and spread the pulp over the scalp. The Lahu boil the leaves of *Litsea glutinosa* and make a shampoo; however, it is more than just a shampoo or soap because they claim it effectively controls both dandruff and head lice. The Mien like to boil plants of *Millettia extensa* to make a solution with which they bathe skin infected with scabies and fungi, such as ringworm.

Fungal infections are a particularly bad problem for farmers of irrigated rice fields as they must be continually wet when working. Most treatments involve either infusions of various plant parts or poultices; both are applied topically. Plants which they claim are effective in the treatment of fungal infections are *Brucea mollis, Byttneria pilosa, Dioscorea pentaphylla, Drymaria cordata, Gmelina arborea, Grewia hirsuta, Laportea interrupta, Litsea glutinosa, Millettia extensa, Ocimum basilicum,* and *Verbena officinalis*.

Frequently tribal people, who may not be able to give details about their skin problems, simply say that the skin itches, or that there is a rash or some other skin problem. I have identified an additional 60 species, excluding those listed in this section, which they use to treat itching, rash, warts, and other skin problems (Appendix 2). Again, almost all of these plants are applied topically as either a poultice or infusion.

Burns

Open fires in houses, village areas, and fields constantly pose the threat of burns, and the highlanders use several plants to treat them. The introduced leaf succulent, *Aloe vera*, is found planted in household gardens or in pots in most tribal villages; highlanders, like Westerners, have learned that the juice from the crushed leaves gives considerable relief from a painful burn (Plate 130). Two other leaf succulents, *Bryophyllum pinnatum* and *Kalanchoe* sp., and a native stem succulent, *Euphorbia antiquorum*, are used in the same way (Plate 133).

In most cases burns are treated with poultices or the liquid from a plant part, though some infusions are used to bathe the injury. Poultices are made from *Archidendron clypearia*, *Careya arborea*, *Clausena excavata*, *Clerodendrum paniculatum*, *Gomphostemma wallichii*, *Milletia caerulea*, *Morinda citrifolia*, *Morus alba*, *Mucuna pruriens*, *Mussaenda parva*, *Oroxylum indicum*, *Plectranthus hispidus*, *Polygonum chinense*, *Shuteria vestita*, *Torenia siamensis*, *Uncaria macrophylla*, *U. scandens*, or *Vanda* sp.

The leaves and young stems of *Berchemia floribunda* and *Selaginella helferi* are dried, then powdered in the hands, and applied to the burn. Each of six species, *Dregea volubilis*, *Euonymus sootepensis*, *Jatropha curcas*, *Maesa ramentacea*, *Markhamia stipulata*, and *Scoparia dulcis*, can be made into an infusion by boiling the plant in water; the burned area is bathed with the liquid. One man told me he had recently been badly burned on one arm by boiling water and had been treated by such an infusion; his arm was completely healed and without scars. This is another example of the strong belief and great trust tribal people have in their traditional herbal medicines.

Tonics

I was continually shown medicinal plants that can best be described as tonics. Few tribal informants knew that word (in fact, their languages simply do not have equivalent terms), but they described plants that are used as a strength medicine, for general health, for weakness, or even for a poor appetite. Some are used only by older people, whereas others are only for children. Many of these tonics are for new mothers to help them regain their strength. I tallied more than 110 species dealing with weakness, poor health, slow recovery from illness, and even old age. Almost all are made into teas, which are drunk by the patient for several days, or even months. One such medicine, used by both the Mien and Karen, is made by boiling the leaves of *Desmodium triquetrum*; the Mien have the patient drink the liquid for a "weak heart and general weakness," whereas the Karen use the same liquid to bathe children who are weak. The Akha, Lahu, and Mien all use *Ficus semicordata* for general weakness by boiling the leaves and drinking the infusion.

Only about a dozen of the more than 100 species which are used as tonics are employed by more than one tribe, which further emphasizes the fact that each tribe has its own distinctive pharmacopoeia and rarely shares it with others.

Venereal Disease

Venereal diseases and other genital problems greatly concern the hill tribes, and they have developed several treatments. Infusions, usually from the roots of *Celosia argentea, Ficus fistulosa, Gelsemium elegans, Glochidion eriocarpum, Imperata cylindrica, Sida rhombifolia, Solanum erianthum, Urena lobata,* or *Ziziphus oenoplia,* are drunk by victims of venereal disease. Mien women drink a tea made from *Jasminum nervosum* for excessive vaginal bleeding, and Hmong men bathe sores on the penis with a liquid made from a species of *Lasianthus.*

MEDICINAL PLANTS FOR PREGNANCY AND CHILDBIRTH

One of the most critical periods in the life of a highlander, whether mother or child, is the time of childbirth. It is not surprising, therefore, that the hill tribes have an immense arsenal of medicines related to fertility, pregnancy, parturition, and the critical few weeks immediately following birth.

Most women were unwilling to discuss this subject with me and my male interpreters. Although men know some of the plants used by women during this period, the subject seems to be almost totally in the domain of women. My Thai colleague from Payap University, Duangduen Poocharoen, a woman who speaks excellent northern Thai, and my daughter, Erica, who also speaks Thai and who served as my assistant for a period of time, were able to gain considerable knowledge of this subject which I could not.

Several Lahu Sheh Leh said that plants have been used to prevent conception, but very few are known or used by people in their tribe. In fact, the only abortifacient that I learned about from any tribe is *Oroxylum indicum,* which the Akha claim will induce an immediate miscarriage. They also say that the same infusion of boiled bark will induce labor immediately. The only other plant that might be described as having regulatory effects on fertility is *Plumbago indica,* which the Lahu Sheh Leh say "regulates a woman's menstrual cycle;" therefore, it improves the chance of conception, but it also quite likely to cause a woman to abort if she is already pregnant. The roots are boiled and the liquid drunk. Some Lahu also say that the same tea gives long life to other individuals. These few plants which are known to control fertility are used each month at the "time the cloth is washed." One male Lahu Sheh Leh medicine man was warned about some plants that induce abortion. His father, who taught him about plants many years ago, told him that they were bad to use. Now an old man himself, he has never used them because he is simply "afraid to use them."

Three plants are used to prevent abortion or miscarriage. The Hmong and

Mien make a tea from the roots of *Urena lobata*; the Mien also claim to use *Eleusine indica* and *Sida rhombifolia* for this purpose.

Several plants are claimed to stimulate sexuality and increase potency and fertility. The Lisu cut up the roots of *Cissampelos hispida* and put them in alcohol, which is then drunk to improve fertility. The Karen believe that a tea made from *Spatholobus parviflorus* helps regulate the menstrual cycle and thus improves the chance of conception (Plate 131). The Mien make an infusion from *Justicia glomerulata*, which women drink to strengthen the body and overcome infertility. The Hmong have an array of fertility-enhancing medicines. Root extracts of *Crotalaria pallida*, *Clerodendrum serratum*, *Mirabilis jalapa* (specifically for impotent men), *Ranunculus cantoniensis*, and *Sambucus javanica* are taken to enhance fertility. Young men who have difficulty ejaculating will drink a tea made from the leaves and stem of *Costus speciosus*, and the reddish bulb of *Scleria* sp. is made into an infusion that is drunk by women who are unable to conceive.

The Hmong even have a plant that determines the sex of an unborn child. If the pregnant woman drinks a tea made from the root of an unforked stem of *Ophioglossum costatum* she will have a boy, but if the tea is made from the root of a forked stem she will have a girl (Pake 1986).

Expectant mothers begin to take a variety of tonics in their sixth month of pregnancy. In one such preparation, a chicken is stuffed with certain herbs, cooked, and then eaten. Women will almost always go to medicine men for the correct plants, fearing that they might gather the wrong ones and be poisoned by them.

Women will almost always give birth to their first child in the village. Later, depending on the severity of problems they had with the first child, they will either stay in the village or go to a hospital. The following are examples of childbirth medicines from three tribes.

Lahu Sheh Leh Childbirth

A Lahu Sheh Leh woman, usually assisted by a mid-wife during parturition, holds onto a rope that has been put over a rafter, thus placing herself in the right position. If the birth is difficult and prolonged, both the husband and midwife may help hold the woman in position. A poultice made from *Clerodendrum serratum* may be rubbed on her stomach to facilitate delivery. To reposition a fetus, the midwife wets her hand in water in which rice has been cooked, thus making it slippery and facilitating her reach into the uterus, or she attempts to change the child's position by massage. Occasionally still births result from a fetus being in the wrong position.

Following parturition the Lahu Sheh Leh midwife massages the new mother to facilitate expulsion of the placenta. If the delivery has been easy, the father and relatives perform a simple ceremony to thank the midwife, presenting her with two wax candles, some rice, and a little money. If the birth has been difficult, the elders are called to the house to help perform a "Calling of the Spirits Ceremony." They take glutinous rice patties and pop them like popcorn over a fire. The rice and some beeswax candles are put on a winnowing tray and placed on the porch to ask the good spirits to return and protect the mother and new child, who both have spirit cords tied on their wrists.

The new mother is given a tea made with ginger (*Zingiber officinale*) for well-being immediately following birth. A bit of ginger rhizome is also pounded up and rubbed on the baby's head.

There are first-day, second-day, and third-day medicines following the birth of a child. The first-day concoction is said to restore peace and calmness to the new mother, and to provide a sense of well-being; the second-day and third-day concoctions help restore lost blood and regain strength. All these concoctions are different. The plants used are *Chloranthus elatior, Desmodium renifolium, Ficus semicordata, Sauropus androgynus,* and *Uraria cordifolia*. The new mother may also be given a steam bath made by boiling the leaves and bark of *Cinnamomum tamala*. Some claim this steam bath prepares her to face the many rigors of life outside the house, such as wind and heat, giving her resistance and protection.

For twelve days after birth the new Lahu Sheh Leh mother and child stay close to the house and bed, with the mother taking tonics frequently to regain strength. She is forbidden to eat beef or the meat of light-colored chickens, but pork and the meat of dark-colored (melanotic) chickens are thought to have medicinal effects so are very good for her. The only time that the woman leaves the house during this period is to relieve herself.

Hmong Childbirth

The Hmong also employ an array of herbal medicines before and during childbirth. The expectant mother, who is accompanied only by older females during the birth process, perches herself on the edge of a low stool, close to the ground or dirt floor of the house. The mother is given a tea made from *Impatiens balsamina* (only plants bearing pink or purple flowers may be used), two different species of *Kalanchoe*, and a species of *Tradescantia* (Plate 134). When the baby is delivered, the umbilical cord is cut and one end tied to the mother's toe while all await the expulsion of the placenta. As soon as this occurs, a medicine is made by boiling three eggs in water in which a sack containing chilies (*Capsicum annuum*) has been placed. The extract from the sack is mixed with the eggs, which are then applied as a poultice to the abdomen to stop bleeding. Hot stones are also gently placed on the stomach.

Akha Childbirth

The Akha also use herbal medicines following childbirth. For example, they give *Albizia chinensis* to new mothers to help them get rid of "bad blood." The leaves of *Blumea balsamifera, Careya arborea, Clerodendrum colebrookeanum,* or *Croton oblongifolius,* are mixed together and heated; the new mother then sits on them to "make her end heal more quickly."

One of the greatest concerns of the hill tribes is that the new mother have sufficient breast milk for her child. This concern is possibly one of the factors that led to the Akha concept of a "terrible birth" in which twins and deformed babies are immediately put to death. The Lahu, on the other hand, feel that the birth of twins is a great blessing for the family and village.

Because of their concern that new mothers have sufficient milk, the highlanders make medicines from more than 20 different plants to stimulate the flow of breast milk. The Akha make infusions of *Argyreia wallichii*, *Codonopsis javanica*, *Elephantopus scaber*, or *Zygostelma benthamii*. Both the Hmong and Lahu boil root shavings of *Dregea volubilis* with chicken and feed it to new mothers to stimulate milk production. The Hmong will sometimes mix parts of two species of *Euphorbia*, *E. antiquorum* and *E. hirta*, boil them in water, and have the woman drink the liquid. The Karen feel that ginger (*Zingiber* spp.) stimulates milk production, as do the fruits of *Aesculus assamica* if they are rubbed on the breasts. Lahu breast milk stimulants include *Embelia stricta*, *Ficus hispida*, *Raphistemma pulchellum* (which is often cooked with chicken and fed to the new mother), *Sterculia lanceolata*, and *Syzygium cumini*. Lisu medicines to stimulate lactation contain *Limnophila rugosa* and a species of *Solanum*. A very important medicine for Mien women is made by chopping up and boiling the woody trunk of the vine *Embelia*; the liquid is then drunk not only to stimulate lactation, but also to relieve menstrual cramps and help in delivery. New Mien mothers will also eat the fruits of a species of *Ficus* to increase milk flow.

The tribal medicine chest contains numerous other plant-derived drugs to combat fever, a variety of pains, nausea, nosebleed, earache, dizziness, paralysis, and even hypertension. Though not sophisticated in medical terminology or the names of specific diseases, the highlanders have medicines to treat virtually every ailment with which they are afflicted. Few of these medicines have been analyzed scientifically and it is difficult to evaluate their effectiveness, but there are no questions in the minds of the tribal people. They cannot be simply dismissed as nonsense. Researchers should undertake a systematic study of those plants widely used among the tribal people and for which there is empirical evidence that they do improve the status of the patient, whether by affecting the immune system, reducing pain and swelling, or simply by killing the infecting organisms. The medicinal knowledge of the tribal people is a treasure that must not be lost, for within that body of information may be cures for some of the worst diseases that plague the human race.

Plate 1. Nestled in the remote mountains of northern Thailand, the Akha village of Li Pha (at 850 meters or 2790 feet elevation) is typical of more than 3000 villages inhabited by the hill tribes.

Plate 2. Elephants, like these at the Karen village of Ruam Mit, were once a major part of forest logging. Today a more lucrative use of elephants is in the tourism industry.

Plate 3. Compared to their married counterparts, unmarried Karen females wear simple clothing—a white shift with a few colored lines woven into it and perhaps some tassels.

Plate 4. Married Karen women wear colorful blouses with distinctive designs.

Plate 5. Blue Hmong woman at a communal maize grinder in Pa Kluoy.

Plate 6. Blue Hmong mother and daughter in the traditional black blouse, batiked skirts, and leggings.

Plate 7. Blue Hmong women sewing in the village of Mae Tho.

Plate 8. Lahu Sheh Leh in the village of Pak Thang. Coffee beans (*Coffea arabica*) are drying on the porch floor.

Plate 9. Lahu Na wedding group.

Plate 10. Mien mother and child in a rice field near Pha Daeng clothed in everyday dress.

Plate 11. Elderly Mien women with silver jewelry in Ban Yao Huai Nam Sai.

Plate 12. Akha woman in village of Ba Go Akha (Ba Kluay Akha) making bundles of broom grass (*Thysanolaena latifolia*). Among the hill tribes, the Akha have the most distinctive and elaborate clothing.

Plate 13. Akha with ornate headdresses in Ba Go Akha.

Plate 14. Although the Lisu prefer to locate their villages on ridges in high mountains, many communities, such as Ba Bo Ngar at 650 meters (2130 feet), have been forced to move to lower elevations and even into lowland valleys.

Plate 15. Lisu woman with material of *Artemisia atrovirens* at the village of Huay Kong.

Plate 16. Lisu women in the village of Nong Khaem.

Plate 17. Lisu village of Nong Khaem at 550 meters (1800 feet) elevation, with deciduous forest in background.

Plate 18. Opium poppies (*Papaver somniferum*) and mountains near the Lahu village of Pak Thang showing Hill Evergreen Forest.

Plate 19. Lahu Sheh Leh village of Ban Doi Mok at 950 meters (3100 feet) elevation.

Plate 20. Old opium fields and Akha village west of Chiang Rai at approximately 1200 meters (3900 feet) elevation. At one time these mountains were covered with Hill Evergreen Forests.

Plate 21. Savanna area in Doi Inthanon National Park at 1300 meters (4265 feet).

Plate 22. Reforestation project with pines east of Phrao.

Plate 23. Reforestation project with *Eucalyptus* north of Chiang Mai. This species is fast growing but of limited commercial value.

Plate 24. Fields cut and ready to burn west of Chiang Rai.

Plate 25. Burning fields near the Akha village of Huay Mae Liam fill the air with smoke for many days.

Plate 26. A recently burned field near Huay Mae Liam appears to be devoid of all life.

Plate 27. New fields near the Akha village of Huay Mae Liam.

Plate 28. Swidden fields near the Lahu village of Pa Kluay.

Plate 29. Fields cut and ready to burn west of Chiang Rai.

Plate 30. Burning of fields east of Ngao.

Plate 31. Burned fields and firewood west of Chiang Rai.

Plate 32. Akha swidden field near Li Pha village showing numerous tree trunks left after clearing and burning the area.

Plate 33. Opium poppies (*Papaver somniferum*) growing in a field cleared by the slash-and-burn technique. This crop was preceded by maize (*Zea mays*).

Plate 34. The Akha digging hoe is used to cultivate the soil and to weed between plants.

Plate 35. The Karen dibble stick, a simple tool consisting of a long bamboo pole and a curved iron blade, is used to make a hole in the ground for planting seed.

Plate 36. Swidden fields west of Wa Wee in Chiang Rai Province in which many of the trees and larger shrubs have been left, thus enabling the vegetation to return more rapidly during the fallow period.

Plate 37. The remarkably sophisticated terraced and irrigated fields near Huey Nam Yen are typical of Karen farmers living in valleys.

Plate 38. Karen rice terraces near Boh Kaew.

Plate 39. Tribal gardens in and around the village, like this garden in the Mien village of Pha Deng, contain an amazing number of different plant varieties.

Plate 40. Ears of maize (*Zea mays*) hanging up to dry from the rafters of a Lisu home at Ba Bo Ngar.

Plate 41. Akha woman removing maize kernels from cobs in Ba Go Akha. When dried and ground, the maize will be fed to the family's animals.

Plate 42. Akha woman in Ba Go Akha spreading maize kernels on a mat where they will continue drying before they are sold or stored in sacks.

Plate 43. Swidden field with diverse crops near the Hmong village of Ban Huay Nam Chang.

Plate 44. Coffee (*Coffea arabica*), ginger (*Zingiber officinale*), and wheat (*Triticum aestivum*) in a Karen field near Boh Kaew.

Plate 45. Coffee and maize near the Karen village of Mae Ka Pu Luang.

Plate 46. Hmong woman and pineapples (*Ananas comosus*) at Ban Tung Sai.

Plate 47. Hmong growing cabbages (*Brassica oleracea* var. *capitata*) near Pa Kluoy.

Plate 48. Drying chili (*Capsicum annuum*) in the Akha village of Ba Go Akha.

Plate 49. Taro (*Colocasia esculenta*) field near the Karen village of Pa Kia Nok.

Plate 50. Bean (*Phaseolus vulgaris*) fields near the Mien village of Pa Daeng.

Plate 51. Sesame (*Sesamum indicum*) seeds at the Lahu village of La Bah are consumed locally or sold as a source of oil.

Plate 52. Potato (*Solanum tuberosum*) field near the White Hmong village of Cheng Meng.

Plate 53. *Sorghum bicolor* drying in the Lisu village of Pa Daeng. Rarely eaten by humans, the grains are fed to livestock.

Plate 54. Hmong woman in field of ginger (*Zingiber officinale*) near Pha Kia Nai.

Plate 55. Akha woman harvesting rice (*Oryza sativa*) near Ba Go Akha.

Plate 56. Mature hill rice near the Hmong village of Cheng Meng.

Plate 57. Irrigated Karen rice fields near Mae Ka Pu Luang.

Plate 58. Rice seedlings in Karen nursery field on Doi Inthanon.

Plate 59. Preparation of Karen irrigated terraces on Doi Inthanon.

Plate 60. Lahu irrigated rice fields near Goshen on Doi Tung.

Plate 61. Akha hill rice fields and field huts near the Burma border on Doi Tung.

Plate 62. Lisu planting rice seeds with a dibble stick near Mae Moo. Note papaya and pineapple (center rear) are also in the field.

Plate 63. Mien rice harvesting knife (*gyip*).

Plate 64. Mien woman harvesting hill rice near Pa Daeng.

Plate 65. Mien rice drying rack near Pa Daeng.

Plate 66. Northern Thai threshing rice with a large bamboo basket near Chiang Saen.

Plate 67. Lisu woman threshing rice.

Plate 68. Karen woman pouring winnowed paddy into a bamboo basket.

Plate 69. Glutinous (bottom) and nonglutinous (top) rice varieties grown in the Lahu Sheh Leh village of Ban Doi Mok.

Plate 70. Lahu Sheh Leh women pounding rice in the village of Lang Muang.

Plate 71. Pounded rice in Lang Muang.

Plate 72. Akha woman winnowing pounded rice.

Plate 73. Lisu woman filling a rice steamer.

Plate 74. Lisu rice steamer.

Plate 75. Lahu Na woman pounding rice with hand pounder.

Plate 76. Son of dead Akha elder standing by sacrificed water buffalo covered with paddy rice in Ba Go Akha.

Plate 77. Akha medicine man standing by a rice spirit house in a field near Li Pha.

Plate 78. Abnormal rice plant near Ba Go Akha. In contrast to a normal plant (see Plate 56), the panicle is compacted and deformed.

Plate 79. Akha men are "feeding" rice liquor to the spirit of the abnormal rice plant.

Plate 80. Covering the abnormal rice plant, the last step in a ceremony to appease the evil spirits.

Plate 81. Rice spirit house near a granary in Li Pha village.

Plate 82. Consisting of a section of bamboo stem, this ancestral altar hangs on the women's side of an Akha house in Ba Go Akha.

Plate 83. Broom grass (*Thysanolaena latifolia*), grows commonly in open areas of the mountains.

Plate 84. Akha woman in Ba Go Akha preparing broom grass (*Thysanolaena latifolia*) for sale.

Plate 85. *Solanum erianthum* (#5295) near the Lahu village of La Bah is one ingredient in gunpowder.

Plate 86. *Gelsemium elegans* (#6031) from near the Lahu village of Mae Poon Luang is very poisonous.

Plate 87. Wild banana (*Musa acuminata*) (#5572) from near Chiang Kham. The species is important medicinally and as a source of food for the Akha, but the Lahu believe the fruits are poisonous.

Plate 88. Rattan (*Calamus* sp.) scabbard and thatch made of *Imperata cylindrica*.

Plate 89. Akha women with firewood near the village of Thoet Thai.

Plate 90. Bamboo "star" at the edge of Ba Go Akha warns evil spirits to stay out of the village.

Plate 91. Bamboo water system in the Lahu Sheh Leh village of Ban Doi Mok.

Plate 92. Bamboo chicken cage in Ban Doi Mok.

Plate 93. Akha woman removing lard from a section of *Dendrocalamus strictus*.

Plate 94. Akha animal trap made of bamboo.

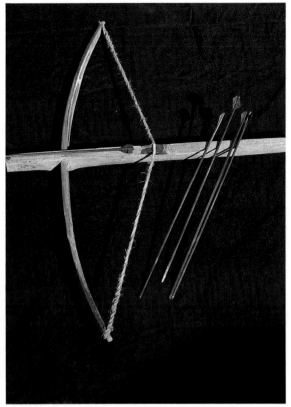

Plate 95. Akha crossbow with bamboo "spring" and arrows.

Plate 96. Akha man weaving a low table with strips of bamboo.

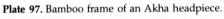

Plate 97. Bamboo frame of an Akha headpiece.

Plate 98. Tribal pony eating bamboo leaves.

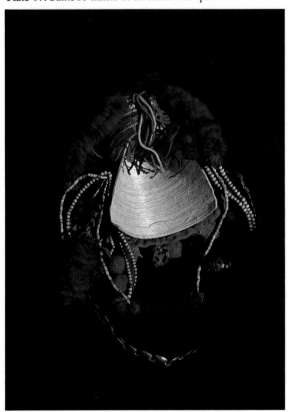

Plate 99. This Akha marriage house in the village of A Hai is made entirely of bamboo.

Plate 100. Bamboo knife for cutting the umbilicus of a newborn baby.

Plate 101. Bamboo blinder, post of *Schima wallichii*, and sacrificial water buffalo tied with the vine of *Bauhinia bracteata* in an Akha funeral ceremony.

Plate 102. Akha coffin with bamboo tube going out of the top (to release gases) and another piece coming out of the bottom (to drain off liquids).

Plate 103. *Dendrocalamus hamiltonii* (bamboo) growing near the Hmong village of Cheng Meng.

Plate 104. *Dendrocalamus giganteus* growing near Ban Tha Khai.

Plate 105. Culms and culm sheaths of *Bambusa vulgaris*.

Plate 106. Woven bamboo basket and knife handle made of *Cephalostachyum virgatum* in Ba Go Akha.

Plate 107. Akha rice fermenting container made of *Dendrocalamus giganteus*.

Plate 108. Akha woman splitting culms of *Dendrocalamus hamiltonii* (bamboo).

Plate 109. Bridge made primarily of *Dendrocalamus hamiltonii* in the Karen village of Mae Cha Ta.

Plate 110. Karen woman carrying water in sections of *Dendrocalamus membranaceus* (bamboo).

Plate 111. Opium poppies (*Papaver somniferum*) growing near the Hmong village of Mae Tho.

Plate 112. White-flowered opium poppies and old maize stems (lower center) in a field near Mae Sa.

Plate 113. Close-up view of *Papaver somniferum* plants with fringed, white flowers.

Plate 114. Special knife used to tap the opium poppy fruits.

Plate 115. Hmong woman tapping opium poppies.

Plate 116. Tapping the poppy capsule to allow the latex to flow.

Plate 117. Opium poppy capsule with latex.

Plate 118. Scraping the congealed latex off the poppy capsule.

Plate 119. Poppy capsules with scars where the latex has been removed.

Plate 120. Congealed opium latex and Hmong scraper.

Plate 121. Raw opium wrapped in paper made from *Broussonetia papyrifera*.

Plate 122. Hmong opium smoker heating a ball of latex over an open flame.

Plate 123. Hmong man smoking an opium pipe.

Plate 124. Akha woman and some common medicinal plants used in Ba Go Akha.

Plate 125. Lahu Sheh Leh medicine man examining medicinal plants.

Plate 126. Hmong exorcist with medicinal plants.

Plate 127. Lisu medicine woman with an array of medicinal plants.

Plate 128. Akha medicine man pounding up material of *Morinda* sp. for use as a poultice.

Plate 129. Medicine man applying a poultice to a sick child.

Plate 130. *Aloe vera* is used by all hill tribes to treat burns.

Plate 131. The bark of *Spatholobus parviflorus* (#5553) is used to treat broken bones and boiled in a tea that a woman drinks to regulate her menstrual cycle.

Plate 132. This Lisu medicine woman will use *Clausena excavata* to treat snake bites.

Plate 133. *Kalanchoe* sp. (#6085), grown in the Lahu village of Goshen, is used to treat burns.

Plate 134. Hmong woman with *Impatiens balsamina* (#5953), one of several plants made into a tea that is drunk by new mothers.

Plate 135. Lahu house in Nong Khieu

Plate 136. Porch of a Lahu house in Ban Doi Mok with a log stairway. Note the pictures of the Thai royalty on the split bamboo wall.

Plate 137. Karen and elephant assisting Akha men drag logs to Maw La Baw Soe.

Plate 138. Lisu man trimming a post of *Gluta usitata*. Known for its durability and strength, the plant can cause poison-ivy-like dematitus.

Plate 139. Mature plant of *Livistona speciosa* (center) growing near Pong Klang Nam. The trunks are used for house posts.

Plate 140. The foliage of *Livistona speciosa* is used to make thatch.

Plate 141. Hmong women bringing *Imperata cylindrica* from the field to make thatch.

Plate 142. Akha woman making thatch "shingles" from *Imperata cylindrica.*

Plate 143. Stack of completed thatch shingles made from *Imperata cylindrica.*

Plate 144. Leaves of *Dipterocarpus tuberculatus* drying before being made into roofing "shingles."

Plate 145. Karen man making "shingles" from dried leaves of *Dipterocarpus tuberculatus* in the village of Moung Phaem.

Plate 146. Karen woman making roofing "shingles" from the leaves of *Phrynium capitatum* (#5664).

Plate 147. Roof of a Mien house in Huay Gaew made from culm sections of *Dendrocalamus* sp.

Plate 148. Akha house in Huay Mae Liam. Note the typically masssive roof structure.

Plate 149. Ornate decorations at the gables of an Akha house in Huay Mae Liam.

Plate 150. Construction of an Akha house in Ba Go Akha involves tightly lashing the superstructure with bamboo or rattan ties.

Plate 151. Elevated Karen houses in the village of Mae Tia.

Plate 152. Karen rice storage container in Moung Phaem.

Plate 153. Lahu Na roof made from *Imperata cylindrica* in which the stems project into the house.

Plate 154. Lahu fireplace and metal tripod on a packed dirt hearth constructed on raised floor.

Plate 155. Hemp or marijuana (*Cannabis sativa*) is grown by the Hmong for its strong fibers which are stripped from the stem of the plant and made into cloth.

Plate 156. Close up of hemp foliage.

Plate 157. Drying hemp (*Cannabis sativa*) stems in the Hmong village of Pha Kia Nai.

Plate 158. Hmong woman creating pattern on hemp cloth as first step in batiking process.

Plate 159. Hmong woman dyeing hemp cloth in container of indigo.

Plate 160. Blue Hmong woman with batiked skirt made of hemp cloth.

Plate 161. Tree cotton (*Gossypium arboreum*) in the Lahu village of Nong Khieu. This tree reaches a height of 3 meters (10 feet) or more and is both wild and cultivated in Asia.

Plate 162. Lahu Na woman making thread with a spinning wheel in the village of Goshen.

Plate 163. Bamboo container to hold cotton spun into thread by Akha women.

Plate 164. Akha woman making thread using a spindle rolled along her thigh.

Plate 165. Like other Akha swings, this one in the village of A Hai is used only at special times during the year.

Plate 166. Four long tree trunks form the frame of an Akha swing, and rattan (*Calamus* spp.), if available, is used for the rope.

Plate 167. Headman of the village of Ba Kha Akha removing the outer green bark of *Pueraria phaseoloides* with his knife to get the fibers used to make the distinctive Akha string shoulder bags.

Plate 168. Twisting the fibers of *Pueraria phaseoloides* on the thigh to make the string.

Plate 169. Karen woman weaving with the back-strap loom.

Plate 170. Akha woman weaving with a bamboo foot-treadle loom.

Plate 171. Newly dyed indigo cloth drying in Ba Go Akha.

Plate 172. Akha woman's headpiece with silver, beads, and seeds of *Coix lachryma-jobi* (Job's tears).

Plate 173. Spirit altar at the sacred tree near the Akha village of Li Pha.

Plate 174. Century plant, *Agave americana*, believed by the Karen to be a deterrent to the bad spirits.

Plate 175. *Opuntia dillenii* (center) and *Jatropha curcas* (tree behind) in the Akha village of Huay Mae Liam. The former is believed to ward off evil spirits; oil from seeds of the latter is used as a purgative and as fuel for lamps.

Plate 176. *Sansevieria trifasciata* is believed to be a "good luck plant."

Plate 177. Stems of *Euphorbia antiquorum* placed over the doorway of a Lahu house keep weretigers from entering.

Plate 178. The seeds of Job's tears (*Coix lachryma-jobi*) are one of the most common seeds used for decoration.

Plate 179. The seeds of Job's tears (*Coix lachryma-jobi*) come in a variety of shapes and sizes.

Plate 180. The gigantic fruit of *Entada rheedii* produces seeds that are used medicinally, for making necklaces, and for playing games.

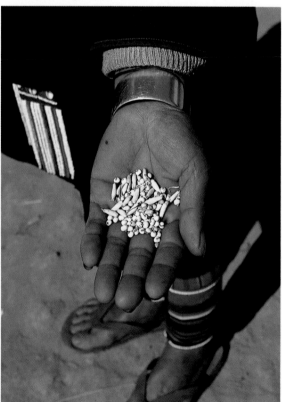

Plate 181. Akha necklace made from seeds of *Caryota mitis,* cowrie shells, and fruits of the small bottle gourd (*Lagenaria siceraria*).

Plate 182. Leech lime (*Citrus hystrix*) is a source of shampoo.

Plate 183. Akha man with loofa (*Luffa aegyptiaca*), used widely in bathing and dishwashing.

Plate 184. The teeth of this Lisu man are stained from chewing betel nut (*Areca catechu*).

Plate 185. Betel nut supplies and container.

Plate 186. Betel nut (*Areca catechu*) (lower and right), pepper leaf (*Piper betle*) (above), and lime (pinkish material on leaves).

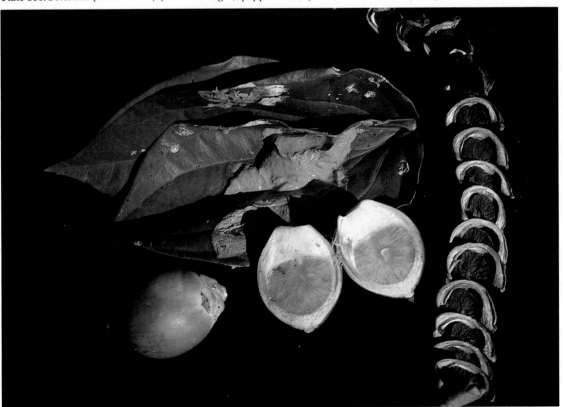

Plate 187. Akha elder smoking an elaborate bamboo pipe.

Plate 188. Akha woman smoking a pipe made of bamboo pieces.

Plate 189. Tobacco plants (*Nicotiana tabacum*) are found in almost every village garden, such as this one in the Lahu village of Pak Thang.

Plate 190. Tobacco drying in the Hmong village of Buak Jan.

Plate 191. Lisu woman shredding dried tobacco leaves.

Plate 192. Bamboo water pipe used to smoke tobacco.

Plate 193. Large clay urns are used to ferment grain, the first step in making whiskey.

Plate 194. A Lisu still.

Plate 195. Hmong boy with a heavy wooden top carved of very hard forest wood.

Plate 196. Hmong boy with a "pea shooter" made of small sections of bamboo.

Plate 197. A Lahu gourd pipe is a musical instrument combining bamboo and *Lagenaria siceraria*.

Plate 198. The Hmong mouth organ may have up to six bamboo pipes.

Plate 199. In the Karen bamboo dance, performers dance within rhythmically moving bamboo poles.

Plate 200. The intricately carved mouth or lovers' harp is made from the bamboo *Dendrocalamus strictus*.

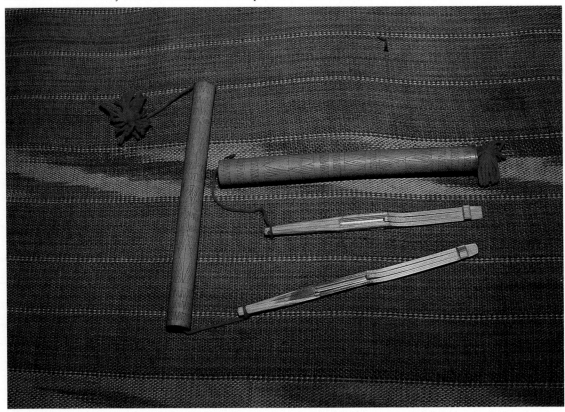

9

Houses From the Forest

When one first sees a hill tribe house, one cannot help but feel that it seems very temporary. Can it be a home? Can it provide these people with a sense of security and possession? Yes, these relatively simple houses structures, made by hand from materials brought from the nearby forest, are indeed homes to the hill tribes. These houses provide security from wild animals; comfort from the rains, blazing sun, or cold nights; a place in which families may store, prepare, and eat their foods; a place to make love and to have children; a place to rest; a place to entertain or mourn; and most importantly, a place to put their ancestral altar. Indeed, these grass, bamboo, and wood structures are their possession in perhaps an even stronger way than our houses are to us, for they have personally gathered the materials from the nearby forest and have constructed their houses by hand, piece by piece, along with neighbors from the village, so that it is really a major part of their existence (Plates 135–136).

MATERIALS FOR HOUSE CONSTRUCTION

Lumber

One of the most essential elements of any hill tribe house is the superstructure, which is almost always made of wood cut and brought from the forest. These beams, posts, logs, and poles make up the framework upon which the walls and roofing materials are attached. Without doubt, the heaviest labor and greatest danger in constructing a house comes from obtaining this lumber, for, from the initial cutting of the tree until the timber is secured in place, there is continual threat of serious injury to the men involved.

Many forest trees have extensive buttresses, which are of little value as straight wood for construction. Therefore, the highlander must usually erect a platform or simply lean a board against the tree trunk 2 or 3 meters (6.5–10 feet) above the ground from which he then chops down the tree with his axe. Once the selected tree has been felled, all usable parts are removed, whether for posts, beams, or fire-wood. The cutting is done by hatchet, machete, or axe. The trunk, if too large for a main post or beam, is split with wedges. Getting the lumber back to the village requires the labor of many men, unless the highlanders have been able to hire an elephant from a nearby Karen village (Plate 137). This is less common now because so many elephants are involved throughout the year in the tourist industry.

The selection of trees to fell is influenced by several factors: accessibility, straightness, durability, and whether a tree will bring good or bad luck. There are dozens of species of trees in the forests of northern Thailand, some of which are limited to lower elevations, others to the higher mountains.

Members of the Dipterocarpaceae (Philippine mahogany family) and Fagaceae (oak family) are used extensively in house construction as the wood is excellent and easy to work, and the trees are relatively common at most elevations. One of the most common species at lower elevations is *Dipterocarpus tuberculatus*, an excellent source of lumber that also provides leaves for covering the roof. Other species of the Philippine mahogany family are also used: *Dipterocarpus alatus, D. costatus, D. obtusifolius, Shorea obtusa, S. roxburghii,* and *S. siamensis.*

Members of the Fagaceae that provide important lumber for construction are *Castanopsis armata,* which occurs at higher elevations, and *C. diversifolia,* which occurs at much lower altitudes. The latter is a very popular tree for construction with all the tribes, as it is both durable and the termites seem to leave it alone. Besides, the prickly fruits are excellent to eat when roasted. Other species of *Castanopsis,* such as *C. acuminatissima, C. indica,* and *C. tribuloides,* are also used.

Lithocarpus elegans is another important member of the oak family; perhaps the most valued of all forest trees for house construction, this species produces excellent wood which will last 20–30 years. It is fairly common in the forests, relatively easy to cut, and can be brought to the village fairly easily. Other members of the Fagaceae that are used for construction include *L. polystachyus, L. thomsonii, Quercus aliena, Q. incana, Q. kerrii,* and *Q. semiserrata.*

The following species of other families are good or bad for one reason or another and some must be used only in particular ways:

Caryota mitis (**Arecaceae**). This palm is cultivated in the villages and used for a variety of purposes, including house construction.

Cassia fistula (**Caesalpinaceae**). The Akha particularly like to use this wood for house posts because they believe it protects them from hurricanes.

Dillenia indica (**Dilleniaceae**). The Akha particularly like to use this wood; however, a related species, *D. obovata,* has wood that is too hard, so the Akha use it only for firewood.

Eurya acuminata (**Theaceae**). A very sacred tree to the Akha, the wood of this tree can only be used for the upper parts of a house, never for under the floor. That would just not be right to them.

Ficus auriculata (**Moraceae**). One of many figs within Thai forests, the Lahu particularly noted this species as a good source of wood. The fruits are also edible.

Gluta usitata (**Anacardiaceae**). This tall tree occurs at lower elevations and produces a deep red wood that is very hard and durable (Plate 138). Tribal people like to use it, though some get serious dermatitis from it.

Holarrhena pubescens (**Apocynaceae**). The light-colored wood of this big tree is particularly treasured by the Karen for a variety of uses, including construction of furniture. Its leaves and stems are used medicinally.

Litsea monopetala (**Lauraceae**). The Lahu refer to this as the "coffin tree," as the lumber is highly desired for making coffins. However, it is difficult to find trees large enough for this purpose. The tree makes excellent house posts as termites do not seem to eat it, and the Lahu make insect-proof boxes from its wood.

Livistona speciosa (**Arecaceae**). Mature plants of this species provide excellent posts and the leaves can be used for thatch (Plates 139–140).

Macaranga denticulata (**Euphorbiaceae**). The wood is used for house construction by the Hmong.

Mitragyna spp. (**Rubiaceae**). Several species of this genus provide lumber that the Karen like to use.

Parinari anamensis (**Rosaceae**). The Akha use the wood for construction and as firewood.

Pinanga sp. (**Arecaceae**). Some of the tribal people do not bother with wood from this palm because of its small diameter. However, the Akha will use it because the pieces may be up to 4 meters (13 feet) long, thus serving as excellent smaller beams.

Pinus kesiya and *P. merkusii* (**Pinaceae**). The only native species of pine in northern Thailand, these and some of the introduced pines are used for house construction, even though their durability is not good.

Polyalthia simiarum (**Annonaceae**). The Akha believe this is one of the finest woods for house construction. They also eat the fruits and make a sweet-tasting tea from the bark.

Schima wallichii (**Theaceae**). The Lahu believe this is a good wood for house construction.

Tectona grandis (**Verbenaceae**). The hill tribes, such as the Karen and Lisu, who live at lower elevations in the teak forests, very much like to use this tree because of its beauty and durability. Even smaller teak trees make excellent posts for houses, but tribal people are now prohibited from cutting it.

Xylia xylocarpa (**Mimosaceae**). The wood can be used on any part of a house except for posts. Apparently termites readily eat it where it comes in contact with the ground.

Bamboos

The many species of bamboo and their uses by the hill tribes are discussed in Chapter 6. However, because of their great importance in house construction, it is appropriate to at least list the 15 species that have been identified as used in construction. They are *Bambusa polymorpha, B. tulda, B. tuldoides, B. vulgaris, Cephalostachyum pergracile, C. virgatum, Dendrocalamus brandisii, D. giganteus, D. hamiltonii, D. latiflorus, D. membranaceus, D. strictus, Gigantochloa latifolia, G. wrayi,* and *Thyrsostachys oliveri.*

Roofing Materials

Roofing materials, like the superstructure and other basic parts of the house, is of great importance for the protection of the household, but the materials used vary somewhat, depending on the elevation and the available plants. Six species have been identified as used for thatch, exclusive of teak and pine which are occasionally made into wood shingles and used by the hills tribes.

Imperata cylindrica (**Poaceae**). By far the most common thatching material is cogon grass, *Imperata cylindrica*, a species that is almost certainly introduced but one that is now widespread in Thailand, especially in badly disturbed or deteriorated swidden areas where it forms extensive savannas (Plate 21). In fact, these so-called savannas of the highlands consist almost completely of this hearty grass, which makes an almost-solid mass of rhizomes at or just beneath the level of the soil. Land that has been taken over by *Imperata* is almost worthless for future agriculture or reforestation.

The grass is harvested and carried to the village (Plate 141) where the women construct a basic "shingle" of thatch, about a meter (3.5 feet) long and half a meter (1.5 feet) wide by tying folded wet grass around a piece of split bamboo (Plates 142–143). These "shingles" are then tied to the bamboo rafters with strips of bamboo that have been soaked in water to make them soft; the ends of the strip are simply twisted together, shrinking and tightening as they dry (Grunfeld 1982). Approximately 300 shingles are used on a house, assembled so that they overlap one another by 12–15 centimeters (5–6 inches). This basic pattern can also be found in Yunnan, China and is used by the Thai as well. Cogon grass thatch lasts about three or four years, sometimes up to seven or eight if an especially thick layer is used. A roof may be patched several times, usually starting in the second or third year, before it must be completely replaced.

Dipterocarpus tuberculatus (**Dipterocarpaceae**). Because *Imperata cylindrica* does not grow at lower elevations, tribal people living there must use different roofing materials. The large, leathery leaves of *Dipterocarpus tuberculatus*, a common lowland tree, make a good, durable roof. The leaves are harvested, placed in thick stacks, covered, and allowed to thoroughly dry (Plate 144). They are tied into "shingles" about the same size as those made of cogon grass (Plate 145), and then lashed into place in an overlapping fashion to make the roof. Dipterocarp roofs last only about three years and cannot be patched as easily as those of cogon grass. The leaves of *Macaranga denticulata* (Euphorbiaceae) are sometimes used in a similar fashion by the Akha living at low elevations.

Phrynium capitatum (**Marantaceae**). A Karen village at over 1000 meters (3300 feet) elevation used the leaves of another "broad-leaved" plant, *Phrynium capitatum*, a swamp plant. The women collected and used these leaves like dipterocarp leaves because there were no suitable trees of that type at that elevation and the *Imperata* grass was too far away (Plate 146).

Livistona speciosa (**Arecaceae**). A palm, *Livistona speciosa*, is fairly widespread at moderate elevations in northwestern Thailand and is also cultivated in many tribal villages (Plates 139–140). Fronds are made into "shingles" similar to those of cogon grass and used for roofing. In some villages roofs are made from a combination of *Livistona* and *Imperata*. Occasionally, bamboo leaves are used like the fronds of *Livistona*.

Bamboo. In many villages the roofs of smaller buildings, such as pigpens and rice storage huts, are made of split bamboo, usually species of *Dendrocalamus*, in which the outside, rounded half lies over the space between two inverted halves in an arrangement like that of a Spanish tile roof (Plate 147).

CEREMONIES AND RITUALS RELATED TO HOUSE CONSTRUCTION

The village site is of great importance, for there must be an adequate water supply, areas nearby for farming, and a good forested area to provide the residents with many items necessary for their survival. The ceremonial head of the village is usually responsible for selecting the site. A common practice followed at the possible location of the new village is to make a small depression in the ground; several grains of rice are placed carefully in it so that they radiate outward from the center. A bowl is then inverted over the depression and left for a specified period of time; later the grains are examined to see if they have been disturbed by the spirits of the ground. If they have not changed, the site is considered satisfactory.

Some tribes also follow additional rituals. The Akha, for example, drop an egg in a special, cleared area to see if the spirits, their ancestors, and the local "Lords of Land and Water" approve. The spirits are displeased should the egg not break when dropped, and a new location is sought (Lewis and Lewis 1984).

Most tribes also follow carefully prescribed rituals in selecting a house site within a village and later moving into it. The Hmong, for example, select a location they feel is acceptable to their ancestors. They make an offering of paper "money" to learn if the ancestors approve of their choice. All building materials are gathered together; on a day they believe is auspicious, the first two posts are installed. At that time the owner of the new house announces to the evil spirits that he is now living there and they must stay away. A sacrifice is made when the house is completed.

The Mien strongly believe that a house is sacred, for it contains the ancestral altar. Therefore, they carefully consider all omens and dreams that occur when they select the house site, hoping that the earth and forest spirits that are disturbed will not take offense.

When an Akha house is completed, items must be taken inside in a prescribed sequence. First, the ancestral altar and the paraphernalia associated with it are placed in the proper position on the women's side of the house; the three-legged metal tripod that is placed over the fire and on which rice is cooked is brought in with these items. After that, other family and personal items may be carried into the new house.

There are times when an entire village must move. The Karen, for example, relocate when the village priest dies, but nowadays they usually move only a short distance so that they do not have to abandon terraced fields, water systems, and so forth. Moreover, the Thai government restricts tribal people from moving to new sites. Also, there simply may be no available new locations to which the highlanders can move even if they get permission.

The need to move a village, coupled with the inability to do so because of government prohibitions or lack of suitable nearby sites, has created some interesting ceremonial compromises. For example, when the spirit man or *pi ma* of Huoe Eurn, an Akha village consisting of 21 houses in western Chiang Rai Province near the Burma border, decided that the village had to move, there was no place to which it could move. Therefore, the people tore down the village and reconstructed it in exactly the same location as before. This carefully prescribed activity was divided into four days. On the first, or Monkey Day, all the houses were torn down. On the second, or Chicken Day, the spirit man's house was built. The third day, Dog Day, the people rested. On the final day, or Pig Day, they began construction of the other houses, a process that actually took several days to complete, even though all supplies were already there. Few new posts and timbers were needed, but many of the bamboo poles and all the thatch was replaced. In addition, each household needed new chop sticks made of split bamboo, as well as a new spirit house. The ancestral items and other paraphernalia involved with the various spirits were carefully packed away and stored during this period of transition. In one temporary hut the family placed their things in a beautiful covered basket made of woven bamboo about a meter high and nearly as large in diameter. Some of these items, such as cloth, bells, some rice, a knife, and a small hatchet, were over 20 generations old. When the house was finished, an egg was cooked; each male of the household ate a bit of it. Next, the basket and metal cooking tripod were carried in first, followed by everything else. The new house required a new ancestral altar, consisting of a section of bamboo that was hung on the women's side of the central partition, which would then be replaced every year at the end of harvest (Plate 82).

COMMUNITY PARTICIPATION IN HOUSE CONSTRUCTION

Houses are usually constructed and repaired during the dry season following the completion of rice harvest. The erection of the timbers of a house requires several strong men, so frequently, as was the case with barn construction in rural United States a few generations ago, the building of a house becomes a community event in which several families assist each other. It is also an opportunity for young people to meet and for elders to work with one another.

In the first week of February several dwellings were rebuilt or newly constructed in the lowland Lisu community of Ba Bo Ngar. Ten men and several boys constructed a new house for a recent widow, who had just come to the village to be near her family.

The main or center posts were beautiful hand-hewn pieces of *tcuh tsuh tzuh* wood, a beautiful red wood of *Gluta usitata* in the Anacardiaceae; the tree is the lac tree, known for its durability and strength but also a plant that can cause severe poison-ivy-like dermatitis (Plate 138). Throughout the construction process the men talked, argued, and joked with one another, while groups of women in nearby houses sat and sewed, but kept close track of the construction process.

At about one o'clock the men took a break. A woman brought a bottle of whiskey, some small porcelain cups, and a huge basket of rice. The men relaxed, ate, drank, and smoked cigarettes rolled with newspaper.

Once the main posts were set, a long hand-hewn ridge beam of *Dipterocarpus tuberculatus* was lifted into place and set in the notched ends of the center posts. In less than 5 hours the main structure of the house had been assembled.

The final steps involved repair of an old roof structure from a house that had been torn down recently, the addition of some new rafters (it was not quite long enough for the new house), the placing of split bamboo sections to make the walls, and, finally, the thatching of the roof. Quickly the sections were set into place and secured by a long section of bamboo along the middle using nails. In a matter of minutes the walls were nearly in place, and other men, some retying rafters and crosspieces, began the thatching of the roof. The roof was thatched with cogon grass, even though large numbers of *Dipterocarpus tuberculatus* trees grew nearby and were an important source of timber. However, the grass is easier to work with and lasts a bit longer than do the broad, flat dipterocarp leaves.

By dark the house was essentially finished, with only interior work, such as construction of door, sleeping platforms, and fireplace, to be done in the days to come. But soon the widow would have her own little house near that of her relatives, all made possible by the community's assistance and participation.

DESIGNS AND CONSTRUCTION OF TRIBAL HOUSES

The six tribes construct their houses in distinctive ways, some built on the ground, others raised on posts. Thatching methods, floor plans, food storage systems, and sleeping arrangements all differ from group to group.

Akha

Perhaps the most distinct of all tribal houses, the Akha house appears at first glance to be nothing but roof. Indeed, the roofs are massive structures, always of cogon thatch, starting at a high peak and coming down nearly to the level of the floor, which is constructed on posts usually about 1.5 meters (5 feet) above the ground (Plate 148). There are two small openings in the roof located at each end of the peak, often with intricately carved wooden decorations at the gables (Plate 149); from both these and the thatch itself comes the smoke from the two fireplaces. The superstructure of the house is of carefully fitted, pegged and notched timbers, upon which have been laid rafters, supports for the walls, and floor joists, all usually of bamboo (Plate 150). These parts are tightly lashed with bamboo or rattan ties. There are no windows, so the interior of the house is always dark—and smoky; however, even in the heat of the day it is also pleasantly cool.

The house consists of three basic parts: the open deck or porch, the men's side, and the women's side. Usually there is a small covered porch at each end of the house under the eave of the roof and actually part of the basic rectangular floor plan. The interior floor plan is simple: the house is divided into two equal halves, the men's (and guests') side and the women's side, divided by a split bamboo partition extending upward about 2 meters (6 feet) and completely dividing the interior except for a passageway on one side. The family's ancestral altar is hung on the women's side of the partition (Plate 82). Fire pits are located in the center of the house on each side of the partition, and, in some houses, a third is found in a corner of the women's section for the cooking of pig food.

Beneath the house are simple corrals in which the family's livestock is kept at night, such as pigs, water buffalo, ponies, and even cattle.

Hmong

The Hmong house is remarkably unlike that of the Akha, being placed on the ground and frequently with the walls made of sawn boards rather than bamboo. However, like the Akha, the Hmong house has no windows. The structure is rectangular or modified-rectangular in shape, approximately 5×10 meters (16×33 feet) in size, and with either one or two doors to the outside. The main door is along the front side and usually under the extended eave of the roof, which makes an open, covered porch. A second door is at the end of the house near the stove on which the pig food is cooked. White Hmong usually have two doors, whereas the Blue Hmong may have only a single one.

Almost all Hmong houses have both a stove and a fireplace. The stove is located in one corner of the house, usually at quite a distance from the main entrance, and used only for the cooking of pig food. It is made of concrete or hard, baked mud. The fireplace, on the other hand, is usually on the hard-packed floor in the middle of the house or near the main door and, at most, may be lined with rocks. Pots and kettles are placed on a metal tripod. Smaller homes may have a single fireplace upon which both human and animal food is cooked.

Along the inside of the main front wall there are a series of rooms, each with a door (or curtain) and a raised wooden or split bamboo platform for sleeping.

The roof is supported by two main posts of wood, and from this are strung the rafters, usually of bamboo but sometimes of wood. Lower beams to support the walls are usually of wood. The roof commonly is thatched with cogon grass, though often the Hmong living at lower elevations have adopted corrugated metal or concrete. The walls may be split bamboo or sawn wood, the latter usually being vertically oriented.

Karen

Karen houses are built on posts, but the height may vary from 2 to 3 meters (6.5 to 10 feet), depending on whether the household owns an elephant. If the family owns an elephant, then the higher house makes it easier to load and unload the animal, as well as store the harness (Plate 151). If the household owns no elephant, then the house is only about 2 meters (6.5 feet) off the ground. The house always has a covered porch or veranda and usually consists of a single room with a central fireplace. Again, the basic plan is that of a rectangle.

The main part of the house is divided into three sections: kitchen, family bedroom, and central area. This latter part is open onto the porch and from it are doors leading into the kitchen and bedroom. The kitchen extends the entire width of the house, but the bedroom only about three-fourths of the way. The other fourth is for storage of tools and other items. The fireplace is in the kitchen and consists of a wooden framework filled with packed mud.

Rice may be stored in separate structures or in huge woven bamboo baskets, which are often sealed with mud. These baskets may be 2 meters (6.5 feet) high and more than one meter (3 feet) in diameter. They are located either next to the kitchen or on the covered porch (Plate 152).

Lahu

In general, Lahu houses are quite simple in design, often consisting of but a single room, and varying somewhat from group to group. Others have a central partition separating the kitchen-living area from the sleeping area. Non-Christian houses have a spirit altar or sacred closet in the bedroom. Lahu houses are built on posts usually about one meter (3.5 feet) high, so there is insufficient space beneath for livestock, which are kept in separate pens nearby. All Lahu houses have an uncovered porch, reached by a notched log or bamboo ladder and from which one enters the house through the single door (Plates 135–136). The posts and main beams are of timbers, with the secondary beams, rafters, studs, and floor joists of bamboo.

Roofs are almost always made of cogon grass, but some of the Lahu, particularly the Lahu Na, prepare it differently from the other tribes. Rather than binding equal lengths of the grass around a piece of split bamboo to make "shingles," the Lahu Na actually carry the grass stems up to the roof where they are tied to split bamboo cross pieces that have been tied to the rafters. The end that goes down into the house is carefully trimmed so that the ceiling consists of row upon row of projecting stems (Plate 153). The effect from inside the house is striking and beautiful. Lowland Lahu roofs may have a combination of cogon grass and dipterocarp leaf, with the latter being placed along the peak. Another distinctive characteristic of a Lahu roof is that often a series of bamboo poles are laid and secured parallel to the rafters but on top of the thatch to help hold it down. Walls and floors are usually of split bamboo, although some houses may have sawn wood floors.

There is usually a single fireplace or stove constructed in the usual manner by building a framework of wood and filling it with packed mud on which to have the fire. Kettles and pots are placed on metal tripods over the fire (Plate 154).

Lisu

Lisu houses are usually constructed on the ground and have very similar floor plans; however, occasionally they will build their houses off the ground when living at lower elevations, for these structures will be cooler in the hot season. There is a single entrance from the main side, which leads into the living area;

usually the ancestral altar is directly across from the door. If the house is built on a hill, the altar is always on the upper side and the door on the lower. The kitchen area and cooking fireplace are in a corner. The other corner at that end of the house is the sleeping platform for the children, usually constructed of split bamboo. The other side of the house consists of two sleeping areas with platforms of split bamboo.

The basic structure of the house consists of hand-hewn posts, which have been placed in holes in the ground and carefully notched to take hand-hewn cross beams. Historically, the posts and beams were secured to each other by notches, reinforced with lashings of rattan (*Calamus* spp.), which would last up to 20 years; however, the rattan palm is now so rare that bamboo ties, which last only about 10 years, must be used. Nails are now readily available, so most beams and even the split bamboo walls are fastened by them. The roof consists entirely of bamboo rafters, crosspieces, and ties; thatch of cogon grass is then secured to that. The Lisu often arrange their thatch in the same way the Lahu Na do, and they claim a roof will last about six years before it needs to be replaced. The walls are made of vertically oriented split bamboo or sawn lumber.

Livestock are kept in corrals and pens near the house, the pig pens about 5 meters (16 feet) in diameter and with meter-high (3-feet-high) walls made of woven split bamboo. Other stockades are constructed of wood.

Mien

Like the Hmong and Lisu, the Mien prefer to have their houses on the ground, though sometimes they are built on posts. Again, the house normally is rectangular in shape and with a fairly standard floor plan. The main posts and beams may be either timbers or sawn lumber; the walls usually are of sawn wood oriented vertically. The fairly steep roofs usually have rafters of bamboo, and they are covered with thatch, wooden shingles, or corrugated metal or concrete.

Many Mien houses have three entrances: the women's door at one end which enters immediately into an area where maize and other animal food is stored and a stove located that is dedicated to cooking pig food; the men's door at the other end of the house leads to the living and guest area; and the main or "big" door, which is used for ceremonies. Upon entering the main door one immediately sees a series of three doors, each leading to a sleeping room, which consists of a raised platform of split bamboo. To the right of the main entrance is the kitchen area with the second door at the end of the building. Often the kitchen area also contains a massive rice pounder (Lewis and Lewis 1984). An altar is located in the center of the living area and directly opposite the "big" door from which the family's direct view of the village's spirit shrines may not be restricted.

The main beams and corner posts are of hand-hewn lumber; secondary beams and rafters, most of which are of bamboo, are lashed with a fiber obtained from a huge leguminous vine, probably *Spatholobus parviflorus*, found in the higher forests. The roof is made of a combination of palm fronds and cogon grass, but not in the usual form of shingles. Rather, it is laid down on the rafters and then bamboo battens are secured on top. Interestingly, the thatch on the overhang along the main side of the house is put on backwards, with the cut end of the grass pointing down. Not only is the grass more manageable and looks better, but with the stiffer ends down the wind coming up the side of the hill will not pick it up and blow it away. A grass roof may have to be replaced in just two or three years, but it is cooler in the summer and warmer in the winter than are metal or concrete roofs.

Animals are kept in wooden or bamboo pens near the house. One such pen in Huay Gaew was raised about 1.5 meters (5 feet) above the ground and had a "tile" roof made of sections of bamboo, one inverted over the next.

MODERN HOUSE CONSTRUCTION

With the continuing influence of the lowland Thai and the infusion of more wealth to at least some tribal households, there is a steady change in house construction. Many tribal villages at lower elevations are now a mixture of thatched roof houses with split bamboo walls and those with wooden walls and corrugated roofs, either of metal or concrete. Some tribal people like the newer style roofs because they do not need to be replaced every few years and there is not the constant threat of fire, a phenomenon that may cause the destruction of parts of entire villages during the dry season. More and more lumber and sawn posts are purchased from Thai who often have illegally cut the trees. The wooden walls and floors are more permanent than those of bamboo, so as the tribal community slowly converts to a cash economy and seeks more ways to earn money, time not spent on constructing or repairing a house can be used for such things as sewing and growing cash crops.

Despite the relentless movement of modern ways into the hills of northern Thailand, one can still see in the more isolated mountains of northern Thailand the traditional tribal house, which is the same as that built and lived in by many previous generations, whether in Burma, Laos, or China. Tribal houses give a fascinating glimpse of how these people can create comfort and protection from what materials the forests offer them.

10

Plant Fibers and Dyes

The small Blue Hmong village of Pha Kia Nai is located in a lovely valley in the mountains west of Chiang Mai at an elevation of just over 1000 meters (3300 feet). This village is famous for growing and weaving hemp fibers into cloth on which the villagers do their traditional batik work.

In late November the edges of some swidden fields had extensive rows of hemp plants (*Cannabis sativa*) nearly 3 meters (10 feet) high (Plates 155–156). When I visited the village in July, the long hemp stems had been harvested, stripped of their leaves, and stacked to dry against fences and the wooden walls of several of the houses (Plate 157). In an open porch, an older woman squatted by a low table on which lay a length of hand-woven, bleached hemp cloth about 4 meters (13 feet) long and 30 centimeters (12 inches) wide. Carefully she drew a design on the cloth in charcoal (Plate 158). Later, using a small instrument which the Indonesians call a *tjanting* and which is made of copper tubing and bamboo, she will carefully apply lines of melted beeswax.

Another woman with darkly stained hands appeared from the back of the house. Gathering up a piece of material that had already been waxed, she returned to the dyeing area. There she deftly immersed the cloth into a hollowed-out log filled with indigo dye (Plate 159). The process of coloring the unwaxed parts of the cloth requires many hours of soaking in the dye solution, but the final product, once the cloth is boiled in water to remove the wax, is a beautifully batiked fabric that will be eventually crafted into a heavily pleated skirt also bearing panels of distinctive Hmong needlework (Plates 7, 160).

Few Blue Hmong now spend the time or effort to create these works of art, preferring to buy ready-made cloth in the lowland markets, some of which even have machine-printed fake batik designs. And, even if fabrics are woven by hand within the villages, many tribal people now use synthetic dyes rather than the natural dyes (in this case, indigo) found in the forest or grown in village gardens.

The art of gathering and preparing fiber plants, weaving, and then dyeing the cloth with natural dyes is a rapidly disappearing activity among the tribal people of northern Thailand, for the pressures of a cash economy and the need to make

157

handcrafts for tourists mean that the highlanders simply can no longer spend the time and effort making and dyeing cloth in the traditional ways. Ropes and cords made by hand from plant fibers are also being replaced by market products.

FIBERS AND WEAVING

Fibers serve a myriad of uses within a highland village. Even though many types of plastic cords and ropes are now available, materials from the forest are nonetheless still essential and most tribal people remain knowledgeable of their sources. Fibers are used for clothing, footwear, blankets, nets and snares, traps, padding and pillows, bags and baskets, ropes and cordage, crossbow strings, and even for decoration. Two plants, cotton and hemp, provide most of the fibers used for clothing by the hill tribes today, although many forest plants are still extensively used in other ways.

Cotton

Cotton (*Gossypium* spp.) is clearly the most widely used of the fiber plants available to the hill tribes, especially for clothing. Tree cotton (*G. arboreum*) (Plate 161) is generally preferred to the small, bushy type (*G. barbadense*) because the fibers are of better quality and more plentiful. Both types of cotton are planted from seed and propagated around the villages. *Gossypium arboreum* often lines the narrow streets of the villages, reaching a height of 4 or 5 meters (12–15 feet). *Gossypium barbadense* is grown in fields near villages, usually as an annual crop. Both plants, however, produce a long-staple fiber that can be made into a high-quality thread.

The mature cotton is harvested and ginned to remove the seeds with simple handmade instruments usually consisting of two rollers. The Akha have a special bowlike structure made of bamboo that is used to "shoot" or fluff the cotton fibers into tufts up to 10 centimeters (4 inches) long (Campbell et al. 1981); this tiny bow is pushed into a small pile of cotton and plucked to produce the long clusters of fibers (Grunfeld 1982). Most of the tribal women make their thread with hand- or foot-operated spinning wheels in a manner very similar to that of Western women (Plate 162). However, Akha women use a different method, one that I have also observed among some groups of South American Indians who make wool thread. Akha women carry long tufts of cotton in small bamboo baskets attached to their waists (Plate 163). As they walk along, usually to and from the fields, they spin these tufts into thread using small weighted spindles which are rapidly spun by rolling them on their bare thighs which have been exposed by hitching up their miniskirts or perhaps even tucking a bit of the hem into their waists (Plate 164). The tuft of cotton is twisted with two fingers of the left hand as the right hand twirls the spindle on the thigh. The thread is attached to the spindle, which twirls rapidly in the air, its weight and rotation being carefully controlled to provide the proper twist to the thread that forms. The thread is then wound onto the shaft of the spindle, with the process being repeated until the shaft is full or all the tufts of cotton are removed from the container and made into thread; finally, the finished product is rolled into a ball and stored in a shoulder bag. In the village the thread is unrolled, starched in water that has been used to soak rice, and then transferred to long I-shaped frames made of bamboo. When the thread is dry, it is ready for weaving.

Hemp

The Hmong have traditionally made most of their cloth from hemp (*Cannabis sativa*), a plant having strong, durable bast fibers (Plates 155–157). Hemp is harvested when 2–3 meters (6–9 feet) high; the leaves are removed, and the stem soaked in water and allowed to dry. The woody parts are then stripped away by breaking the stems with a long thumbnail which the women grow specifically for this activity. Breaking the stems exposes the coarse, yellowish fibers. The women fray the ends of the fibers and fold together three small clusters of fibers, which create a long chain that is spun into thread with a foot-operated spinning wheel. The finished threads are boiled in ashes, which serve as a bleach, and rinsed two or three times until the desired light color is obtained. The threads are then ready to be woven into cloth.

The Lisu still occasionally grow *Cannabis sativa* for fiber but, like the Hmong, usually do not smoke or eat it for its hallucinogenic effects. They do, however, eat the leaves as greens, and they use the coarse, strong fibers to make crossbow strings and to weave heavy blankets. Years ago the Lisu made much of their clothing from this type of fiber, creating the threads and weaving the cloth in a manner much like the Hmong. They then either dyed or repeatedly washed the cloth to make it almost white. They never batiked it. Today they buy cotton or polyester-blend cloth in lowland markets and spend their time sewing intricate patterns and stripes onto it, thus creating the distinctive Lisu products.

Kapok

Ceiba pentandra, the kapok or silk-cotton tree, is commonly grown along streets in tribal villages. The fibers of this tree, as well as those of *Bombax ceiba*, the red silk cotton tree, are used for making pillows and mattresses, although some tribal people prefer to sleep on woven bamboo or reed mats.

Fibers for Ropes and Other Purposes

A number of forest plants provide fibers that are still used by the hill tribes for a variety of purposes, though cordage and ropes purchased in lowland markets are replacing them, again, because of the time and work involved in harvesting, preparing, and weaving or twisting the natural product into the item desired. The fibers described below are still used by tribal people, especially in more isolated areas.

The palm, *Caryota mitis,* is a source of fiber for strings in bird traps. The fiber is also used as tinder when striking a fire. Another palm, *Wallichia caryotoides,* is an important source of fibers for making ropes and cordage.

The rattan palms, *Calamus* spp. (also described in Chapter 5), are used for many purposes because the fibers are so strong and flexible. Rattan fibers are made into string and rope, first by twisting three fibers together, holding one end by the big toe, to make a thin string. A thicker, stronger piece is made by adding three more fibers at a time to the thin string. A wooden fork is also used when a larger number of fibers are added in order to keep them from getting tangled. Ropes can also be made by braiding three to twelve threads, and a flat, plaited rope is often constructed by using up to twenty thin threads.

One of the most interesting uses of rattan is for making rope for the gigantic Akha swings which are used in village celebrations each August (Plates 165–166). However, rattan palms are now almost impossible to find in many areas of northern Thailand because of overcollecting and the destruction of forests having the rich, moist soils in which they are found. These palms are still common in the Shan states of Burma, so often Akha groups will go there to collect it.

A somewhat satisfactory substitute for the Akha rope swing is the vine *Mucuna macrocarpa,* which the Akha call *a beu ci ni.* Its woody stems are cut into lengths of several meters, the bark removed, and the strips of fiber extracted and braided into a long piece of rope more than 10 meters (33 feet) long. Unfortunately, this substitute rope is not as strong as rattan rope and there have been some serious accidents when it has broken while people were swinging.

The bamboo, *Thyrsostachys siamensis,* is sometimes used as a substitute for rattan, but it does not have the flexibility and durability of the palm.

The bark of the common fern, *Lygodium flexuosum,* is used to weave bags and baskets and to make cordage.

Pueraria phaseoloides is a vine whose fibers the Akha use to make open, string shoulder bags. They shave the stem of the vine with a knife to remove the outer green bark, then strip off the inner fibers and make a long string by twisting the fibers on the thigh (Plates 167–168).

Smilax ovalifolia is a fairly common woody vine whose bark is peeled off to remove the inner fibers, which the women weave into necklaces or bracelets.

Several forest trees provide fibers from which ropes and cords are made. The bark of the shrubby tree, *Hibiscus tiliaceus,* is used to make rope and cordage, as is the much smaller plant, *Rhynchotechum obovatum. Bauhinia bracteata,* a small tree, is used to make the rope by which the Akha tie the water buffalo that is sacrificed in funeral ceremonies (Plate 101). The young bark of the plant known to the Hmong as *tcia* (*Maoutia puya*) is stripped off to make rope and a type of cloth. The bark is carefully made into thread, which is then woven or twisted. This material is particularly good for crossbow strings.

One of the mulberries, *Broussonetia papyrifera,* is commonly used by the Lahu to make cord or rope. They make it from the roots and also from the outer, very thin layer of the bark; the latter makes a particularly good cord. The fibers from this same plant are also used by the Shan people to make "Shan paper" or what many Westerners incorrectly refer to as "rice paper." In fact, raw opium is wrapped in this paper, the only kind of paper used for this purpose (Plate 121). This species of mulberry, as well as two others which are used for food and medicine, are often grown in swidden areas and in the villages.

The Lahu use *Sterculia pexa* for making different kinds of bindings for various things, especially in Burma. The bark is cut into strips, which are then used for tying things. The bark of *Syzygium cumini*, known to the Lahu as *pfuh hkaw ceh*, is pulled off, twisted, and made into cord or rope. *Urena lobata* is a small herbaceous plant that is also used as a source of fiber, particularly by the Lahu.

Gnetum sp. makes a very fine, durable fiber that is used to tie on the crossbow arrow stabilizer. A type of paper is made from the bark of *Broussonetia papyrifera*, which the Akha will gather and sell for as much as 10 Baht (U.S.$ 0.40) per kilogram (2.2 pounds). The Akha occasionally use the leaves of *Bauhinia ornata* as cigarette papers.

The fiber pad which the Karen place beneath the "saddle" on the elephant comes from the bark of *Antiaris toxicaria* (Moraceae). This very tall emergent tree, known as *k'aw* to the Lahu, is also the source of a very potent poison (see Chapter 5). Seen near Mae Hong Son, this tree is rare in Thailand but much more common in Burma. Phengklai and Khamsai (1985) state that elephant pads are also made from the bark of *Lannea coromandelica*.

The bamboos (see Chapter 6) also provide fibers for a multitude of uses. Traditionally, the Hmong made a special ceremonial paper from bamboo by chopping up and boiling young shoots, which were then pounded and further boiled until the cellulose fibers were separated. This pulp was spread on loose cotton cloth stretched on a wooden frame and allowed to dry (Bernatzik 1970).

The Akha cultivate *Abelmoschus moschatus* as a source of both medicine (a root poultice is applied to infected sores or wounds) and fibers, which are obtained from the roots and bark by soaking and then scrubbing the plant parts to separate out the fibers. They are later dyed and used in the women's headpieces (Katherine Bragg, herbarium label #20).

Weaving Techniques

Although hill tribe people now purchase hand-woven cloth from the Shan of Burma or machine-made fabrics in lowland Thai markets, weaving is still practiced in many villages. Most women (and some men) prefer to wear the traditional costume made of handmade fabrics on special occasions. Moreover, the handmade cloth upon which the distinctive tribal patterns have been sewn can sometimes bring in considerable cash income through selected tourist markets. Weaving is done in the winter months following the harvest of rice and is strictly women's work.

The threads that have been spun or purchased in the lowland markets must first be carefully prepared for the loom. Lahu women take long skeins of thread to open areas near their houses and string them between pieces of bamboo pounded into the ground about 10 meters (33 feet) apart; they also construct bamboo frames on which to stretch the skeins. Dozens of threads are thus stretched out. In some cases the threads may first be soaked in rice water and then dried, thus stiffening them for easier handling. Just before the threads are gathered to make the warp or lengthwise pieces of thread on the loom, they may be sprinkled with fine wood ash to absorb any excess starch and to assure that each thread is thoroughly dry (Grunfeld 1982).

Three types of looms are used by the hill tribes to weave cloth from cotton or hemp fibers. The Karen use the horizontal **back-strap loom** for all their weaving, and the Lahu and Lisu use it for certain types of articles, such as shoulder bags (Plate 169). One end of the warp is attached to a pole, often a house post, and the other to a small wooden frame or "cloth stick," which is then secured to a wide leather strap that is put around the small of a woman's back. Thus, the weaver literally becomes part of the loom. She sits nearly straight-legged on a woven mat on the floor so that she can push with her feet against a piece of bamboo or wood that is anchored to the floor or wedged against the house post to which the warp threads are attached, thus enabling her to provide the proper tension as she weaves. The width of the woven piece more or less corresponds to the width of the woman's hips, usually about 40 centimeters (16 inches), because she must be able to move the shuttle containing the weft threads from side to side while in her stationary position. The warp threads are changed with a "shed sword" and "heddle rod" as the weaver leans against the leather backstrap. Each time she changes them, she rapidly pushes the shuttle from one side to the other, adding the weft and beating it with a stick inserted in the warp.

The second type of loom is the more familiar **foot-treadle loom,** usually constructed of a large wooden or bamboo frame in which four posts nearly 2 meters (6 feet) high are firmly sunk into the ground (Plate 170). The upper ends of the posts are held rigid by a framework of tightly lashed smaller bamboo pieces. The massive wooden shedding mechanism with fiber heddles in wooden frames is hung by ropes from the framework above; dangling below are two bamboo treadles attached by cords which function as the lamms to raise and lower the heddle frames during the actual weaving process. A comblike reed keeps the warp threads evenly spaced and separated, and a wooden paddle is used as the "beater" to pack the weft or cross threads tightly together.

The Akha feel that *Saccharum spontaneum* is the only plant that can be used to make the reeds for weaving. Bamboo, broom grass (*Thysanolaena latifolia*), or other plants cannot be used. Each day the weaver sets up and takes down her loom, leaving only the bamboo frame. In less than five minutes she can attach the already-woven fabric to the front of the loom and the warp threads to the back. She then hangs the shedding mechanism and reed portion to the frame and is ready to again take up the weaving process, standing to one side of the heddle frames in a position that enables her to easily reach each side of it and the treadles. Quickly she pushes the massive shuttle containing the weft threads wound on a long spool through the open space, changes treadles to raise one heddle harness and lower the other, and pushes the shuttle back to the original side. Proper tension on the warp and newly woven material is maintained by keeping the rolled fabric tight on the cloth beam at the front of the loom; a piece of bamboo is inserted through the beam and tied to the frame. The Akha typically weave cloth no more than 25 centimeters (10 inches) wide, but sometimes the Lahu may weave pieces up to 50 centimeters (20 inches) wide. An experienced weaver makes 6 or 7 meters (20–30 feet) of new cloth in a long day's work.

The third type of loom is used by the Hmong and is a combination of the above two types. The loom is a bamboo frame similar to that of the foot-treadle type; at one end there is a wooden bench and at the other the warp is firmly attached to the back beam, which is part of the frame. The weaver then attaches the other end of the warp to a cloth stick and wide backstrap, putting it around her back to provide the proper tension as with the back-strap loom. However, in this case she sits on the bench while weaving and changes the warp threads with foot treadles and a simple shedding mechanism.

DYES AND DYEING

Once a piece of cloth is woven, it is usually dyed, and traditionally the sources of these colors and various mordants or setting agents were plants from the garden or forest. After extracting dyes that come from trees, the highlanders carefully cover the cuts with mud to help them heal. Thus, this harvesting of natural forest products does not seriously affect the vegetation. However, in recent years more and more tribal people are buying synthetic coal tar dyes from lowland Thai markets. The resultant colors, in my opinion, are much more garish and certainly less typical of the subtle and beautiful hues they formerly obtained from natural dyes, but the tribal women feel these synthetic dyes are more permanent, more sellable to tourists, and certainly less work to prepare. Nonetheless, there are still many tribal women who continue to use—or at least still remember how to use— the natural dyes obtained from plants.

Indigo

The most widely used color of the hill tribes is a black or dark blue, such as that obtained with indigo (*Indigofera tinctoria*). The plant is grown in village gardens and the young branches and leaves harvested when it is time to do the dyeing. The collected material is placed in a vessel of water and allowed to soak and decay for two to six days. Occasionally the branches are turned over to make sure they soak evenly. During this time the color goes from the plant into solution in the water. Lime is added to absorb (fix) the color and act as a medium for the dye. In a day or two the color and the lime settle to the bottom of the container. The plant material and water are poured off, leaving the thicker, heavier slurry at the bottom of the vessel. A special sieve is prepared with the leaves of *Osbeckia stellata* and the bark of *Schima wallichii*; charcoal ash is placed on this plant mixture. First water is slowly poured through the sieve, followed by the slurry containing the dye. Great care is taken to prevent the ash from passing through.

A variation of this procedure is to simply mix the ashes with water to create an alkaloid solution. When the ashes settle, the water is poured off and added to the dye slurry containing the lime, which is now distinctly blue in color. Some tribes add a bit of rice whiskey and let the solution sit for another two or three days. This indigo dye is now ready to use and will produce a dark blue color (Campbell et al. 1981).

The actual dyeing process is accomplished by immersing the cloth into the container with the dye and leaving it for about half an hour (Plate 159). It is then removed, hung up, and allowed to dry. This procedure is repeated two to four times each day until a fabric with the desired darkness is obtained; in some cases the fabric is dipped up to 30 times (Campbell et al. 1981). The dyed material is washed two or three times, and after that the dyeing process is finished.

The Lahu use *Artemisia austroyunnanensis* and *A. dubia* as mordants to fix the indigo dyes; the plants are dried and burned, with the ash then being boiled in the water in which the dyed fabric is immersed.

Black or Dark Blue Dyes

The hill tribes have found some dye plants that work like indigo, producing dark blue or black colors: *Chloranthus elatior* and *Pseuderanthemum andersonii*. The dyes are obtained from the leaves by soaking them in water for about a week. The plant material is taken out of the water and put in a large jar or container and allowed to settle over a period of several days; everything but the sediment is then disposed of. Ash is mixed with water and screened through a sieve, consisting mostly of *Albizia chinensis*, to leave a brownish red liquid, which is poured into the container with the lime and plant material. The solution is carefully mixed and almost immediately turns blue. It is stirred again just before dying a piece of cloth.

Three species of *Strobilanthes: S. anfractuosa, S. lanceifolius,* and *S. pentstemonoides* make a black dye using the same method of preparation as described above for the indigo dyes, but no other plant is added in the preparation.

Baphicacanthus cusia is another source of blue dye, very similar to indigo, as is *Marsdenia tinctoria,* which the Akha actually refer to as wild indigo (*ni myah*). These dyes are prepared in the same way as true indigo.

Dunbaria longeracemosa apparently is used as a substitute for indigo by the Lahu.

The fruits of *Diospyros ehretioides* produce a black dye (Sukthumrong et al. 1979).

Red, Yellow, and Green Dyes

All the plants considered up to this point produce either blue or black dyes, but other colors are available.

A big tree known to the Akha as *k'oe toe* is the source of a red dye, but villagers are not sure if it still grows in Thai forests. If so, it occurs very close to the Burma border and I have not yet been able to find and collect it for identification. The Akha remove the bark in thin strips, which are then boiled in water, to make the red dye. This same bark can also be used as a substitute for betel.

Macaranga gigantea is called *hpa to* by the Lahu and *lm pya* by the Akha. Both tribes are well aware that when the plant is cut, a sticky, mucilaginous material comes out, which is the source of a red dye. The plant parts are boiled to release the dye-containing gum, which is used mainly as a paint rather than as a dye.

The Lahu remove the bark of *Nothaphoebe umbelliflora*, dry it, and mix it with other ingredients to make a paint for their houses.

A very important source of an orange-red dye for the Karen is *Morinda angustifolia*, which they know as *kho*. A large amount of root material is gathered, the best parts pulverized, and then boiled in water. A small amount of pork fat or vegetable oil is added, as well as a small amount of citrus juice (citric acid). The solution is watched carefully as it boils. A piece of cloth is put in the solution, which continues to boil until the cloth gets to be the right color. The cloth is then taken out and allowed to drip dry; it is never wrung out. The color is best described as a brick- or orange-red. The brightly colored red clothing of the Karen today has unfortunately been colored in synthetic dyes purchased in the market. Few Karen even know how to use the roots of *Morinda*. The Akha also say that they use the leaves of *Morinda* for dying cloth red.

Buddleja asiatica is used to color rice; its flowers are boiled to make a yellow liquid, which is then poured over the rice. *Curcuma domestica*, known as turmeric, is the source of a yellow dye. The rhizome is pounded and boiled in water; the plant material is then removed and the colored water used for dyeing.

A few additional native and introduced plants yield dyes for the hill tribes: *Bixa orellana*, introduced from South America, has seeds that are a source of red dye (Katherine Bragg, herbarium label #112); *Caesalpinia sappan* yields a red dye from the bark or wood and a yellow one from the roots (Phengklai and Khamsai 1985); the heartwood of *Garcinia dulcis* produces a green dye (Sukthumrong et al. 1979); and *Terminalia bellirica* yields a yellowish green dye from the bark and fruits (Sukthumrong et al. 1979).

Dyeing Techniques

The Akha Technique. The Akha, who are masters of dyeing fabrics dark blue or black, have developed a very elaborate process unique to their tribe, which involves the following steps:

1. One or more of the dye plants, which are in the Acanthaceae (e.g., *Phlogacanthus curviflorus, Strobilanthes lanceifolius*, and another member that has not yet been positively identified, but which is probably another species of *Strobilanthes*) are placed in water in a wooden barrel or trough. The plants are allowed to ferment or rot for two or three days (it can be as long as five or six days) with occasional stirring. When the process is complete, the leaves are squeezed dry and removed.

2. Lime is added to the liquid and the mixture gently stirred by dipping a ladle in and out of the solution.

3. The mixture is allowed to settle for about 24 hours. During this time the dye settles into a slurry at the bottom. The clear fluid on top is poured off and thrown away.

4. A container consisting of two sections is now set up. Leaves of *Osbeckia stellata* and *Schima wallichii*, which are softened over the fire to serve as a sieve, are put at the bottom of the upper section. Ash made from the

bark of any *Lithocarpus* species is placed on top of this sieve. Water is added and allowed to pass through the sieve of ash and leaves into the lower section.

5. The dye slurry is put into this lower section with the solution that has just been filtered through the sieve. The two are mixed very carefully and slowly so that no sediment from the ash solution gets in. Stirring results in a slight foam, and at this point the dye is ready to be used. The proportion of dye to ash solution is determined by the color of the liquid, but most of the mixture is ash solution. More dye slurry probably will be added as the cloth absorbs the dye mixture and takes on color.

6. Cloth is placed into the dye once a day for one to two hours, taken out, and allowed to dry (Plate 171). The Akha believe that sunlight and heat are very important in getting a permanent, dark color. On a very hot, sunny day the cloth may be dyed twice. The degree of darkness (from blue to black) is determined by both the strength of the solution and the number of times the cloth is dyed. It may be dyed five or six times, but a strong dye solution will produce a dark blue color in just one dyeing.

7. The *yeh* plant (*Albizia chinensis*) is then used as a mordant to make the dye fast by boiling the leaves in water and then putting the dyed cloth into the liquid to soak for a period of time.

The Hmong Technique. The Hmong are the only tribal people in northern Thailand who do batik work. Described earlier in this chapter, the technique is done only on cloth made from hemp fibers with beeswax gathered in the forest and applied with a *tjanting* usually made of copper and bamboo. The Hmong claim there are only seven basic batik designs, but each woman combines them in different ways to produce a wide variety of patterns (Plate 158). Interestingly, no curves are ever part of the design; only straight lines and dots are used (Campbell et al. 1981). Once the pattern has been drawn on the cloth, a woman carefully applies the beeswax, taking about two hours to do a piece of cloth about 15 centimeters (8 inches) square. The waxed material is dyed in indigo and then boiled in water to remove the wax, which is carefully collected to be reused in later batiking. The resulting fabric has an intricate white design on a dark blue background wherever the wax had been applied. It may take up to six months to make sufficient material for a woman's skirt; once the main, batiked portion is finished it is immersed in rice starch and pleated by pressing it with large flat stones. The pleats are then basted to hold them permanently (Campbell et al. 1981). Hmong women believe that only the heavy pleated skirts made of hemp fibers sway correctly when they walk, so they take great pride in making these beautiful pieces of clothing and wearing them on special occasions.

The Karen Technique. Karen women practice the technique of *ikat* or tie-dyeing, in which they color some of the warp threads prior to weaving. Each woman loops the undyed threads around a bamboo pole and then ties them in a distinctive pattern using grass stems that have been soaked until they are flexible. Another method is to wrap the threads around a piece of bamboo that has been cut in the shape of a hand; the grass ties are then put on the thread between the "fingers" of bamboo. The tied threads on the pole are immersed in a red dye to produce the distinctive pattern. Most women jealously guard their distinctive methods of *ikat* dyeing, even doing it hidden in the forest because they believe it is associated with spirit worship (Campbell et al. 1981).

THE INFLUENCE OF FIBERS AND DYES ON TRIBAL DRESS

The different weaving and dyeing techniques of the highlanders are clearly reflected in each tribe's distinctive style of dress. Each group also employs different combinations of decorations, such as seeds, embroidery, and applique, to make their clothes.

Almost all cloth used by the Akha is dyed black or dark blue, and decorations are mostly red or white. Men wear plain black trousers and shirts. Women wear short skirts, halters, black jackets, and leggings; the skirts are rarely decorated but the other items are usually covered by exquisite patterns of applique, embroidery, and bead and seed designs of many colors (Plate 12). The magnificent headpieces contain silver coins, buttons, beads, tassels, dyed feathers, and gibbon fur (Plates 13, 172).

Some Blue Hmong still weave hemp cloth for making skirts and baby-carrying cloths, but otherwise they now purchase almost all the cloth they use, as well as the colorful threads with which they create their elaborate needlework patterns (Plates 6, 134).

Traditionally, the skirts of White Hmong women were woven from white hemp and without ornamentation; however, these are now worn only on special occasions and the usual mode of dress for women is loose black trousers and jackets, the latter of which are trimmed with their distinctive embroidery. The cloth of both men's and women's outfits is cotton, frequently velvet or satin. Now other fabrics are used, such as polyester and rayon.

The Karen are perhaps the most skillful weavers among the hill tribes, and will use either embroidered or woven decorations depending on the geographical area and their group's tradition. Many Sgaw Karen, for example, embroider patterns almost exclusively with Job's tears (*Coix lachryma-jobi*) (Elaine Lewis, personal communication). Shirts and dresses are created from two pieces of woven cloth that have been partially sewn up the sides and on one end with threads which often differ in color from the woven fabric, thus leaving head and arm holes. Unmarried girls wear undyed, white dresses in which some patterns have been woven or stitched. Married women, on the other hand, wear predominantly red-colored sarongs with colorful, woven horizontal designs. Their blouses may contain elaborate woven geometric designs of many colors, but most often either red or black (Plates 4, 169).

Though not strictly part of the dyeing process, it is interesting to note that the coloring of the skin tatoo still widely seen among tribal men, especially the Karen, is obtained by taking the fluid from the gall of a water buffalo or cow and mixing it with pulverized, select, completely powdered charcoal, usually obtained from trees of the Fagaceae. The concoction is then boiled to a sticky consistency (a very important step). The point of the metal tattooing instrument is then dipped in the solution to make the impression in the skin and the design.

Few Lahu women now weave fabric for their clothing, although they continue to make material on both foot-treadle and back-strap looms for their distinctive shoulder bags and other articles that are worn at the New Year and other celebrations; they also sell these articles to tourists. However, the women still do much handwork in sewing patterns and designs on purchased cloth.

Likewise, the Lisu now weave very little, nor do the Mien, who prefer to buy "Shan cloth" and then spend their available time making the marvelous cross-stitch embroidered panels on the fronts of their trousers and turbans.

Hill tribe handwork is distinctive and an important part of the tribal cultures. In recent years it has become an important source of income. However, with the financial stimulus to create large quantities of their work, many women have resorted to short-cuts and modern techniques, abandoning their traditional use of plant fibers and dyes. Sewing machines are commonly used even in remote villages, as are threads, dyes, and fabrics purchased in lowland markets. Slowly the tribal knowledge of natural fibers and dyes is being lost, but for those women who continue to create fabrics and designs in the traditional way, there will always be some who will appreciate and respect them, such as textile collectors and "back to mother-earth" people. Many will also continue to value the magnificent baskets and other articles woven from rattan and bamboo. Hopefully, the highlanders will continue to create these products, for they represent an important aspect of their distinct cultures.

11

Plants and the Spirit World

Most of the hill tribe people are animists, aware of a world full of spirits, with whom they must constantly deal. They have learned ways of protecting themselves through ceremonies, sacrifices, potions, and behavior. Plants play a significant role in these continuing encounters with the spirit world.

SPIRITS OF THE FOREST

There are special places in the forests where spirits dwell; these areas and the sacred trees within them are to be respected. Sometimes the sacred place is considered an especially auspicious place for burials, so hunting is usually prohibited in the area right around the large sacred tree or trees. If a tree is large, it is often considered a likely place for the spirits to dwell. Very tall trees are important to the Hmong because it is then easy for the "spirits on high" to reach the villages.

The Akha almost always have a sacred tree or a small area near the village that is set aside and revered. No plant may be cut, so everything grows large and has an aura of great power. The Akha visit this area when there is a need to appease or call upon the assistance of the owner-lords of the region, which may include several villages in the vicinity. The people therefore can enlist the help of not only the lords of land and water and other local forces, but also those of the other villages, especially when there is a special need, such as a serious disease or some other disaster. At the base of the sacred tree they construct a spirit altar with four posts, three walls, a roof, and a ladder going up to the small floor within (Plate 173). There must be nine steps on the ladder. No particular species of tree is necessarily chosen. The shaman must do the honors at the altar, making offerings and sacrifices of pigs, male and female chickens, sesame, chili, rice, and maize. Parts of the pig, such as the tail, ears, hooves, and so forth, are put onto the altar. Almost anything that is edible can be offered in small amounts. Small containers with rice liquor and water are also placed in the altar.

Some of the hill tribes, as well as the Shan, especially revere species of the genus *Ficus*. Individual banyan trees (*F. benghalensis*) or other massive trees of the

169

forest are often considered holy, and spirit houses or altars are constructed at their bases. Perhaps some of this respect is due to the influence of the bo tree (*F. religiosa*) in Buddhism. However, the Lahu are afraid of the spirits of *F. religiosa*, probably derived from the Thai, who also fear it.

Every village has a tree renovation ceremony at the beginning of the New Year. This idea, borrowed from the Shan, is practiced by the Lahu, Karen, Lisu, and Akha. The tree selected for this ceremony is a jungle tree but it must be one that produces fruit at some time. Each year the men go to this sacred tree located on a hill above the village and make an offering usually "every third sabbath" unless the village has problems and more frequent offerings are necessary. When a person departs permanently or arrives to live in the village, he or she also makes an offering at this tree, usually including a pig or chicken and a variety of plants and other items that are important in daily living: tea, rice, and chili. The Lahu Sheh Leh almost always have such sacred forest areas where they go to make offerings and they absolutely prohibit the shedding of blood in this area. Sometimes medicinal plants can be taken from the area, but if so, the spirits must be properly notified. There is a platform on or near the tree that is supported by a post upon which the offering is placed. Only the headman and medicine men can approach this altar on behalf of the village. All firmly believe that honesty and integrity are necessary virtues when the village leaders, medicine men, and even the villagers themselves deal with the spirits. As deforestation continues, the size of the sacred forest area becomes smaller and smaller so that sometimes it may consist of only one large tree.

Plants themselves also seem to take on almost magical properties for the people of the hills, who often feel they can even control the forces of nature with certain plants. For example, the Akha believe *Iris collettii* protects them from thunder, so it is planted by their huts in the rice fields to protect the farmers when they stay there. However, for the Lahu, *Albizia chinensis* has almost the opposite effect; it is called the "lightning tree" and, if possible, is not used in house construction because it "attracts lightning." The Lahu claim that if this tree and another type of tree are standing together in a storm, this is the one that inevitably will be struck by lightning. *Cassia fistula* (*k'a ceh*) is a small forest tree, and the Akha claim that it is a very important protection against "hurricanes" or high winds. They burn the dried fruits (legumes) because the smell seems to take the high winds away from the house or village. The Lahu say that if you eat *Centella asiatica* at every meal of every day for three years, then you become invulnerable. You cannot be shot, cut, or hurt in any other way.

At times plants can protect one from animals—or even fear of them. The Lahu sometimes plant *Impatiens balsamina* near ponds to keep snakes away. The flower buds supposedly produce something that is toxic to snakes. *Rubus blepharoneurus* is an important plant used as a charm to overcome fear, used in a pile with cowrie shells in a sort of snare. A stack of nine layers of this species is made and then placed so it can be flipped over by a special kind of bird snare. If the pile of leaves flips over properly, then that means the Lahu need not be afraid any more.

The spirits of the forest also dictate how and where trees may be cut. Often the Akha ask permission from the "spirit-owner" of a large tree by hacking the tree with a large machete which they leave in it overnight. If they return the next day and find the machete has fallen to the ground, they believe that the "spirit-owner" has not given them permission to cut down the tree.

If a tree were to be cut down too near the village gate, their water source, the graveyard, or their sacred tree, the Akha believe that tigers will kill some of their livestock and villagers will become ill (Paul Lewis, personal communication).

EVIL SPIRITS

Plants exert great powers over the spirits, so the hill people have developed a large arsenal of herbs to protect themselves and influence the spirits' actions with the village.

Tribal people work fervently to keep evil spirits away from members of their families and their villages. Several common plants are employed in this way by both priests and family members when the need occurs. For example, peach wood (*Prunus persica*) is burned by the Lahu to keep evil spirits away, and even rice (*Oryza sativa*) has great importance in this respect. One common practice is to sprinkle around the house pounded rice that has been sanctified by a medicine man. A further step involves placing a sickle under the pillow of a person especially in need of protection, such as a pregnant woman. If the person thus guarded perceives the odor of burning cotton, it means the evil spirit has gotten into the house. Quickly the sickle is snatched up and swung wildly through the room to get rid of the spirit.

The Karen believe that the introduced Century Plant, *Agave americana*, bearing spines along the margins and at the tip of its large succulent leaves, is an important deterrent to the bad spirits (Plate 174). If bad spirits bother anyone in a Karen village, a piece of *Agave* is placed above the doorway. A plant in the yard nearby is also good protection. *Opuntia dillenii*, a cactus introduced from the New World, has taken on a similar significance because of its spines (Plate 175). Several tribes plant it in their villages because they believe its prickly nature wards off evil spirits. Likewise, the Karen use *Smilax ovalifolia* when a person is often sick, believing these thorny stems hung on the door will scare away the evil spirits.

To appease the powerful spirit *Bga*, the Karen may use many different kinds of seeds. *Bga* is a spirit which may affect the general well-being or living conditions of a person or family, so continued offerings are important. Roasted seeds are even placed with the body of a dead person in the coffin so *Bga* has no question that the person is dead. The Karen use seeds of watermelon (*Citrullus lanatus*), peanut (*Arachis hypogaea*), and even maize (*Zea mays*). They believe that pairs or even numbers of seeds are bad; ideally, one should have three, five, or seven different kinds of seeds.

The Akha employ a wide variety of plants in their constant battle against a world possessed by evil spirits. *Smilax ovalifolia* is a very important plant when someone is possessed by an evil spirit. The vine is woven into three circles or rings to which three cords made with white cotton string are added. This large, hoop-like structure is then passed from top to bottom over the victim's body to rid him or her of the evil spirit. The Akha then take a large piece of the herb called *meh*

(*Alpinia* sp.), and use it as a whip, hitting the person on the back to help get the evil spirit out. This same procedure can also be used if a person has had a run-in with the police and has been badly scared by them. Another plant used in the continuing battle against evil spirits is *Vitex trifolia*. A piece of this vine is dipped in water and then used to hit the affected person under the arms, behind the knees, and on the toes to "chase the spirit." The leaves are also bundled in the shaman's headdress while he consults the spirits. Old men may even wear leaves on their ears when praying in the swing ceremony. *Diploclisia glaucescens* is another vine that keeps a person safe from evil spirits. Sometimes it is put over the door of the house and sometimes it is worn on the ear or on the body to provide safety. The fruit of *Sapium discolor* is used to kill a greatly feared and powerful spirit (Pake 1986).

The beautiful leaf succulent, *Sansevieria trifasciata*, is often planted by the Akha in a front yard because it is a "good luck plant" (Plate 176). The villagers believe it helps prevent jealousy, which could bring bad luck. It also helps promote friendship, even preventing evil spirits from coming.

Once a year the Akha have a festival in which they "drive" the evil spirits out of the village. Everyone carves a sword or gun from wood, except for one person who dresses up as an evil spirit and is then chased symbolically from the village. The wooden guns, swords, and spears are placed on a wooden rack at the main entrance to the village by the gate as a warning to the evil spirit not to return. Numerous other plants provide additional "weapons" in this on-going battle against spirits. Hunters often carry a small amount of the gymnosperm, *Gnetum*, into the forest. A small piece of the plant is burned in the campfire at night to ward off evil spirits and to protect the hunters from other dangers lurking in the forest.

Evil spirits are particularly feared during a period of illness when someone is weakened or when a child is born in the house. It is at these times that the inhabitants of the house are particularly vulnerable. The Akha place *Acacia megaladena, Cassia* sp., *Costus speciosus, Gnetum* sp., *Rubus blepharoneurus,* and *Smilax perfoliata* over the entry way to the yard, over the doorway, along an interior wall of the house, or along the partition dividing the men's from the women's side. None of these plants are eaten or drunk, and, if possible, at least three of these plants should be used together to ward off the evil spirits.

Frequently evil spirits take the form of an animal or even a human being. One of the most feared forms is the weretiger (called *taw* by the Lahu), a human that is possessed by an evil spirit. One kind of evil spirit is satisfied by eating chickens (chicken taw), but the other type eats people. It has the power to become not only an apelike creature, but it can also become catlike. The Lahu believe it most frequently looks like an ape. This much-feared creature can eat babies, the flesh of dead people, or that of sick people. A household with a new baby can be protected by putting some stems of the succulent plant *Euphorbia antiquorum* over the doorway. It is such a powerful plant that even the weretiger will not enter the house (Plate 177). Several other plants protect villagers from weretigers. The wood of *Schima wallichii* is burned in houses where there is the fear of weretigers because the smoke, which is very offensive and burns the eyes, keeps the weretigers at bay. The roots of *Dianella ensifolia* are boiled and the liquid drunk when a person has been bitten by a weretiger.

For an Akha couple (or a village) to have twins is one of the most serious calamities that can befall them. The father must immediately suffocate both children with rice husks and ashes. The dead bodies are wrapped in the leaves of *Molineria capitulata* and placed in a section of split bamboo, which is bound closed, and buried in the forest in a location where people will never go again. The metal digging tool that was used to dig the hole is left on top of the grave. The parents implore the infant spirits, if asked in the spirit world to name their parents, to say "your mother was the bamboo container and your father the digging tool." The family then must burn everything they have—including their house and clothing but excluding their money—spend two nights and three days with the local shaman or witch doctor in the forest, and live outside the gate (e.g., the village) for one year. As further evidence of the seriousness of a "terrible birth," the woman must remove her headdress, something that is almost never done. The couple must travel on separate paths to the fields and stay completely apart from other people. While living in this isolated way, they use *Molineria capitulata* (*a lah jah ma*) for clothing.

Leaves of *Alocasia macrorrhiza* are sometimes used for spirit incantations when there has been a "terrible birth." *Kaempferia* sp., another very important Akha plant in "warding away the spirits," is also employed when twins are born. The villagers make a long string of this plant, which they call *meh* (a name they use for other Zingiberaceae), and put it all the way around the village to give protection from further incursion.

ILLNESS, INJURY, AND THE SPIRITS

The dividing line between the effects of being possessed by evil spirits and illness caused by microbes and viruses is very blurred in the world of the hill tribes. To many, disease simply is caused by evil spirits, so exorcism is a major activity by which someone is cured (see Chapter 8). Just as plants play a role in healing the sick, plants again play an important role in exorcism events.

The Hmong believe that a spirit, having entered the body of a person and caused shock or fainting, must be exorcised. To do this, they boil the roots of the sensitive plant, *Mimosa pudica*, and have the victim drink the liquid. The stems of this plant are also placed in a small bag and hung around the victim's neck as an amulet for protection. A branch of the *tse ma mo* tree (*Antidesma sootepense*) is used by the doorway in their exorcism event. Traditionally the paper money used in exorcism ceremonies was made from the bark of *Boehmeria sidaefolia*. The Akha use *Homalomena occulta* and *Piper pedicellatum* together when a witch doctor has cast a spell on someone, causing illness. These plants, which are soaked in alcohol to be drunk by the victim, help expel the illness.

The tribal pharmacopoeia is immense (see Chapter 8), and plants often are used for both the treatment of disease and the exorcism of evil spirits. This intimate association between health and the spirit world has resulted in many ceremonies having to do with the collection, preparation, and utilization of medicinal plants. For example, the Lahu perform a small ceremony when medicinal plants are collected in the forest. When the area with the desired plants is located, the medicine man then walks indirectly toward the plant to be collected and prays to the spirits to help him heal the person. He then sprinkles the plant and area around it with hulled rice. Having done this, he harvests the plant. If some pieces of certain medicinal plants remain once the patient has been healed, they must be taken out of the house and village for disposal.

The Akha also carry out certain prescribed activities at the time of collection, especially when the disease requiring treatment is very serious. When the plant is collected, it is necessary to make an offering to the spirits, usually money or some silver scrapings. If scrapings are offered, the village priest is careful to see that none of this "gift" to the "medicine spirit" actually falls onto the ground. The priest only acts as if he is scraping silver off the bracelet or other silver article as he goes through the motion three times (Paul Lewis, personal communication). In this ceremony the "owner" of the plant is being "paid off."

The Hmong believe that certain plants are especially powerful and their presence in a mixture of herbs makes almost any medicine more effective. *Kaempferia* sp. is one such example. It is used with other medicinal plants for many types of ailments.

Lahu medicinal practitioners undergo certain privations or perhaps eat certain things to insure that their treatment of various ailments is effective. The civet cat, rhesus monkey, wild goat (goral), and bear meat should not be eaten by practitioners in their houses or even their villages, but it is permissible to eat them in the forest. If these meats were to be eaten in the house (or even village), the household spirits would become very upset. However, the villagers always give the medicine man the best cuts of meat from any other slaughtered or shot animals. Several common plants which grow within the village (*Carica papaya, Cucurbita* spp., and *Sansevieria trifasciata*) are often mixed together by the Lahu to make a medicine to treat internal pain. However, to make it work the person who is making this medicine must eat the seeds of lettuce (*Lactuca sativa*) or parsley *(Petroselinum crispum)* because this will help the spirits "make the medicine work."

There are other strange procedures which must sometimes be done to make the medicine "work." One such example is the Lahu method of preparing medicine for a person having trouble urinating. The fruit of *Pandanus furcatus* is boiled along with the pitch of *Pinus merkusii*, but it will only work if three rocks from under a bridge are in the container!

The Akha place the leaves of *Scleria terrestris* near the body of a person with high fever, believing that even its presence in the room will help heal.

Costus speciosus is used by the Lisu in a somewhat strange way to treat infected ears. A length of the plant about the same height as the person with the infection is cut, brought into the house, and placed near a source of heat such as the fire. The plant is then kept near the patient and allowed to dry to help bring about the cure. The Akha take the stem and leaves of a species of *Amomum*, wrap it around into a circle, and use it to strike a sick person to promote healing.

BIRTH, DEATH, AND THE SPIRITS

Pregnancy is not a time of rest and little work for hill tribe women, for fields must be tended, other children cared for, animals fed, clothing made, and meals prepared. However, certain precautions are followed to allay the disfavor of the spirits and to avoid the "bad" effects of some plants. For example, the Karen prohibit expectant women from eating jackfruit (*Artocarpus heterophyllus*) because they believe it will cause the baby to be born with a skin disease.

A number of procedures are followed at the time of birth, somewhat different from tribe to tribe. Almost all tribes cut the umbilicus of a new baby with a bamboo knife (Plate 100). The Karen take the placenta, wrap it in cloth, and place it in a bamboo culm. The culm is then hung in a tree just outside the village or buried at the foot of the steps leading up to the veranda. A bamboo basket is overturned and a knife placed on top to protect the mother and her baby from evil spirits. The Hmong are very aware that "purple fever," which is caused by the spirits, follows childbirth. The new mother actually develops a fever, which can be cured with *Agastache rugosa*. The flowers of this plant are put in a cloth or handkerchief and tied around the mother's neck for protection. A baby whose fever is caused by a "spirit bite" is bathed with liquid made by soaking the fruit of *Leucaena leucocephala* in cold water.

Death, an inevitable fact of life, but greatly feared by the tribal people, involves much superstition and ceremony. I witnessed an Akha funeral ceremony, which illustrates the importance of plants for both family and village during this critical event. To appease the spirit of death so that it will not return to the house of the deceased, *Maesa montana* (*leh nyoe*) is used with a chicken sacrifice. After cooking the chicken, the leaves of the plant are added to the dish, then eaten. This sacrifice "says" to the spirit of the dead person: "Stay away; we don't want you any more." Another important plant is *Saurauia roxburghii*; it is part of the "separation meal" after death. The shaman "gives" the plant to the person who has died and says: "We divide this with you. Now you are in the land of the spirits. This is for you." The plant is used for no other purpose for it is believed to be a "food for the spirits." *Eranthemum tetragonum* is also used in spirit incantations when a person dies.

The funeral ceremony lasts for several days, but the most important event is the sacrifice of a water buffalo, followed by a meal within the house of the deceased in the presence of the coffin. As the water buffalo ceremony begins, the shaman or Pima places a branch of *Debregeasia velutina* before him while reciting the funeral ceremony. The animal is then ritually killed in the yard by an elder who spears it through the heart (Plate 101). As soon as it falls, villagers leap upon the dying beast and pour water down its throat to prevent it from making any sound. If the dying animal were to make a noise, it would bring bad luck to both the deceased and his or her family. After the animal is dead, some of the fern, *Dryopteris cochleata*, is mixed with the animal's blood and an incantation or prayer is given by the spirit priest. The water buffalo is then carefully washed (amidst considerable laughter and the throwing of water) and a frond of *Pteris biaurita* placed on the head with cowrie shells on top. Paddy rice is poured over the head and upper body until they are completely covered (Plate 76). Sometimes the frond of *Araiostegia pulchra* is also placed on the head of the sacrificed animal; this plant is said to direct the dead person's spirit to the ancestor's home. The cowrie shells and fern can also be used when a pig or some other animal is sacrificed, because the type of animal killed depends on the wealth of the deceased's family. The leaf of *Callicarpa arborea* is placed in the house by the guests as they enter for the funeral meal. Finally, the bark of *Litsea monopetala* is rubbed on the casket before burial.

Cryptostegia grandiflora is known as the "widow plant," for widows believe that if they wear the flowers of this plant, they will marry again.

Death of a relative is not final, especially for the Akha, for several times each year offerings are made to the ancestors. Each Akha house contains an ancestral altar consisting of a section of bamboo, hung on the women's side of the partition separating the men's and women's halves of the house (Plate 82). One of the most sacred Akha plants is *Eurya acuminata,* so three tips of the compound leaves are used in the special offerings to the ancestors nine times a year. The tips of the shoots and other things, including rice, are placed in the ancestral altar. The wood of *Eurya acuminata* is also very sacred, so much so that it cannot be used *under* anything; it can only be used on the upper part or on top of the house in its construction. Special drinks or teas are made with *Zingiber officinale* to give to the ancestors and for the people to drink when they make offerings. In fact, the Akha believe that if *any* spirit offering is made, it must have ginger in it because it is very important in making the spirits happy. A little *Chlorophytum orchidastrum* is also put in the ancestor offering.

LEGENDS OF PLANTS AND THE SPIRITS

Several legends emphasize the significant role plants play in the spiritual and ceremonial aspects of the various hill tribe cultures. Traditionally, when someone wanted to cause harm to another person, he or she would cut up tiger's whiskers and put them in the victim's food to cause indigestion and other serious problems in the digestive system. *Mukia javanica* is called *la meh tzeu* or "tiger's whiskers" in Lisu. This plant was used to induce vomiting to purge the body of the tiger's whiskers. It continues to be used, but only when bad food or something poisonous has been eaten.

An old legend, probably originating in China, has long existed about *Plumeria rubra*. This plant, used by the religious leader to protect the Lahu from evil spirits, controls a variety of spirits, especially those affecting young people. The shaman may be asked to use this plant in a love affair, for example.

The Karen place the seeds of *Acacia concinna* in water as a good symbol. If the seeds and water are given to a person, it both shows respect and is a means of apology. In other words, the seeds are a symbol of great respect and relationship. The face and hands are washed with the water, often at New Year's. If someone has bad thoughts about ancestors or has said something wrong, then he or she bathes in the water to get relief from this guilt. When a new mother has no money to pay a woman who has helped deliver her child, then she can give her this water as thanks and respect.

If the Karen make a musical flute or horn from the wood of the tree *Rhus chinensis*, then they do not blow it when the moon is full, for that brings bad luck. The Hmong believe that squash (*Cucurbita* spp.) is an antiaphrodisiac, so they feed it to soldiers—and to elephants.

The Akha believe that *Rhus chinensis* is important when choosing a safe place to store rice. When a granary is constructed for rice storage, a rock is put down and then covered with nine leaves of this plant. The rice is then put on top of this.

Phoebe lanceolata is a small tree that grows in the undisturbed forest. The Akha pound up the leaves and apply them as a poultice for a scar that they believe has become swollen because of the odor of whiskey.

One of the interesting tribal legends involves *Clerodendrum serratum*, which is an important medicinal plant to many tribes. Lahu legend says that this plant was made available by God to a sick person centuries ago. The person offered one "hock" or portion of a cow if this plant would cure him. Thus, the plant is known to the Lahu as *nu te keu*, or "one half of the cow." It is used to treat malaria, as well as abdominal pains and problems (Paul Lewis, personal communication).

The Lahu also use *Cleistanthus hirsutulus* to treat infected wounds. Their legend is about a female water buffalo who had an infected genital tract in which maggots had started to eat away the organ. The gods gave the Lahu this plant as medicine to cure this bad infection. It worked so well, however, that the water buffalo then had difficulty urinating. This medicine is use on both humans and animals (Paul Lewis, personal communication).

The Akha use *Toddalia asiatica* (*ho ca la sah*) for treating the disease caused by the burning of rodents, which has been known to kill people. Known as *ho ca ho dzah*, the disease results in swellings and red spots. The Akha claim a mouse may actually come out of the sore, and that those who smell the smoke of burning rats or mice are susceptible to the disease. The plant is mixed with red and white mouse droppings that come from the house and applied as a poultice on the sore.

Spirits pervade every aspect of the hill tribe environment, and plants are a critical ingredient by which the highlanders can deal with the spirits. Whether it be crops, forest trees, disease, birth, or death, plants are used in various spirit ceremonies and rituals as part of this difficult and uneasy relationship.

12

Plants for Beauty and Pleasure

The old man spit and then smiled as he answered my questions about medicinal plants. My thoughts were momentarily distracted by his darkly stained irregular teeth. A sign of beauty to many of the hill tribes, betel nut–stained teeth are often a bit shocking to outsiders. Many of the older people in this Lahu Sheh Leh village chewed betel, for it is to them one of the real pleasures of the day. A stimulant, the seed of the betel nut palm (*Areca catechu*) is widely used throughout Asia and is particularly common among the Karen, Lisu, Akha, and Lahu tribes. It is just one of a number of plants that provide beauty and give pleasure.

The highlanders of northern Thailand enjoy beautiful things and have developed a number of activities that bring pleasure to their rigorous lives. Among the first things evident in even an isolated mountain village are flowers, sometimes a few orchids hanging in simple baskets from the corners of porches, or perhaps marigolds and rock roses near the house. Possibly the family garden may have some day lilies or roses mixed with the lemon grass (*Cymbopogon citratus*), chili (*Capsicum annuum*), sugar cane (*Saccharum officinarum*), and roselle (*Hibiscus sabdariffa*). Chantaboon Sutthi (1989a) has listed nearly 200 native and exotic ornamental plants found in tribal villages; I have not included the vast majority of these plants in Appendix 1 unless they serve a function other than as ornamentals. I believe that virtually every plant that has been adopted as an ornamental by the Thai has now found its way into one or more hill tribe villages.

CLOTHING DECORATIONS

The tribal women of northern Thailand have strikingly beautiful clothing, not only because of the fine weaving or sewing involved, but also because of the magnificent ornamentation that usually accompanies the outfit. Whether it be the headpiece or tunic of an Akha woman, the beads of a Karen, or the silver work of a

179

Mien, all are items of beauty, for, like women almost everywhere, they are desirous of attracting attention, especially that of the males within their societies.

The hill tribes are rightly famous for their remarkable silver work, as shown so beautifully in Lewis and Lewis (1984). Likewise, the old whiteheart and glass beads that probably came over the ancient silk route from Venice hundreds of years ago are treasures of great beauty. Yet, some of the tribal people have been able to create considerable beauty for their clothing simply by using seeds and other items from nearby fields and forests.

Undoubtedly the most widespread plant product used for the decoration of clothing is the seed of Job's tears (*Coix lachryma-jobi*), which adorn the blouses of every married Karen woman and the clothing and shoulder bags of Akha women (Plates 178–179). One of the most attractive features of these seeds is the variation of both color and shape. For example, an Akha woman's tunic may be festooned with several forms of these seeds; some may be white and elongated, while others are nearly round and either white, gray, or a beautiful gray-brown. The size of individual seeds may also vary from only 1–2 millimeters (0.04–0.08 inch) in diameter up to nearly a centimeter (0.4 inch). The plant may be found in the forest but is also grown as a crop, for the seeds can be eaten and occasionally are sold for cash.

Another plant, *Elaeocarpus sphaericus*, produces seeds that Akha women use for making necklaces and decorating their clothes. A huge woody vine, *Entada rheedii*, grows in the mountains and produces gigantic legumes (fruits) up to one meter (3 feet) long (Plate 180). The large seeds are used both medicinally and for necklaces. Another leguminous plant, *Mucuna macrocarpa*, also produces large fruits and seeds, which are used for decoration, but usually after being cut in two.

One of the most beautiful seeds used for clothing decoration is that of the legume, *Adenanthera pavonina*, a striking red seed that is somewhat oddly shaped and which varies in size from 0.5 to more than one centimeter (0.25–0.5 inch) in diameter. This seed is often combined with others, particularly those of Job's tears. Children like to play games with these colorful seeds, and men use them as bait in snares and traps.

Several members of the Monocotyledonae also provide seeds for decoration. One is the wild banana, *Musa acuminata*, which produces typical bananalike fruits that are full of seeds, each one nearly one centimeter (0.4 inch) in diameter (Plate 87). The seeds are used for decoration by members of several tribes, whereas the seeds of a species of *Rhaphidophora* in the Araceae apparently are used only by the Akha. The Akha also enjoy using the seeds of *Caryota mitis*, the fishtail palm, that is often grown in villages. One centimeter in diameter (0.4 inch), these seeds are made into necklaces, which also contain several small bottle gourds (*Lagenaria siceraria*). This interesting combination is worn by young Akha girls to indicate that they are unmarried and available (Plate 181). Occasionally, the seeds of a fourth monocot, *Sorghum bicolor*, are used for decorating clothing (Plate 53). Bracelets and simple necklaces are also woven from fibers obtained from the common vine, *Smilax ovalifolia*.

Tribal women—and occasionally men—adorn themselves with flowers of various plants, especially on special occasions, such as weddings and New Year celebrations. Examples of such flowers are *Quisqualis indica, Disporum calcaratum* (young Lisu girls like to put the flowers in their pierced ears and hair), and *Mussaenda parva*. Of course, many of the introduced ornamental flowers are also used in the same way.

COSMETICS AND TOILETRY ITEMS

Only a few plants have been identified that play a role in a man or woman's daily toilet. One plant, *Citrus hystrix*, the leech lime, is a source of shampoo (Plate 182); the fruit is cut up, boiled, and mixed with sesame (*Sesamum indicum*) leaf to make a foamy liquid. Two species of the Euphorbiaceae, *Jatropha curcas* and *Ricinus communis*, provide a hair oil. The Akha claim they make combs out of the stiff leaves of *Pandanus furcatus*, whereas the Lahu make them from the small bamboo, *Thyrsostachys siamensis*. The loofa or vegetable sponge, *Luffa aegyptiaca*, is widely used for bathing (and dishwashing), as it is in many other cultures (Plate 183). Finally, Karen women use the bark extract of *Millingtonia hortensis*, along with parts from some other plants, to make a beauty cream which they apply to their faces.

MASTICATORIES AND FUMATORIES

Tribal people enjoy chewing and smoking several plants, as do many people in Southeast Asia. By far the most common are betel nut and tobacco.

Betel Nut

The striking features of betel nut chewers are their darkly stained teeth and the blood-red saliva that they frequently spit. Though perhaps repulsive to us, the dark teeth and red mouth are attractive to many Southeast Asians, including some tribal people (Plate 184).

Tribal people claim they chew betel nut for several reasons. First, they enjoy it; second, it protects their teeth from decay; and third, it sweetens their breath.

This plant has been chewed by Asians for at least two thousand years, and it was first described by Heroditus in 340 B.C. (Hill 1952). Marco Polo commented on its use in 1298 A.D., when he traveled through Asia, and it is even mentioned in Hindu mythology. As much as one-tenth of the world's population chews it, especially those people in south and Southeast Asia (Janick et al. 1981). The seed of *Areca catechu*, a palm native to the Malay peninsula, contains several alkaloids, as well as tannins, fats, carbohydrates, and proteins. Its basic effect is to mildly stimulate.

Some betel nut chewers among the hill tribes have handsome woven bamboo containers for their chewing supplies, whereas others simply keep them in plastic bags (Plate 185). The one essential ingredient is the fruit of the betel nut palm which is almost always bought in the lowland markets and may be used either fresh or dried. The second necessary item is the leaf of the pepper, *Piper betle*, which is sometimes grown in the tribal villages, but more often is also purchased at a lowland market. The third item is lime, usually in the form of a paste, which is rubbed onto the fresh pepper leaf (Plate 186). The three ingredients are rolled into a wad and gently chewed. Many tribal people also add other items, the most common being tobacco. If the pepper leaf cannot be obtained, the leaves of *Rubus blepharoneurus*, a common shrub of the Rosaceae within the mountains, is sometimes substituted. Occasionally, the Akha will chew a bit of the rhizome of *Aspidistra sutepensis* with the betel. Or they may chew a bit of bark of *Callicarpa arborea* or a species of *Shorea*; even the bark from the roots or stems of *Streblus asper* may be added.

Phyllanthus emblica

The Akha have used another plant, known as *ci ca a baw*, as a masticatory and to blacken their teeth. The plant, *Phyllanthus emblica* of the Euphorbiaceae, has distinctive bipinnately compound leaves and a fruit (1–2 centimeters or 0.4–0.8 inches in diameter) that tastes something like a tamarind (*Tamarindus indica*). The fruit is placed above the fireplace for three days to "ripen." In the meantime the person who wants to stain his or her teeth takes the sour leaf of another plant (undetermined) and keeps it in his or her mouth to "soften the teeth." The fruit is then placed in the mouth to make the teeth black. I was told by a village headman that few young people now want to use this plant, so apparently this style is dying out.

Tobacco

Tobacco (*Nicotiana tabacum*) is a widely used plant that brings much pleasure to the hill tribes. Though native to the Western Hemisphere and not introduced into Southeast Asia until the late fifteenth or early sixteenth century, the highlanders of Thailand assume that they have always had it. Not only do they enjoy it privately, but they also find it an important communal activity, whether walking to the fields or sitting around a fire at night (Plates 187–188). Many both smoke tobacco and chew betel, simply alternating them as the situation warrants.

Tobacco plants are seen in most village gardens (Plate 189), which supply each household with most of its needs. Leaves are removed as they mature and allowed to dry or cure, usually by hanging them under the eaves or above the fireplace in the house (Plate 190). However, the Karen will usually wrap the freshly cut tobacco leaves in leaves of *Alocasia macrorrhiza* to allow them to ferment slightly before drying. Once the leaves are dry, they are cut up and placed in plastic bags or woven bamboo containers (Plate 191). A tobacco plant will produce leaves virtually throughout the year in this subtropical climate.

Most tribal people smoke tobacco, although, as noted in a previous section, many also chew it with their betel nut. The forms of smoking are many, but historically the most common method has been the pipe. Some groups, such as the

Sgaw Karen, still prefer the pipe; each individual, whether man or woman, is very proud of his or her pipe, which is often covered with silver. To the young Karen woman, her pipe is one of her most treasured possessions. In fact, a Karen couple may exchange tobacco pipes to symbolize their engagement, just as Westerners use an engagement ring (Lewis and Lewis 1984). Old men may often be seen in villages smoking pipes they have had for decades. An Akha village elder proudly showed me a beautiful pipe made of the Buddha's belly bamboo, *Bambusa ventricosa*, which had come from China more than a generation ago. On the other hand, one may also see a very simple pipe constructed of a piece or two of bamboo (Plate 188).

Smoking cigarettes is much more common now as lowland products make their way into the hills. Some highlanders roll their own cigarettes, using corn husks, very thin bamboo sheaths, or waste paper.

The water pipe is, in a sense, a combination of the cigarette and pipe, for usually a cigarette is used as the source of tobacco. The water pipe, constructed of two sizes of bamboo (the larger part usually of *Dendrocalamus membranaceus* or *D. strictus* and the smaller part of *Bambusa arundinacea*), consists of a long portion up to about 80 centimeters (30 inches) long and approximately 6 to 8 centimeters (2.4 to 3.2 inches) in diameter, with a much smaller piece of bamboo inserted into the side of the larger piece about one-fourth of the way up (Plate 192). The tip of this smaller piece is made into a small bowl for loose tobacco or is modified to hold a cigarette. Water is poured into the large portion to a height above where the smaller piece is attached. The smoker then inserts some tobacco or a cigarette into the end of the small piece, lights it, and puffs mightily on the end of the large piece while keeping it sealed with his cheek and mouth. The vacuum pulls air through the lighted cigarette, bringing the smoke down the small bamboo, through the water as bubbles, and up the large bamboo into the mouth of the smoker. Almost every tribal home will have a water pipe leaning against the wall within easy reach of a person sitting by the fireplace. Individuals often share the same water pipe, thus making it a distinctly social activity.

There appear to be few substitutes for tobacco, although some Akha will smoke the leaves of *Rhynchotechum obovatum* if they run out of tobacco. They will also smoke a combination of opium and the leaves of *Telosma pallida*.

LIQUOR

Compared to tribal people in other parts of the world, especially the New World, the highlanders of northern Thailand use very few mind-altering substances. Some smoke opium (*Papaver somniferum*) for the pleasurable experience it induces, though, as stated in Chapter 8, most use it primarily as a medicine. Likewise, few tribal people smoke or eat marijuana (*Cannabis sativa*). However, when it comes to the alcohol-containing substances, these people are like those almost everywhere in the world. Indeed, their local home brew is essential to daily life, as well as to ceremonial events.

Methods of making whiskey vary slightly from village to village and from tribe to tribe, but, nonetheless, all use the same basic technique. The Thai government now prohibits the making of whiskey, so the situation among the tribal people is similar to that of the moonshiners in the Ozarks of the United States in that they must hide their manufacture of it. The process is therefore carried out in a remote part of the village out of sight of strangers or officials. However, I was willingly shown the process by a Lisu friend.

The process is carried on throughout much of the year, so long as ingredients are available. Large clay urns containing perhaps 40 liters (10 gallons) are used for the fermentation process (Plate 193). Each urn, filled about three-fourths full of rice and/or maize grains, is placed in a cool, shady place. The source of the fermenting organisms varies widely; some simply add yeast that has been purchased in a lowland market, whereas others use an interesting combination of local plants for the purpose. They make a fine white powder consisting of finely crushed rice mixed with the dried and pulverized parts of more than a dozen different local plants. Only a few of these plants could be identified: *Thunbergia similis, Spilanthes paniculata*, and *Helicteres plebeja*. The brewers claim the plants give the whiskey "flavor," but this almost certainly is the source of the natural yeasts, about ten tablespoons of which are put in each urn. No water is added and the urn is covered tightly. Fermentation times vary from a few days up to a month, but the Lisu say that, although it can be drunk after ten days, it is best left for a month.

The distillation process is an example of amazing ingenuity. The still, consisting of a 208-liter (55 gallon) drum that is about one-fourth full of water, is placed over a fire. A tightly woven bamboo sieve is then wedged into the drum just above the water level, and onto this is carefully emptied the contents of an urn, avoiding contact of this mass of fermented material and the water below. A specially constructed unit shaped like a wooden paddle is inserted into the top of the barrel so that its handle protrudes through a small hole in the side; a counterweight insures that it stays properly oriented (Plate 194). This paddle is slightly concave on top and has a small groove cut the entire length of the upper part of the handle. Lastly, a large metal dish pan filled with cold water is placed on the top of the barrel, thus sealing it. As the fire heats the contents, some of the liquid from the urn's mixture drips down to mix with the water below, which is slowly brought to a boil. Over a period of time the alcohol, which has a slightly lower point of vaporization than does the water, begins to evaporate, as does some water. The vapors of the two substances rise within the barrel, come in contact with the cold bottom of the dish pan, condense, and drip onto the concave paddle, which is so oriented that the liquid then runs down the groove in the handle and into a container sitting beside the drum. This freshly-brewed "white fire" has a high alcohol content, varying between 30 and 50 percent ethyl alcohol. It may be transferred to other bottles, which are often corked with corn cobs, so each participant can take his share home for use in ceremonies, meals, or simply pleasure.

GAMES

Hill tribe children play games as do children anywhere in the world. However, their toys are almost always derived from products of the forest, whether they be coasters made of wood and bamboo, seeds with which they play marbles, or stilts and hoops made from bamboo.

One of the most widespread games played with great skill by tribal men and boys is the spinning of a large, heavy top by a string wrapped about its base (Plate 195). The game consists of making a large circle in the dirt; the first person starts his top spinning within the circle; the others then attempt to cast their newly spinning tops to knock the other top out of the ring while theirs stay in. Each top is larger and more solid than those found in the West, usually carved of very hard forest wood (often *Shorea obtusa, Stereospermum neuranthum,* or *Terminalia bellirica*) and up to 10 or 12 centimeters (4 or 5 inches) in diameter. Top spinning is a popular sport especially during the New Year's celebration.

Lahu women play a game similar to the one the men play with tops, but they use the large seeds of *Mucuna macrocarpa,* a leguminous vine with fruits up to 2 meters (6 feet) long and seeds up to 5 centimeters (2 inches) in diameter. The women set their seeds in a row and then knock them down according to rules established before the game starts. They may shoot the seeds as one does marbles; run with them on one foot, then kick them at their opponents' seeds to try to knock them out of the row; or even let the seeds roll off their skirts to knock them down (Elaine Lewis, personal communication). Children of both sexes may play this game, but men never do.

Another leguminous vine, *Entada rheedii,* produces similar-sized seeds in meter- or yard-long fruits (Plate 180). These seeds are used by children for playing marbles. Similarly, children play with the red seeds of *Adenanthera pavonina.*

Akha boys create simple but effective coasters for riding down steep paths within their villages. They make the framework mostly of bamboo and then carve wheels from hard forest wood. The wheels are held onto the simple axle by a bamboo pin. The front axle can even be steered by means of a pivot in the frame.

Boys in both Hmong and Lisu villages fashion "pea shooters" out of small sections of bamboo (Plate 196). Inserting "peas," which are often the small fruits of *Microcos paniculata,* into the tube, they forcefully shove a plunger through the tube, thus throwing the small fruit some distance and creating a loud pop.

MUSICAL INSTRUMENTS

The hill tribes have a great love for music, which they now perform regularly for tourists. Whether it be youngsters at night in the Akha courting area, adults singing and dancing at New Year's celebrations in Hmong or Lahu villages, or Karen wedding processions, an interesting variety of musical instruments has been constructed primarily from products of the forest.

Perhaps the most widely used instrument is a simple bamboo flute, which is played much like the recorder in the West. The Lisu flute, for example, is 40–45 centimeters (15–18 inches) long, 2 centimeters (0.8 inches) in diameter, and with six finger holes. The reed and air hole are about 5 centimeters (2 inches) from the end.

Frederic Grunfeld describes a remarkable Akha musical instrument, which he calls a "bagless bagpipe" (Grunfeld 1982). Made from a very thin section of bamboo containing a tiny bamboo reed, this instrument is put far into the player's mouth with only about 8 centimeters (3 inches) sticking out. By fingering the three holes and puffing out one's cheeks, a series of interesting sounds are created, suggestive of a Scottish bagpipe.

A much more elaborate instrument is the gourd pipe, a combination of bamboo pipes (primarily made from *Bambusa arundinacea* and *Cephalostachyum virgatum*) and reeds, which pass through holes cut in a gourd (*Lagenaria siceraria*) and sealed with beeswax (Plate 197). The very thin bamboo reeds are made from *Thyrsostachys siamensis* and are so attached by the beeswax that they act like valves, vibrating as the air passes from the gourd out into the pipes. The player blows into a hole cut in the small end of the gourd, thus making an even flow of air through all the pipes and enabling the player to get various combinations of tones. When he blows into the pipe there is one set of tones, and when the air is sucked out there is a different set. Then as the player puts his thumb on the various holes in the back of the gourd (there are five of them) and moves his index finger on the holes on the sides of the bamboo pipes, a remarkable variety of sounds is produced.

Some of these gourd pipe instruments are less than half a meter (20 inches) in length, whereas others of the Hmong and Lahu tribes may be nearly 2 meters (6 feet) in length. The smaller Lahu gourd pipe has five pipes protruding through the gourd, each with a small hole or slit; the pipes vary from 19 to 37 centimeters (7.5 to 15 inches) in length, with the longest two equal in length. One of these pipes has a small hole about one-third of the way up, whereas the other has a 0.5 by 4 centimeter (0.2 by 1.5 inch) slit about two-thirds of the way towards the end. The instrument is tuned to the pentatonic scale, and the end of the pipe by the gourd must be covered to make a sound. The large Lahu gourd pipe may be up to a meter (3 feet) long, and the longest pipe may be topped with a second gourd which makes the drone tone sounding chamber; Lahu Sheh Leh believe this very low "drone tone" sound goes all the way up to God. Gourd pipes are played only by men except for the Lahu Nyi, who have a small, high pitched instrument that is played by the women (Campbell et al. 1981).

The Hmong mouth organ (*kaen*) has up to six pipes of different lengths, most nearly a meter (yard) long (Plate 198). The bamboo or wooden part which one blows into is also more than half a meter (1.5 feet) in length. Great skill is required to play these large instruments, and the musician, always a man, dances, turns, dips, and sways rhythmically as he blows into it.

One of the most renowned Karen instruments is the Karen harp (*t'na*), with either five or six strings. Consisting of a carved wooden body or sounding chamber about 10 by 26 centimeters (4 by 10 inches) in diameter, the top is covered with a carefully shaped piece of metal that is nailed to the body and to which the steel strings are attached. At the other end, the strings are attached to tuning pegs, which in turn are attached to a carved wooden piece, about 45 centimeters (18 inches) long, and which curves gracefully from the body. Traditionally, the strings were made of cotton, but now they are almost always of steel; likewise, the top of the body once was covered with a piece of barking deer skin rather than with metal (Renard et al. 1991).

Another Karen instrument, the *haw tu*, is very similar to a guitar. Carved from wood, this instrument may be over 90 centimeters (35 inches) long. The body is nearly round, about 22 centimeters (8.5 inches) in diameter and 6 centimeters (2 inches) deep. Traditionally, it had only three strings, but the instrument I examined had four strings and they were steel.

One of the most striking stringed instruments is the Lisu lute or mandolin (*cue-bue*). The body or sound box is of carved wood about the size and shape of a tin can. A piece of snake skin is stretched and securely fastened over the box. The finger board is then nearly 60 centimeters (24 inches) long, and there are three steel or brass strings.

A very simple child's instrument is made from a bamboo section in which a portion of the wall is removed except for several thin pieces which serve as "strings" that can be plucked (Campbell et al. 1981).

Drums are also important in tribal music. Carved wood drums are of various sizes and shapes, depending on the tribe. Karen drums are in the shape of large, footed urns. The tops are covered with animal skins, which are tightly secured with cotton twine and rope, and which in turn are made taut by wooden pegs driven underneath the rope. Each has a fiber shoulder strap made of rattan (*Calamus* spp.) or bamboo. Mien drums are of a distinctly different shape, being only half as high as in diameter. The body of the Mien drum is carved from a solid piece of wood, and both top and bottom are covered with animal skin. Three series of half pegs to which the skin is secured by bamboo pegs are then fastened around the body of the drum. Akha drums are similar to Karen and Mien drums, though they do not have the half pegs distinctive of the Mien. In addition, the Akha make drums from large sections of bamboo, particularly *Dendrocalamus giganteus*. They also create another simple percussion instrument: bamboo sections of different length are simply hit in rhythm against an overturned pig trough. One often hears the beat of these drums after a successful hunt.

Bamboo plays another role in tribal music. The Karen, as well as the Thai, enjoy performing a bamboo dance in which there are eight long sections of bamboo (usually *Bambusa vulgaris* or *Thyrsostachys oliveri*) moved rhythmically to music; the performers then dance within the moving poles (Plate 199).

The Karen have a xylophone, which is made from 11 sections of green bamboo. They are played with small bamboo hammers, making a sound something like the hitting of empty bottles (Renard et al. 1991).

Finally, there are the small, intricately carved bamboo mouth or lovers' harps, which several of the tribes possess (Plate 200). Each tiny instrument, about 14 centimeters (5.5 inches) long, is usually made from the culms of *Dendrocalamus strictus* and attached by a colorful string to a bamboo container made from *Cephalostachyum pergracile*. Whereas the Akha mouth harp is a single piece of carefully fashioned bamboo, the packet of the Lahu consists of two or three tuned instruments attached to one another by a string. The Karen mouth harp, on the other hand, is a much larger instrument, often measuring up to 30 centimeters (12 inches) in length. This instrument is played by placing it in the mouth and twanging the end to make it vibrate; the mouth acts as the sound box. These instruments are commonly used in courting activities; the low-toned harps are referred to as "male," while those with a higher pitch are called "female" harps (Campbell et al. 1981). Boys will "talk" to the girls with their instruments, twanging monosyllables that mimic love songs (Grunfeld 1982).

Despite their rigorous and often difficult lives, the hill tribes enjoy beauty and do a number of things for pleasure. Whether it be adornments for their clothing, a drink of rice whiskey, a smoke with a friend, or a chew of betel nut, the highlanders make their lives more pleasant through the frequent use of these plant-derived items.

13

The Future: Will the Cultures and the Plants They Use Survive?

The Mien couple talked with me for more than an hour. Several times we were interrupted by the arrival of tourists anxious to see the hill tribes in their remote setting. In fact, this particular village was located near a paved road less than an hour's drive from Chiang Rai. Sweaty tourists descended upon the village from the road above, walking quickly down the main street and taking photographs of the people and animals. The woman with whom I was talking jumped to her feet and went to a table arranged with tribal crafts which she hoped to sell to the visitors during their brief stop. One tourist came up to her, pointed his camera in her face, and took a couple of pictures, not bothering to ask her permission. He picked up an item or two from the table and asked the price of each. When she held up four fingers, meaning forty Baht, he grimaced and said it was too much. The two bargained in sign language and a few words of English; then the tourist tossed down the handmade items and walked away with not a word of thanks. Other tourists stopped briefly at the stand, two bought small items, and then the group departed as quickly as it had come. A half an hour later another van arrived and the process was repeated.

I asked my host what he thought about the almost continuous stream of foreign and Thai tourists. He smiled and said that most visitors were nice, but that their presence had greatly changed the village's way of life. Of course villagers welcomed the extra income from the sale of tribal crafts, yet it was clear in his expression that he and his wife longed for the uninterrupted life of the past.

I have seen dramatic changes in northern Thailand during the 15 years that my family and I have lived and visited there. Chiang Mai is no longer a quaint, picturesque town, but rather a bustling, dirty, overcrowded city, with tourists on almost every street. In 1976 only one company offered treks into the mountains to see the hill tribes. Today literally hundreds of such groups offer trips to almost

189

every tribal village in the north. New and improved roads have made previously inaccessible regions only a drive of a few hours. In short, there are virtually no hill tribe villages that do not have almost daily contact with the outside world, either in the form of Thai traders and foreign tourists, or through radio and television. The tribal people and their cultures will never be the same.

What can these villagers expect in the future? What will happen to the distinctive cultures of the hill tribes? Will their knowledge of the forests and the plants within the mountains be lost? Will the forests themselves disappear forever?

One elder who has lived in a village west of Chiang Rai for 40 years lamented the loss of the forests and the very different lives the people were forced to live. He said that many different types of trees and bamboo have disappeared, as have numerous edible foods from the forest. The people can no longer cut wood to build their houses but often must buy lumber from the lowland Thai (who very possibly cut the lumber illegally in the very same forests). Most of the old forests are now gone and the Royal Forest Department prohibits the cutting of any trees that remain. The village has set aside small protected areas containing a few precious trees and some large stands of bamboo. They have learned to cultivate bamboo within the village to ensure a continual supply of that essential commodity. But clearly the future, though not necessarily bleak for all the tribal people in a strictly economic sense, will be different from what they have known and in how they have lived for generations.

The loss of these cultures would be a serious catastrophe, for the people have developed an intimate and significant relationship with their surroundings. Their cultures are an irreplaceable treasure containing a knowledge of the plants of the region that cannot afford to be lost. All humans may benefit from what they know.

FACTORS LEADING TO THE LOSS OF THE TRIBAL CULTURES

The hill tribes and their distinctive cultures are in peril because of the following problems.

Thai Government Policy

The Thai government had a policy of noninterference regarding the hill tribes until the late 1950s, but since that time it has steadily become more and more involved with them (Bhruksasri 1989a). The reasons for this change were threefold. First, the government became aware of the ever-increasing destruction of forests in the north, due in part to tribal slash-and-burn agriculture, which, in the opinion of some forestry people, was not only despoiling the land, but was also affecting major watersheds. The second reason was that some of the tribal people were cultivating opium poppies, which the government had recently (1955) declared illegal. Moreover, foreign governments were also exerting pressure on the Thai to do something about the flow of drugs from their country, which greatly increased during the Vietnam conflict. Finally, the Thai government was increasingly concerned about the security of its borders with Laos and Burma, especially as the war in Vietnam was heating up and many feared the domino effect of Communist expansion in Southeast Asia, and wanted to make sure that the hill tribes would not support Communist infiltrators or be involved in insurgency.

In more recent years a fourth factor has been added: assimilation. The Thai government has consciously attempted to bring the hill tribes into the lowland Thai culture through the construction of numerous new road systems and schools, through visits by government officials to villages, and by permitting development programs to be carried out by both Thai and international organizations. The pace of acculturation is now much greater than what it was only a decade or so ago, and much of the increase is through the intentional policy of the government. It wants the tribal people to be a part of Thailand.

Tourism

The presence of tourists in tribal villages has exploded from a few intrepid trekkers of a decade or two ago to literally thousands a day. Their presence in once-remote villages clearly transforms both the appearance and behavior of the villagers (Cohen 1983). Without doubt, the least that happens when tourists visit is that villagers lose their dignity, for most tourists come to a tribal village in much the same way they visit a zoo. They want to look at, but not interact with, the "different" people.

Many basic cultural mores are affected by the economic exploitation of tourism. For example, the elders of a Christian Karen village in Chiang Rai Province, who felt that the people should continue to observe Sunday as a day of rest and worship, were pitted against owners of elephants who, because of the severe pressure exerted on them by tour operators, wanted to provide their services every day of the week. The traditional hospitality of the tribal people is also being lost through growing tourism, for commercial exploitation works both directions. Today visitors are asked to pay for services, which formerly were considered basic tribal hospitality and for which there would never have been a charge.

Population Growth

Birth Rate. Health programs and better medical care have reached many tribal villages. Whereas previously highlanders suffered from terribly high infant mortality and a life span of 50 years or less, many are now in better health and more infants are surviving. In most villages the birth rate is now two or three times the death rate (Task Force 1987). One of the most notable health advances has been in malaria control; it is claimed that nearly 75 percent of the tribal villages in northern Thailand have been provided with protection from this debilitating and deadly disease (Task Force 1987). In addition, more than half of the tribal villages have access to family planning services (Task Force 1987). The availability of medical care in many villages has greatly improved the lives of the tribal people, but better survival and longer lives have led to a rapid population growth, which has had serious consequences on the environment and the prospects for future agricultural success.

Refugees. The continuing oppression of tribal people in Burma and Laos has added to the problem of population growth. The population of hill tribes has quadrupled in the past 35 years, much of it due to immigration of refugees (Lewis and Lewis 1984). Over half the tribal people now living in northern Thailand have arrived there since 1978 (Task Force 1987). So long as the military dictatorship in Burma continues its policy of persecution and genocide, the hill tribes will continue to seek security and peace in Thailand.

Needless to say, this continued immigration creates problems for the Thai government, as well as for the overpopulated mountains in the north. The Thai government issued a decree in 1976 that anyone immigrating into Thailand would be subject to repatriation. Nothing was done until 1987, when suddenly officials forcibly returned some tribal people (John McKinnon, personal communication). For example, more than 2500 Akha were transported to the Burmese border and told not to return to Thailand. Many of them were, in fact, legal residents of Thailand, but the paramilitary personnel carrying out the task failed to check for documents. The Akhas' villages were burned and the people lost their possessions, including money, most personal belongings, livestock, and a nearly mature rice crop (Paul Lewis, personal communication). Several of those who were "repatriated" died from malaria soon after entering the area of Burma to which they were released, known to be one of the most heavily infected malaria areas in Southeast Asia.

A group of Hmong were repatriated to Laos at nearly the same time; more than 30 died at the hands of Laotian government troops.

The Thai government has also attempted relocation programs, but most have failed because of poor planning, corruption, and lack of suitable places to relocate the hill tribes.

Despite the government's limited efforts to restrict immigration of refugees (the government officially refers to them as "illegal immigrants") (Kammerer 1988), tribal people continue to move across the borders to escape the terrors and hardships in Burma and Laos. So long as conditions in those countries (which, ironically, have vast areas of mountainous land that are suitable for the tribal people) remain as they are, there seems to be little that the Thai government can do to stop this flow of humanity.

Land Shortages

Both population growth and immigration are significant factors in causing the serious shortage of land available to the highlanders. Coupled with reforestation programs and up-mountain migration of lowland Thai, the shortage of land is critical. Tribal farmers are no longer able to rotate their fields in the traditional and ecologically sound way they have followed for generations. Many highlanders simply do not have enough land to let any lie fallow.

The deterioration of land, destruction of further watershed and forested areas, and insufficient crop yields also make the future survival of the hill tribes very precarious. Moreover, the legal owner of most of the mountainous land of the north is the state, as administered by the Royal Forest Department (McKinnon 1989). Thus, most tribal people do not own the land on which they farm and dwell, and the securing of land rights, though often promised by government officials, is infrequent and sometimes impossibly expensive because of bureaucratic "delays." Nearly 70 percent of all tribal people do not have a legal right to the land they work (Task Force 1987).

Lack of Citizenship

Lack of citizenship prevents children from continuing their schooling past the fourth year, ill people from receiving hospitalization, and workers from getting jobs or appropriate compensation even if they do have work. Many tribal people are legally entitled to citizenship, but often obstacles are created to deny them this coveted status. Two requirements are official house registration documents, which only about half the tribal people have, and individual registration documents (ID cards), which slightly more than a third of them possess (Task Force 1987). Frequently the hill tribes cannot obtain these documents because perhaps they cannot prove where they were born, or their birth was not registered soon enough after the event, or they cannot prove how long they have lived in Thailand (they must have lived there since 1985 if from Laos, or since 1986 if from Burma). Citizenship may even be blocked by officials who demand exorbitant payments for completing the process.

Disease and Malnutrition

Serious epidemics and illnesses are widespread among the hill tribes, despite major efforts by the United Nations, church, and government to provide adequate health care. Significant advances have been made in controlling malaria and tuberculosis, although over 20 percent of the hill tribes are plagued by at least one of these diseases, and improved sanitation in many villages has reduced intestinal disorders (Task Force 1987). However, malnutrition, especially protein deficiency, continues to be a serious problem (Maneeprasert 1989) and will probably get worse as the land shortage forces the ever-larger population of tribal people to meet their nutritional needs on smaller and more deteriorated plots of land. Moreover, the shrinking forests cannot provide game and wild plant harvests to supplement tribal diets.

Poverty

The hill tribe economy is changing from a subsistence economy to a cash economy in which the people are becoming more and more dependent upon the lowland Thai markets and traveling merchants. The government's continued efforts to prohibit growing opium poppy commercially has removed a traditional source of cash income, though some villages now grow other lucrative crops and are actually better off than before. Numerous projects, some personally overseen by His Majesty the King and Her Majesty the Queen, have been instituted to facilitate this change from opium poppies to suitable substitutes. For those villages included within the project areas, as well as for some others, income has improved. However, many others simply have no way to augment the rice crop which must be grown on all available land for subsistence; consequently they are slowly slipping farther and farther into poverty.

A rapidly increasing market for handicrafts is providing opportunities for many villagers, yet by seeking tourists and overseas markets for their products, tribal artisans are changing the traditional ways in which they do their handwork; hopefully, they will not forget such things as how to use natural dyes or weave with fibers from the forest. Another interesting change is that this activity, which has been almost exclusively for women, now involves men anxious to contribute to the family's cash income.

Some hill tribes have found ways to participate successfully in Thailand's cash economy, but for others, who do not have the land, expertise, or opportunity, the likelihood is that they will slip slowly into an ever-greater poverty.

Clash of Cultures

Most dominant cultures are ethnocentric. This attitude leads to various forms of discrimination between the competing cultures. Sadly, this problem exists in Thailand, where many Thai consider themselves to be more culturally advanced than the tribal people whom they feel are inferior subjects of Thailand. These competing cultures clash frequently and discrimination, sometimes subtle but at other times blatant, results in unfortunate incidents. For example, Thai tourists often ridicule and demean village people. Government officials sometimes assume, mistakenly, that tribal leaders are unable to make good decisions; treating them as inferiors, government officials do not include tribal leaders when making decisions about a village. Some tribal leaders are therefore hesitant to cooperate with government-approved projects. Too often they have been slighted or even insulted by Thai who are unwilling to consult with them regarding important decisions concerning village welfare. Job discrimination is also widespread. One of my tribal interpreters, who was literate in six languages, was unable to find any job except an occasional one helping unload merchandise from trucks at the market. The clash of cultures in Thailand is especially severe because the Thai and hill tribes are competing for limited land and resources. Only by education and mutual understanding will the hill tribes gain the respect they deserve.

Education

Most tribal children now have the opportunity of a Thai education, for schools are being constructed throughout the mountains of northern Thailand and Thai teachers sent to teach at them. This educational opportunity certainly provides an avenue by which tribal young people may integrate into the dominant Thai society; it also means greater chances of technical training, better-paying jobs, and improved health. Yet this opportunity creates one of the most severe challenges to the perpetuation of the tribal cultures and their traditional ways. For example, school children are not allowed to wear tribal clothing, but must wear official school uniforms (white shirts or blouses and blue skirts or trousers). This means that Akha girls are forbidden to wear their headpieces. The students are instructed in the Thai language, which, as happened with Native Americans in the United States who were instructed in English, is a great unifying factor among the tribes. An education provides these children with the opportunity to read and better understand radio and television, but it also pulls them away from their cultures. Indeed, though education of the tribal children is certainly desirable, it is nonetheless a mixed bag and may ultimately be one of the main factors leading to the loss of the tribal cultures.

Forest Destruction

Clearly, one of the most serious problems facing the hill tribes is the disappearance of the forests upon which they are so dependent. Though many Thai blame the tribal people for the loss of forests, other factors have played major roles in the destruction of wooded areas.

There is no question among those who have studied deforestation in Thailand that the main problem is logging, and, although the Thai government presently prohibits it, numerous illegal cutting operations continue, even in protected zones such as national parks. The movement of lowland Thai farmers upward into the hills where they, too, practice slash-and-burn agriculture, has led many experts to state that the Thai themselves now cultivate more swidden land than do the hill tribes. Road construction has been a high government priority in recent years, so dirt or paved routes now go into most previously isolated mountainous areas. These roads permit communication with the tribal people, but their construction has led to terrible erosion and sedimentation problems in many regions of the north. Whether the loss of forests has really changed the climate of the region is open to debate, but there is no question that land degradation has occurred. Sadly, the few remaining teak forests of neighboring Burma are being destroyed as the Burmese military dictatorship, desperate for foreign capital to support its army, is selling off its forest heritage to Thai lumber companies.

Whether this rampant loss of forests can be stopped in either Burma or Thailand is uncertain, but one encouraging sign is the growing conservation movement among the Thai. Wildlife Fund Thailand, for example, has been a strong leader in several conservation projects and has been instrumental in stopping ill-advised programs, such as the construction of the Nam Choam dam that would have seriously affected part of the remaining undisturbed natural region of the country. Hopefully, education and the enforcement of existing laws will enable the Thai to become more effective in stopping the destruction of the remaining forests.

TRIBAL KNOWLEDGE AND CONSERVATION

If it is important to conserve the forests of Thailand, then the hill tribes and their intimate knowledge of the mountains in which they live must also be conserved. Acculturation into the dominant Thai culture will lose this critical resource, and, unfortunately, it is occurring rapidly as a result of better communication, education, tourism, and the government's policy to bring the highlanders into the Thai way of life. The Thai culture—and numerous others, such as Western and Japanese cultures—are significantly changing the world of the hill tribes. Throughout the book I have referred to many of the effects of these foreign cultures on tribal ways: new crops, market fabrics, dyes, construction materials, medicines, and religions. I am not optimistic that many of the tribal ways will persist, and one of my greatest fears is that we will lose the highlanders' great knowledge of the forest.

Of course, the forests, too, are disappearing. While there is little hope that this sad destruction will stop soon, it could be slowed down. There is even less of a chance that forests as they originally occurred in northern Thailand will regrow, for reforestation projects involve mainly exotic trees, such as nonnative pines and *Eucalyptus*. Thus, most of the forests that the tribal people have long known, and from which they have been so ably provided, will soon cease to exist. Is there hope that some patches may be preserved? Are the tribal people aware of ways in which they might conserve even small areas of forest so that they can continue to teach their youth the ways of their forefathers? I believe there may be a way: that is by developing further the concept of the "Sacred or Holy Hill."

Holy Hill Concept

The Chinese botanist Pei Sheng-ji claims that in the region of Xishuangbanna, Yunnan Province, China, there are approximately 400 "Holy Hills" associated with the Dai minority (Pei 1985, 1987). They vary from 10 to 100 hectares (25 to 250 acres) in size and are more or less naturally forested. The Dai people, with whom the hill tribes have long been associated, traditionally were animists closely bound to the natural world. Thus, the forests with their associated plants and animals contained the supernatural forces of nature. The early Dai were encouraged to live in harmony with their surroundings so as to show proper respect for these spirits. Lack of respect led to the displeasure of the gods who punished the Dai with misfortune of one form or another. Even though Buddhism and atheism may have largely replaced the older polytheistic religion of the Dai, the concept of the "Holy Hill" as the residence of the spirits has been retained.

These hills and trees therefore comprise a significant ecological element in the areas of Yunnan in which the Dai reside.

Pei states (1987) that there are two types of Holy Hills. The first, *Nong Man,* usually consists of a forested hill 10–100 hectares (25–250 acres) in size that is worshiped by the people in a single nearby village. The second, called *Nong Meng,* occupies a much larger area and is worshiped by the inhabitants of several villages. The wildlife that lives on a Holy Hill are considered either companions of the spirits or simply living things in the gardens of the spirits. Thus, the Holy Hill is a natural preserve, and because the plants and animals belong to the spirits, they are conserved. Several of these Holy Hills have long been preserved, so the vegetation frequently is pristine. Human activities, such as wood-cutting, hunting, gathering, and cultivating, are strictly prohibited.

Other minorities in Yunnan Province respect the Dai Holy Hills and do not enter the forests, for many believe that violation of the Holy Hill may lead to terrible disasters and misfortune, including fires, floods, diseases, windstorms, earthquakes, insect plagues, and even attacks by wild animals.

Offerings are regularly presented by the Dai to the spirits of the Holy Hills. The most important annual ceremony sometimes lasts as long as three days and includes gifts of animal meat, rice, fruits, flowers, liquor, and beeswax candles.

Tree Renovation Ceremony

Several of the hill tribes of northern Thailand have what is known as a tree renovation ceremony (see Chapter 11). Quite probably this ceremony is derived from the Holy Hills concept of southern China because most of these Thai tribal people have migrated from Yunnan Province where many of their relatives are still present and associating with the Dai people. However, because the tribal people in Thailand have less land available for agriculture and village sites, the sacred area may only be a large forest tree or small patch of vegetation located on a hilltop overlooking the village. It is to this tree or sacred area that the village leaders usually go about every three weeks to make offerings of tea, rice, chili, and wax candles to renovate or pacify the spirits on behalf of the villagers. These gifts are placed on a platform that has been constructed at the base of the tree or within the small patch of forest. Occasionally medicinal plants are removed from the sacred area, but only after the spirits have been properly notified. Blood must never be shed within the area.

Some Akha villages, for example, have a sacred tree or a small area that is set aside as holy. Nothing can be cut in the area, so in time the trees have become quite large. The Akha claim these areas are set aside for a very important purpose: they must be visited when there is a need to appease various spirits of the region, which includes all villages within the area. Since village spirits are shared, when disease or disaster comes to a village, it may call upon the spirits of the whole region to assist it.

The sacred tree or patch has a special spirit altar where the village shaman makes sacrifices and offerings (Plate 173). The area is very sacred and all villages greatly respect it and the surrounding forest.

Two Strategies

Conservationists should study the Holy Hills concept, as well as the tree renovation ceremony, for they might provide a strong cultural force among the hill tribes for the continued preservation or protection of some of the remaining forests in northern Thailand. Village leaders should be encouraged to expand these special areas. When talking to these leaders, conservationists should emphasize the wisdom of the ancients who clearly recognized the value of undisturbed forests and wildlife. Even though today some tribal people feel that the spirits no longer reside in the forest, the phrase "respect and love of wildlife, both plant and animal" could replace the word *spirits*.

Another conservation strategy for northern Thailand is to extensively involve the tribal people in the reforestation and forest preservation projects of the Royal Forest Department. In East Africa, for example, tribal people, who were once poachers and destroyers of the natural wildlife, have been successfully involved in wildlife conservation programs on that continent. Unfortunately, few highlanders have been hired or asked to participate in such programs in Thailand. Their deeply ingrained respect of wildlife and the close ties to the forest make them excellent candidates for active involvement in virtually all kinds of conservation work. Unfortunately, prejudice and discrimination have thus far thwarted the few efforts made in this direction, but clearly, the matter must be pursued with vigor. It is very possible that the future of the forests depends on including the hill tribes in conservation projects.

Extinction of Rare, Threatened, and Endangered Plants

Relatively little is known about the flora of northern Thailand, but there is no doubt that many species are in serious danger of extinction because of the extensive loss of forests. Numerous times in my research hill tribe people told me that certain plants used to grow nearby but are now gone or found only in the forests of Burma.

The forests of northern Thailand are surprisingly rich. For example, J. F. Maxwell (personal communication) has now determined that over 2000 taxa of plants occur on Doi Suthep, a national park and one of the highest mountains in Thailand. It is common knowledge that many orchids are threatened by over-collection throughout Thailand and that Doi Suthep has more than 250 species (Bänziger 1988). I have also seen the rare and beautiful *Sapria himalayana* Griffith, a root parasite of the Rafflesiaceae, on this mountain. Forest destruction can easily obliterate the few known populations of this plant, as well as those of other species. Thus, forest destruction and the loss of the opportunity to learn about the plants of this area of Asia are serious problems. Conservation groups and the Thai government must cooperate in learning about the plants of northern Thailand and preserving those that are left.

CONCLUSION

At first I, too, was just a curious observer of the hill tribes, but as I have come to know them and parts of their remarkable cultures, I can now see clearly that they may very well provide answers on how to deal with the loss of forests and the destruction of the mountains in northern Thailand. The hill tribe people are, indeed, an important cultural resource, and their knowledge can be a valuable tool for conserving what remains of Thailand's natural heritage. So, like the plants and animals that make up the natural setting of the north and should be conserved, the hill tribes are also an integral and critical part of this relationship. They, too, must not be lost or their knowledge destroyed.

The future of Thailand's hill tribes is cloudy, but the results of clouds are rains, and, as seen so clearly at the start of the monsoon season in the hills of northern Thailand, this moisture means new life and new opportunities in a world that appears almost dead and with no future. The greening of the mountain sides means the promise of a better future. From the many clouds and discouraging signs presently facing the hill tribes and their cultures, we must seek ways to both perpetuate and value what they know of the forests in which they dwell. Indeed, some forests will still disappear and many traditional tribal ways may be replaced by modern Thai or even Western ones, but hopefully our respect for and our need of understanding these people will encourage them to retain records and memories of their traditions, to be shared with generations of people to come, not only among the tribal people themselves, but with outsiders as well. The future is certainly cloudy, but these clouds may be bringing rains of survival that result in a new worldwide awareness of the importance of understanding and preserving many parts of the marvelous cultures of the people of the Golden Triangle.

APPENDIX 1

Plants Used by the Hill Tribes

Key to uses:
1 = food
2 = fiber
3 = ceremony, spirit world
4 = medicine
5 = construction
6 = dye, paint
7 = pleasure, masticatory
8 = tinder, firewood, torches
9 = decoration, ornamental

10 = other
11 = cash

Vouchers:
4-digit numbers or EFA = E. F. Anderson
B and digits = K. Bragg
Nelson = R. Nelson
Pake = C. Pake
T and digits = C. Thangthawonsirikhum

Binomial	Family	Uses	Vouchers
Abelmoschus esculentus (L.) Moench	Malvaceae	1, 11	
Abelmoschus moschatus Medic	Malvaceae	2, 4	B20
Acacia concinna (Willd.) DC.	Mimosaceae	1, 3, 4	
Acacia farnesiana (L.) Willd.	Mimosaceae	1	
Acacia megaladena Desv. var. megaladena	Mimosaceae	3, 4, 10	5173, 6006, 6039, 6092
Acacia pennata (L.) Willd. ssp. *insuavis* (Lace) Nielsen	Mimosaceae	1	
Acalypha hispida Burm. f.	Euphorbiaceae	4, 9	
Achyranthes aspera L.	Amaranthaceae	4	5802
Achyrospermum wallichianum (Benth.) Benth. ex Hook. f.	Lamiaceae	1, 4	5416
Acorus gramineus Aiton	Araceae	4	5393
Adenanthera pavonina L.	Mimosaceae	4, 7, 9, 10	
Aeginetia indica Roxb.	Orobanchaceae	4	
Aesculus assamica Griffith	Hippocastanaceae	4	5261
Afzelia xylocarpa (Kurz) Craib	Caesalpinaceae	4, 9	
Aganosma marginata (Roxb.) G. Don	Apocynaceae	4	5447
Aganosma sp.	Apocynaceae	4	5977
Agastache rugosa (Fischer & C. Meyer) Kuntze	Lamiaceae	3, 4	5952

Binomial	Family	Uses	Vouchers
Agave americana L.	Agavaceae	3	
Agave sisalana Perrine	Agavaceae	2, 4	
Ageratum conyzoides L.	Asteraceae	4	5343, 5475, 5634, 5888, 6056, 6138
Aglaonema costatum N. E. Br.	Araceae	1, 4	
Aglaonema sp.	Araceae	4	5790
Agrimonia nepalensis D. Don var. nepalensis	Rosaceae	4	B213
Albizia chinensis (Osborn) Merr.	Mimosaceae	3, 4, 6, 10	5201, 5351
Allium ampeloprasum L.	Liliaceae	1, 11	
Allium cepa L.	Liliaceae	1, 11	
Allium sativum L.	Liliaceae	1, 11	
Allium schoenoprasum L.	Liliaceae	1	
Allium tuberosum Rottler & Sprengel	Liliaceae	1	
Allophyllus cobbe (L.) Räusch	Sapindaceae	4	5638
Alocasia cucullata (Lour.) Schott	Araceae	4	5677, 5812
Alocasia macrorrhiza (L.) Schott	Araceae	1, 3, 4, 9, 10	5763
Alocasia sp.	Araceae	4	6144
Aloe vera L.	Liliaceae	4	5229, 5243, 5694
Alpinia bracteata Roxb.	Zingiberaceae	1, 4	5505, 5816
Alpinia galanga (L.) Willd.	Zingiberaceae	1	B89
Alpinia malaccensis (Burm. f.) Roscoe	Zingiberaceae	1, 4	5482
Alpinia zerumbet (Pers.) B. L. Burtt & R. M. Smith	Zingiberaceae	1, 4	
Alpinia sp.	Zingiberaceae	1, 3	5399
Alpinia sp.	Zingiberaceae	1, 11	B204
Alstonia scholaris (L.) R. Br.	Apocynaceae	4	5348, 5906
Altenanthera sp.	Amaranthaceae	4	5817
Amalocalyx microlobus Pierre ex Spire	Apocynaceae	4	5828
Amaranthus gracilis Desf.	Amaranthaceae	1	
Amaranthus spinosus L.	Amaranthaceae	1	6150
Amaranthus tricolor L.	Amaranthaceae	1, 9, 11	
Amomum megalocheilos (Griffith) Baker	Zingiberaceae	1, 4	
Amomum siamense Craib	Zingiberaceae	1, 4	5502
Amomum sp.	Zingiberaceae	4	5716
Amomum sp.	Zingiberaceae	3	B148
Amorphophallus campanulatus (Roxb.) Blume ex Decne.	Araceae	1, 4	
Anacardium occidentale L.	Anacardiaceae	1, 11	
Ananas comosus (L.) Merr.	Bromeliaceae	1, 4, 11	
Anaphalis margaritacea (L.) Benth. & Hook.	Asteraceae	8	5218
Annona cherimolia Miller	Annonaceae	1	
Annona muricata L.	Annonaceae	1	5242
Annona reticulata L.	Annonaceae	1, 4	
Annona squamosa L.	Annonaceae	1	

Binomial	Family	Uses	Vouchers
Anomianthus dulcis (Dunal) Sincl.	Annonaceae	1	
Anthurium andraeanum Linden	Araceae	4	
Antiaris toxicaria (Pers.) Lesch.	Moraceae	2, 10	
Antidesma acidum Retz.	Euphorbiaceae	4	5435
Antidesma ghaesembilla Gaertner	Euphorbiaceae	4	
Antidesma sootepense Craib	Euphorbiaceae	3	6099
Antidesma sp.	Euphorbiaceae	4	5623
Antidesma sp.	Euphorbiaceae	1, 4	6024
Aporusa villosa (Lindley) Baillon	Euphorbiaceae	5	
Aporusa wallichii Hook. f.	Euphorbiaceae	4	5601
Arachis hypogaea L.	Fabaceae	1, 3, 11	
Araiostegia pulchra (D. Don) Copel.	Davalliaceae	3, 4	B64
Aralia chinensis L.	Araliaceae	1, 4	4088, 5488
Aralia sp.	Araliaceae	4	5848
Araucaria bidwillii Hook.	Araucariaceae	4	
Araucaria cunninghamii D. Don	Araucariaceae	1	
Archidendron clypearia (Jack) Nielsen	Mimosaceae	4	5182
Ardisia crenata Roxb.	Myrsinaceae	4	Pake
Areca catechu L.	Arecaceae	7	
Arenga pinnata (Wurmb) Merr.	Arecaceae	1, 4	
Argyreia capitiformis (Poiret) Ooststr.	Convolvulaceae	4	T273
Argyreia wallichii Choisy	Convolvulaceae	4	B231
Argyreia sp.	Convolvulaceae	4	5346
Artabotrys sp.	Annonaceae	4	
Artemisia atrovirens Hand.-Mazz.	Asteraceae	3, 4, 5	5646, 5985, 6139, 6156
Artemisia austroyunnanensis Ling & Y. R. Ling	Asteraceae	3, 4, 6	5718
Artemisia dubia Wallich ex Besser	Asteraceae	4, 6	5282, 5718
Artocarpus heterophyllus Lam.	Moraceae	1, 3, 11	
Artocarpus hypargyreus Hance ex Benth.	Moraceae	4	5739
Asparagus filicinus Buch.-Ham. ex D. Don	Liliaceae	4	5697
Aspidistra sutepensis K. Larsen	Liliaceae	1, 7	
Averrhoa carambola L.	Oxalidaceae	1, 4	
Azadirachta indica Juss.	Meliaceae	1, 4	
Azolla pinnata R. Br.	Azollaceae	10	
Baccaurea ramiflora Lour.	Euphorbiaceae	1, 4	5175, 5478
Baliospermum montanum Muell. Arg.	Euphorbiaceae	4	5459
Bambusa arundinacea (Retz.) Willd.	Poaceae	1, 4, 7, 10	5445
Bambusa pallida Munro	Poaceae	10	5307
Bambusa polymorpha Munro	Poaceae	1, 5, 10	5251
Bambusa tulda Roxb.	Poaceae	1, 4, 5, 10	5391
Bambusa tuldoides Munro	Poaceae	1, 5, 10	5679, 5960, 5961
Bambusa ventricosa McClure	Poaceae	7, 9	
Bambusa vulgaris Schrader	Poaceae	1, 4, 5, 7, 10	5308, 6174
Baphicacanthus cusia (Nees) Bremek.	Acanthaceae	6	5381, 5382

Binomial	Family	Uses	Vouchers
Barleria strigosa Willd.	Acanthaceae	4	5983
Basella alba L.	Basellaceae	1, 4	5227, 5797
Bauhinia acuminata L.	Caesalpinaceae	1, 4, 8	5589
Bauhinia bracteata (Graham ex Benth.) Baker	Caesalpinaceae	2, 3	
Bauhinia monandra Kurz	Caesalpinaceae	8	
Bauhinia ornata Kurz	Caesalpinaceae	1, 2, 10	5873
Bauhinia purpurea L.	Caesalpinaceae	1, 4, 9	
Bauhinia sp.	Caesalpinaceae	4	B184
Benincasa hispida (Thunb.) Cogn.	Cucurbitaceae	1	
Berchemia floribunda (Wallich) Wallich ex Brongn.	Rhamnaceae	4	B181
Betula alnoides Buch.-Ham.	Betulaceae	4	5205, 5214, 5878, 5940
Bidens pilosa L.	Asteraceae	4	5544, 5640
Bischofia javanica Blume	Euphorbiaceae	4	5998, 6071
Bixa orellana L.	Bixaceae	6	B112
Blumea balsamifera (L.) DC.	Asteraceae	4	5294, 5521, 5548, 5587, 5939, 5976, 5987, 6008, 6117
Blumea fistulosa (Roxb.) Kurz	Asteraceae	4	5972
Blumea lacera (Burm. f.) DC.	Asteraceae	4	B70
Blumea lanceolaria (Roxb.) Druce	Asteraceae	4	5500
Blumea membranacea DC.	Asteraceae	1, 4	5200, 5270, 5647
Boehmeria sidaefolia Wedd.	Urticaceae	2, 3	
Boerhavia chinensis (L.) Asch. & Schwein.	Nyctaginaceae	4	5771
Bombax ceiba L.	Bombacaceae	2	
Borassus flabellifer L.	Arecaceae	1	
Bougainvillea spectabilis Willd.	Nyctaginaceae	4, 9	3768
Brassaia actinophylla Endl.	Araliaceae	4	Pake 85
Brassaiopsis ficifolia Dunn	Araliaceae	4	5486
Brassica juncea (L.) Czerniak	Brassicaceae	1, 11	
Brassica napus L.	Brassicaceae	1	
Brassica oleracea L. var. *botrytis* L.	Brassicaceae	1, 11	
Brassica oleracea L. var. *capitata* L.	Brassicaceae	1, 11	
Brassica rapa L.	Brassicaceae	1, 11	
Broussonetia papyrifera (L.) Vent.	Moraceae	1, 2, 4, 10	5178, 5192a, 5819, 6001
Brucea mollis Wallich ex Kurz	Simaroubaceae	4	6102
Bryophyllum pinnatum (Lam.) Oken	Crassulaceae	4	5221, 5446, 5804
Buddleja asiatica Lour.	Loganiaceae	4, 6, 10	5177, 5293, 5413, 5397
Byttneria pilosa Roxb.	Sterculiaceae	4	5342

Binomial	Family	Uses	Vouchers
Caesalpinia coriaria (Jacq.) Willd.	Caesalpinaceae	10	
Caesalpinia mimosoides Lam.	Caesalpinaceae	1	
Caesalpinia pulcherrima (L.) Sw.	Caesalpinaceae	9	
Caesalpinia sappan L.	Caesalpinaceae	6	
Cajanus cajan (L.) Huth	Fabaceae	1, 11	4071
Caladium × *hortulanum* Birdsey	Araceae	1, 4	5791
Calamus kerrianus Becc.	Arecaceae	1, 2, 5, 11	6180
Calamus rudentum Lour.	Arecaceae	1, 2, 5, 11	
Calathea ornata (Linden) Koern.	Marantaceae	9	
Calathea veitchiana Hook. f.	Marantaceae	9	
Callicarpa arborea Roxb. var. *arborea*	Verbenaceae	3, 4, 7	B161
Calotropis gigantea (L.) Aiton f.	Asclepiadaceae	4	3824
Camellia sinensis (L.) Kuntze var. *assamica* (J. Masters) Kitam.	Theaceae	1, 7, 10, 11	5917
Canarium subulatum Guillaumin	Burseraceae	1, 4	
Canavalia ensiformis (L.) DC.	Fabaceae	1, 10	
Canavalia virosa (Roxb.) Wight & Arn.	Fabaceae	4	B257
Canna indica L.	Cannaceae	9	B28
Cannabis sativa L.	Cannabaceae	1, 2, 7, 11	5678, 6111
Canthium parvifolium Roxb.	Rubiaceae	1, 4	5323, 5592, 5909
Capparis spinosa L.	Capparidaceae	1, 11	
Capsella bursa-pastoris L.	Brassicaceae	4	Pake
Capsicum annuum L.	Solanaceae	1, 3, 4, 11	5255
Capsicum frutescens L.	Solanaceae	1, 10, 11	
Carex sp.	Cyperaceae	4	5292
Careya arborea Roxb.	Lecythidaceae	4	5532
Carica papaya L.	Caricaceae	1, 4, 11	5509
Carthamus tinctorius L.	Asteraceae	3, 6	
Caryota mitis Lour.	Arecaceae	1, 2, 4, 5, 8, 9	5394, 6044
Cassia fistula L.	Caesalpinaceae	3, 4, 5	5422, 5925
Cassia floribunda Cav.	Caesalpinaceae	9	
Cassia garrettiana Craib	Caesalpinaceae	4	
Cassia hirsuta L.	Caesalpinaceae	4	5651
Cassia javanica L.	Caesalpinaceae	9	
Cassia occidentalis L.	Caesalpinaceae	4	5527, 5604, 5691, 6153
Cassia timoriensis DC.	Caesalpinaceae	4	
Cassia sp.	Caesalpinaceae	3	6093
Cassytha filiformis L.	Lauraceae	4	Pake
Castanopsis acuminatissima (Blume) A. DC.	Fagaceae	1, 5, 8	
Castanopsis armata (Roxb.) Spach	Fagaceae	1, 5, 8	
Castanopsis diversifolia (Kurz) King ex Hk.f.	Fagaceae	1, 3, 5	B228
Castanopsis indica (Roxb.) A. DC.	Fagaceae	1, 5, 8	
Castanopsis tribuloides (Smith) A. DC.	Fagaceae	1, 5, 8	
Catharanthus roseus (L.) G. Don	Apocynaceae	9	

Binomial	Family	Uses	Vouchers
Cayratia trifolia (L.) Domin	Vitaceaae	1, 4	5529
Ceiba pentandra (L.) Gaertner	Bombacaceae	2, 11	EFA
Celosia argentea L.	Amaranthaceae	4, 9	3792, 5301
Celosia argentea L. var. *cristata* (L.) Kuntze	Amaranthaceae	3, 4, 9	5540, 6141, 6142, 6169
Celtis tetrandra Roxb.	Ulmaceae	4	5900
Centella asiatica (L.) Urban	Apiaceae	1, 3, 4	5409
Centipeda minima (L.) A. Br. & Asch.	Asteraceae	4	5252
Cephalostachyum pergracile Munro	Poaceae	5, 7, 10	5304, 5318, 5367, 5368
Cephalostachyum virgatum (Munro) Kurz	Poaceae	5, 10	5359
Chloranthus elatior Link	Chloranthaceae	4, 6	5577, 5703, 5723, 5736, 5743, 5744, 5899, 5951, 6101
Chloranthus nervosus Collett & Hemsley	Chloranthaceae	4	6019
Chloranthus sp.	Chloranthaceae	4	6159
Chlorophytum intermedium Craib	Liliaceae	4	5674
Chlorophytum orchidastrum Lindley	Liliaceae	1, 3	5384
Chrysanthemum × *morifolium* Ramat.	Asteraceae	4, 9	
Cicer arietinum L.	Fabaceae	1	
Cinnamomum camphora (L.) J. S. Presl	Lauraceae	4	Pake
Cinnamomum iners Reinw. ex Blume	Lauraceae	10	5277
Cinnamomum tamala Nees & Eberm.	Lauraceae	4	5490
Cinnamomum sp.	Lauraceae	4	B17, B232
Cissampelos hispida Forman	Menispermaceae	4	5714, 5779, 6069, 6152
Cissus cf. *discolor* Blume	Vitaceae	4	5354
Cissus hastata Miq.	Vitaceae	4	5738, 5751
Cissus repanda (Wight & Arn.) Vahl	Vitaceae	4	5820, 5881
Cissus repens Lam.	Vitaceae	4	5583
Citrullus lanatus (Thunb.) Matsum. & Nakai	Cucurbitaceae	1, 3, 11	
Citrus aurantiifolia (Christm.) Swingle	Rutaceae	1, 11	
Citrus hystrix DC.	Rutaceae	1, 10	6034
Citrus maxima (Burman) Merr.	Rutaceae	1, 4, 11	6053
Citrus medica L.	Rutaceae	1	
Citrus reticulata Blanco	Rutaceae	1, 11	
Citrus sinensis (L.) Osbeck	Rutaceae	1, 11	
Claoxylon indicum Hassk.	Euphorbiaceae	4	Pake
Clausena excavata Burm. f.	Rutaceae	1, 4	5184, 5296, 5327, 5550, 5949, 6052
Clausena longipes Craib	Rutaceae	4	5981
Cleidion spiciflorum (Burm. f.) Merr.	Euphorbiaceae	4	5968

Binomial	Family	Uses	Vouchers
Cleistanthus hirsutulus Hook. f.	Euphorbiaceae	3, 4	5915
Clematis siamensis J. R. Drumm. & Craib	Ranunculaceae	4	B237
Clematis subpetala Wallich	Ranunculaceae	4	B76
Cleome gynandra L.	Capparidaceae	1	
Clerodendrum colebrookeanum Walp.	Verbenaceae	4	5333, 5468, 5612, 5832, 6131, 6148
Clerodendrum fragrans Vent.	Verbenaceae	4	5752, 5948
Clerodendrum japonicum (Thunb.) Sweet	Verbenaceae	4	Pake
Clerodendrum paniculatum L.	Verbenaceae	1, 4	5269, 5455, 5568, 5635, 5770
Clerodendrum serratum (L.) Moon	Verbenaceae	3, 4	5169, 5170, 5347, 5439, 5558, 5562, 5658, 5749, 5756, 5778, 5944
Clerodendrum sp.	Verbenaceae	4	6007
Clitoria macrophylla Wall. ex Benth.	Fabaceae	1, 4	
Clitoria ternatea L.	Fabaceae	9	5264
Cnestis palala (Lour.) Merr.	Connaraceae	4	T275
Coccinia grandis (L.) Voigt	Cucurbitaceae	1, 4	
Cocos nucifera L.	Arecaceae	1, 10, 11	
Codonopsis javanica (Blume) Hook. f.	Campanulaceae	1, 4	5209
Coelogyne trinervis Lindley	Orchidaceae	4	5556
Coffea arabica L.	Rubiaceae	1, 11	5253
Coffea canephora Pierre ex Froehner	Rubiaceae	1, 11	
Coffea liberica Bull ex Hiern	Rubiaceae	1, 11	
Coix lachryma-jobi L.	Poaceae	1, 9, 11	5241
Coleus amboinicus Lour.	Lamiaceae	1, 9	
Colocasia esculenta (L.) Schott	Araceae	1, 4, 11	5763
Colocasia sp.	Araceae	4	5283
Combretum deciduum Collett & Hemsley	Combretaceae	4	5528
Commelina diffusa Burm. f.	Commelinaceae	4	5707
Congea tomentosa Roxb.	Verbenaceae	4	B52
Connarus semidecandrus Jack	Connaraceae	4	6062
Conyza sumatrensis (Retz.) Walker	Asteraceae	4	6060
Cordyline fruticosa Goeppert	Agavaceae	1	
Coriandrum sativum L.	Apiaceae	1	4201
Corypha umbraculifera L.	Arecaceae	1	
Cosmos caudatus Kunth	Asteraceae	1, 3, 9	
Cosmos sulphureus Cav.	Asteraceae	1, 3, 9	
Costus speciosus (C. König) Smith	Costaceae	1, 3, 4	5284, 5575, 5641, 5792, 5863, 6089, 6122, 6161

Binomial	Family	Uses	Vouchers
Crateva magna (Lour.) DC.	Capparidaceae	1, 4	
Cratoxylum cochinchinense (Lour.) Blume	Clusiaceae	4	5590, 5926
Cratoxylum formosum (Jack) Benth. & Hook. ex Dyer subsp. *pruniflorum* (Kurz) Cogn.	Clusiaceae	4	5297, 5452, 5471, 5866, 6165
Crinum amabile Donn	Liliaceae	9	
Crinum asiaticum L.	Liliaceae	4, 9	5538, 5786
Crotalaria dubia Graham ex Benth.	Fabaceae	4	5344
Crotalaria pallida Aiton	Fabaceae	4	5206
Crotalaria verrucosa L.	Fabaceae	4	5444
Crotalaria sp.	Fabaceae	4	6155
Croton crassifolius Geiseler	Euphorbiaceae	4	Pake
Croton oblongifolius Roxb.	Euphorbiaceae	4	5513, 5436, 5597
Croton robustus Kurz	Euphorbiaceae	4	5932, 5989
Crypsinus cruciformis (Ching) Tag.	Polypodiaceae	4	5967
Cryptostegia grandiflora R. Br.	Asclepiadaceae	3	
Cucumis melo L.	Cucurbitaceae	1, 11	
Cucumis sativus L.	Cucurbitaceae	1, 11	
Cucurbita maxima Duschesne	Cucurbitaceae	1, 4, 11	
Cucurbita moschata (Duschesne) Poiret	Cucurbitaceae	1, 11	
Cucurbita pepo L.	Cucurbitaceae	1, 4, 10, 11	5510
Curcuma aeruginosa Roxb.	Cucurbitaceae	1, 4	5432
Curcuma domestica Valeton	Zingiberaceae	1, 4, 6	5565
Curcuma parviflora Wallich	Zingiberaceae	4	5230
Curcuma zedoaria Roscoe	Zingiberaceae	4	5431
Curcuma sp.	Zingiberaceae	4	5404, 5684, 5685, 5720, 5735, 6079, 6080
Cuscuta australis R. Br.	Cuscutaceae	4	
Cuscuta reflexa Roxb.	Cuscutaceae	1, 4	5279
Cyathea latebrosa (Wallich ex Hook.) Copel.	Cyatheaceae	1	
Cyathula prostrata (L.) Blume	Amaranthaceae	4	5618
Cycas revoluta Thunb.	Cycadaceae	4	Pake
Cycas siamensis Miq.	Cycadaceae	1, 4, 9	
Cydista aequinoctialis (L.) Miers	Bignoniaceae	1, 4	EFA
Cymbopogon citratus (DC. ex Nees) Stapf	Poaceae	1, 4, 11	5512, 5644a, 6077, 6078
Daemonorops spp.	Arecaceae	5	
Dalbergia cf. *paniculata* Roxb.	Fabaceae	4	5378
Dalbergia stipulacea Roxb.	Fabaceae	6	B116
Datura metel L.	Solanaceae	4, 11	Pake
Daucus carota L.	Apiaceae	1, 11	
Debregeasia velutina Gaudin	Urticaceae	1, 3, 4	5892

Binomial	Family	Uses	Vouchers
Dendrobium sp.	Orchidaceae	4, 9	5557
Dendrocalamus brandisii (Munro) Kurz	Poaceae	5, 10	
Dendrocalamus giganteus (Wallich) Munro	Poaceae	1, 5, 7, 10	5305
Dendrocalamus hamiltonii Nees & Arn. ex Munro	Poaceae	1, 4, 5, 10	5273, 5320, 6176
Dendrocalamus latiflorus Munro	Poaceae	1, 5, 10	6083, 6183
Dendrocalamus membranaceus Munro	Poaceae	1, 4, 5, 7	5364
Dendrocalamus strictus (Roxb.) Nees	Poaceae	1, 5, 7	5267, 5317
Derris elliptica (Wallich) Benth.	Fabaceae	4, 10	5194, 5278, 5595, 5637
Desmodium gangeticum (L.) DC.	Fabaceae	4	5766, 5929
Desmodium heterocarpon (L.) DC.	Fabaceae	4	6061
Desmodium laxiflorum DC.	Fabaceae	4	5629
Desmodium longipes Craib	Fabaceae	4	B74
Desmodium oblongum Wallich ex Benth.	Fabaceae	4	6018
Desmodium renifolium (L.) Schindler	Fabaceae	1, 4	5629, 5896
Desmodium triangulare (Retz.) Merr.	Fabaceae	4	5945, 6128
Desmodium triquetrum (L.) DC.	Fabaceae	4	5349, 5543, 5602, 5762, 5725, 5922, 6059
Desmos sootepense (Craib) J. F. Maxwell	Annonaceae	3, 4	B59
Dianella ensifolia (L.) DC.	Liliaceae	3, 4	6033
Dicliptera roxburghiana Nees	Acanthaceae	4	5811, 6076
Dichrocephala integrifolia (L. f.) Kuntze	Asteraceae	4	5702, 6109
Dicranopteris linearis (Burman) Underw.	Gleicheniaceae	4	5210, 5424
Diefenbachia seguine (Jacq.) Schott	Araceae	9	
Dillenia indica L.	Dilleniaceae	1, 4, 5	
Dillenia obovata (Blume) Hoogl.	Dilleniaceae	8	
Dillenia pentagyna Roxb.	Dilleniaceae	1	
Dimocarpus longan Lour.	Sapindaceae	1, 11	
Dioscorea alata L.	Dioscoreaceae	1, 4, 11	5187
Dioscorea bulbifera L.	Dioscoreaceae	1, 4	5174
Dioscorea esculenta (Lour.) Burkill	Dioscoreaceae	1	
Dioscorea hispida Dennst.	Dioscoreaceae	1, 4	5498
Dioscorea pentaphylla L.	Dioscoreaceae	4	5894
Diospyros ehretioides Wallich	Ebenaceae	6	
Diospyros kaki L. f.	Ebenaceae	1, 11	
Diospyros virginiana	Ebenaceae	1, 11	
Diplazium esculentum (Retz.) Sw.	Aspleniaceae	1, 4	5849
Diploclisia glaucescens (Blume) Diels	Menispermaceae	3	5371
Dipterocarpus alatus Roxb.	Dipterocarpaceae	5	
Dipterocarpus costatus Gaertner f.	Dipterocarpaceae	5	3729

Binomial	Family	Uses	Vouchers
Dipterocarpus obtusifolius Teysmann ex Miq.	Dipterocarpaceae	5	3956, 4126, 4470
Dipterocarpus tuberculatus Roxb. var. *tuberculatus*	Dipterocarpaceae	5	6028
Disporum calcaratum D. Don	Liliaceae	4, 9	5700, 6000
Dolichandrone serrulata (DC.) Seem.	Bignoniaceae	4	5299
Dolichos lablab L.	Fabaceae	1, 4	
Dombeya wallichii (Lindley) Benth.	Sterculiaceae	4	
Dracaena angustifolia Roxb.	Agavaceae	4	
Dracaena fragrans (L.) Ker Gawler	Agavaceae	1, 4	5462
Dracaena loureiri Gagnep.	Agavaceae	4	
Dracaena sp.	Agavaceae	1	
Dregea volubilis Benth. ex Hook.	Asclepiadaceae	1, 4, 6	
Drymaria cordata (L.) Willd. ex Roemer & Schultes	Caryophyllaceae	4	B105
Dryopteris cochleata (D. Don) C. Chr.	Aspleniaceae	3, 4	
Drypetes sp.	Euphorbiaceae	4	5750
Duabanga grandiflora (Roxb. ex DC.) Walp.	Sonneratiaceae	1, 4	5356
Dumasia leiocarpa Benth.	Fabaceae	1, 4	5328
Dunbaria longeracemosa Craib	Fabaceae	6	3952
Dysoxylum andamanicum Kurz	Meliaceae	4	B172
Elaeagnus conferta Roxb.	Elaeagnaceae	1, 3, 4	5643
Elaeocarpus sphaericus (Gaertner) Schumann	Elaeocarpaceae	9	
Eleocharis dulcis (Burm. f.) Trin. ex Henschel	Cyperaceae	1, 11	
Elephantopus scaber L.	Asteraceae	4	5370, 5414, 5534, 5545, 5713, 5776, 5992, 6125
Eleusine coracana (L.) Gaertner	Poaceae	1	
Eleusine indica (L.) Gaertner	Poaceae	3, 4	5620, 6136, 6137
Eleutherine bulbosa (Miller) Urban	Iridaceae	3, 4	5226, 5648, 5667, 5695, 5704, 5711, 6075, 6160
Elsholtzia blanda Keng	Lamiaceae	4	B136
Elsholtzia sp.	Lamiaceae	10	Nelson
Embelia stricta Craib	Myrsinaceae	1, 4	5291
Embelia sp.	Myrsinaceae	4	5837
Emilia sonchifolia (L.) DC. ex Wight	Asteraceae	1	5389
Engelhardia serrata Blume	Juglandaceae	1	5480
Entada rheedii Sprengel	Mimosaceae	4, 7, 9	5358, 6123
Equisetum debile Roxb. ex Vaucher	Equisetaceae	4	6009
Eranthemum tetragonum A. Dietr. ex Nees	Acanthaceae	3, 4	B2
Eryngium foetidum L.	Apiaceae	1, 4	5385

Binomial	Family	Uses	Vouchers
Erythrina subumbrans (Hassk.) Merr.	Fabaceae	4	5617
Erythrina variegata L.	Fabaceae	4	
Etlingera littoralis (König) Giseke	Zingiberaceae	1, 4	5503
Eucalyptus canaldulensis Dehnh.	Myrtaceae	4, 10	5511
Eucalyptus citriodora Hook.	Myrtaceae	4, 10	
Eugenia angkae Craib	Myrtaceae	1	5479
Eugenia clarkeana King	Myrtaceae	1	
Euodia glomerata Craib	Rutaceae	4	5742, 5924
Euodia triphylla DC.	Rutaceae	4	5467
Euodia sp.	Rutaceae	4	5950
Euonymus sootepensis Craib	Celastraceae	4	6105
Eupatorium adenophorum Sprengel	Asteraceae	1, 4	5203, 5271, 5337, 5788
Eupatorium odoratum L.	Asteraceae	4, 8	5268, 5441, 5470, 5551, 5659, 5765, 6013, 6027, 6140
Eupatorium stoechadosum Hance	Asteraceae	4	5224
Euphorbia antiquorum L.	Euphorbiaceae	4	5280, 6035
Euphorbia heterophylla L.	Euphorbiaceae	4	5448, 6157
Euphorbia hirta L.	Euphorbiaceae	4	5523, 6057, 6120
Euphorbia pulcherrima Willd. ex Klotzach	Euphorbiaceae	9	
Euphorbia sp.	Euphorbiaceae	4	5463
Eurya acuminata DC. var. *cerasifolia* (D. Don) H. Keng	Theaceae	3, 4, 5	5276, 5353, 5403
Eurya acuminata DC. var. *wallichiana* Dyer	Theaceae	4	5183
Eurysolen gracilis Prain	Lamiaceae	4	B49
Excoecaria oppositifolia Griff.	Euphorbiaceae	1	
Fagopyrum dibotrys (D. Don) Hara	Polygonaceae	1, 4	5388
Fagopyrum esculentum Moench	Polygonaceae	1, 4	B6
Ficus auriculata Lour.	Moraceae	1, 2, 5	5180
Ficus benghalensis L.	Moraceae	3	
Ficus carica L.	Moraceae	1, 11	
Ficus chartacea Wallich ex King	Moraceae	4	5883
Ficus fistulosa Reinw. ex Blume	Moraceae	1, 4	5188, 5858, 5911, 6177
Ficus hispida L. f.	Moraceae	1, 2, 4, 10	5192b, 5458, 5497, 5933
Ficus racemosa L.	Moraceae	1, 4	
Ficus religiosa L.	Moraceae	3	
Ficus semicordata Buch.-Ham. ex Smith var. *semicordata*	Moraceae	1, 2, 4, 10	5300, 5474, 5603, 5815
Ficus virens Aiton	Moraceae	1	
Ficus sp.	Moraceae	4	5902

Binomial	Family	Uses	Vouchers
Ficus sp.	Moraceae	1	5412
Flemingia macrophylla (Willd.) Kuntze ex Prain	Fabaceae	4	5407
Flemingia paniculata Wallich ex Benth.	Fabaceae	4	B78
Flemingia sootepensis Craib	Fabaceae	4	5633
Foeniculum vulgare Miller	Apiaceae	1, 4, 11	
Fragaria sp.	Rosaceae	1, 11	
Garcinia cowa Roxb.	Clusiaceae	1	
Garcinia dulcis (Roxb.) Kurz	Clusiaceae	6	
Garuga pinnata Roxb.	Burseraceae	1, 4	5352
Gelsemium elegans (Gardner & Champ.) Benth.	Loganiaceae	4, 10	6031
Gendarussa vulgaris Nees	Acanthaceae	4	5535
Gigantochloa latifolia Ridley	Poaceae	5, 10	5266
Gigantochloa wrayi Gamble	Poaceae	5, 10	5361
Girardinia heterophylla Decne.	Urticaceae	4	
Gladiolus × *hortulanus* L. Bailey	Iridaceae	4, 11	
Glochidion eriocarpum Champ.	Euphorbiaceae	4	Pake
Glochidion kerrii Craib	Euphorbiaceae	4	5868
Glochidion sp.	Euphorbiaceae	1	B79
Gluta usitata (Wallich) Ding Hou	Anacardiaceae	5	4160
Glycine max (L.) Merr.	Fabaceae	1, 11	
Gmelina arborea Roxb.	Verbenaceae	4	5313, 5656, 5869
Gnetum montanum Markgraf	Gnetaceae	4	5350
Gnetum sp.	Gnetaceae	2, 4	6012
Gnetum sp.	Gnetaceae	3	6040, 6091
Gomphostemma wallichii Prain	Lamiaceae	4	6015a
Gonocaryum lobbianum (Miers) Pierre	Icacinaceae	4	5734
Gossypium arboreum L.	Malvaceae	2, 11	5642, 6081
Gossypium barbadense L.	Malvaceae	2, 4, 11	5653
Grewia hirsuta (Korth) Kochummen	Tiliaceae	4	5784
Gymnema inodorum Decne.	Asclepiadaceae	4	
Gynostemma pedatum Blume	Cucurbitaceae	4	5285
Gynura pseudochina (L.) DC.	Asteraceae	4	5800
Harrisonia perforata (Blanco) Merr.	Simaroubaceae	4, 6	5518
Hedychium coronarium J. König	Zingiberaceae	4	
Hedyotis elegans Wallich	Rubiaceae	4	5923
Hedyotis capitellata Wallich ex G. Don var. *pubescens* Kurz	Rubiaceae	4	5559
Hedyotis tenelliflora Blume	Rubiaceae	4	Pake 39
Helianthus annuus L.	Asteraceae	1	
Helichrysum bracteatum (Vent.) Andrews	Asteraceae	4, 9	
Heliciopsis terminans (Kurz) Sleumer	Proteaceae	4	6005
Helicteres elongata Wallich ex Bojer	Sterculiaceae	4	5722, 5754
Helicteres hirsuta Lour.	Sterculiaceae	4	B214

Binomial	Family	Uses	Vouchers
Helicteres lanceolata DC.	Sterculiaceae	4	B36
Helicteres plebeja Kurz	Sterculiaceae	4, 10	5341, 5614, 5978
Hemerocallis lilioasphodelus L.	Liliaceae	9, 11	
Hemigraphis glaucescens (Nees) C. B. Clarke	Acanthaceae	4	B120
Hibiscus sabdariffa L.	Malvaceae	1, 2, 11	5254
Hibiscus tiliaceus L.	Malvaceae	2	
Holarrhena pubescens (Buch.-Ham.) Wallich ex G. Don	Apocynaceae	4, 5	5979, 5995
Homalium ceylanicum (Gardner) Benth.	Flacourtiaceae	4	5916
Homalomena lindenii (Rodigas) Ridley	Araceae	9	
Homalomena occulta (Lour.) Schott	Araceae	3, 4	5773, 5874, 5887
Homalomena rebescens (Roxb.) Kunth	Araceae	9	
Homalomena sp.	Araceae	1, 4	5477
Hopea odorata Roxb.	Dipterocarpaceae	5	
Hordeum vulgare L.	Poaceae	1, 11	
Houttuynia cordata Thunb.	Saururaceae	1, 4	5383, 5772
Hydrocotyle javanica Thunb.	Apiaceae	1, 4	B155
Hymenodictyon excelsum Wallich ex Roxb.	Rubiaceae	1, 4	5533
Hypericum japonicum Thunb.	Clusiaceae	4	Pake
Impatiens balsamina L.	Balsaminaceae	3, 4, 9	5953
Imperata cylindrica (L.) P. Beauv.	Poaceae	1, 3, 4, 5, 10	5461, 5891, 6171
Indigofera squalida Prain	Fabaceae	4	6003
Indigofera tinctoria L.	Fabaceae	6	5316, 5379
Inula cappa (Ham. ex D. Don) DC.	Asteraceae	4	5546
Ipomoea aquatica Forsskal	Convolvulaceae	1	B131
Ipomoea batatas (L.) Lam.	Convolvulaceae	1, 4, 11	
Iresine herbstii Hook.	Amaranthaceae	4, 9	5223, 5536
Iris collettii Hook. f.	Iridaceae	3, 4, 9	5810
Ixora cibdela Craib	Rubiaceae	3, 4	B56
Ixora finlaysoniana Wallich ex G. Don	Rubiaceae	4	B259
Jacaranda obtusifolia Humb. & Bonpl.	Bignoniaceae	4	
Jasminum nervosum Lour.	Oleaceae	4	5626
Jasminum sambac (L.) Aiton	Oleaceae	9	
Jatropha curcas L.	Euphorbiaceae	4, 10	5250, 6037, 6049
Justicia glomerulata Benoist	Acanthaceae	4	5836, 5876
Justicia quadrifaria (Wallich ex Nees) T. Anderson	Acanthaceae	4	5665
Justicia spp.	Acanthaceae	1	
Kaempferia pandurata Roxb.	Zingiberaceae	1	
Kaempferia rotunda L.	Zingiberaceae	4	5688, 5753, 5980

Binomial	Family	Uses	Vouchers
Kaempferia sp.	Zingiberaceae	4	6074
Kaempferia sp.	Zingiberaceae	1, 3, 4	5228
Kalanchoe laciniata (L.) DC.	Crassulaceae	4	Pake
Kalanchoe spp.	Crassulaceae	4	5769, 5803, 5956, 6084
Lactuca sativa L.	Asteraceae	1, 3, 11	
Lagenaria siceraria (Molina) Standley	Cucurbitaceae	1, 9, 10, 11	
Lagerstroemia calyculata Kurz	Lythraceae	4	5492
Lagerstroemia speciosa (L.) Pers.	Lythraceae	9	
Lannea coromandelica Merr.	Anacardiaceae	2	
Laportea interrupta (L.) Chew	Urticaceae	4	5785
Lasia spinosa (L.) Thwaites	Araceae	1, 4	
Lasianthus sp.	Rubiaceae	4	5838, 5893
Leea indica (Burm. f.) Merr.	Leeaceae	4	5433, 5466, 5555, 5782, 5794, 5826, 5879
Leea rubra Blume ex Sprengel	Leeaceae	4	5745, 5921
Leea sp.	Leeaceae	4	5326, 6119
Lepisathes rubiginosa (Roxb.) Leenh.	Sapindaceae	4	5605
Lespedeza parviflora Kurz	Fabaceae	4	5930
Leucaena leucocephala (Lam.) de Wit	Mimosaceae	1, 3, 4	5457
Limnophila rugosa (Roth) Merr.	Scrophulariaceae	1, 4	6016, 6072, 6162
Litchi chinensis Sonn.	Sapindaceae	1, 11	
Lithocarpus elegans (Blume) Hatus. ex Soepadmo	Fagaceae	5	5199
Lithocarpus polystachyus (A. DC.) Rehder	Fagaceae	4, 5	5324
Lithocarpus thomsonii (Miq.) Rehder	Fagaceae	5	
Litsea glutinosa (Lour.) C. Robinson	Lauraceae	2, 4	5663, 5732
Litsea monopetala (Roxb.) Pers.	Lauraceae	3, 4, 5, 10	5179, 5357
Livistona speciosa Kurz	Arecaceae	5	6042
Lobelia angulata Forster	Campanulaceae	1, 4	5212, 5339
Lonicera macrantha DC.	Caprifoliaceae	4	Pake
Lophatherum gracile Brongn.	Poaceae	4	Pake
Lophopetalum sp.	Celastraceae	4	5807
Luffa acutangula (L.) Roxb.	Cucurbitaceae	1	3849
Luffa aegyptiaca Miller	Cucurbitaceae	1, 10	
Lycopersicon esculentum Miller	Solanaceae	1, 11	
Lygodium flexuosum (L.) Sw.	Schizaeaceae	1, 2, 4	5438, 5495, 5608, 6021
Lygodium japonicum (Thunb.) Sw.	Schizaeaceae	4	Pake
Macaranga denticulata (Blume) Muell. Arg.	Euphorbiaceae	4, 5	5632
Macaranga gigantea (Reichb. f. & Zoll.) Muell. Arg.	Euphorbiaceae	6, 10	5274
Machilus parviflorus Meissner	Lauraceae	4	B77
Machilus sp.	Lauraceae	10, 11	B126

Binomial	Family	Uses	Vouchers
Machilus sp.	Lauraceae	4	B223
Maclura amboinensis Blume	Moraceae	4	B84
Maclura fruticosa (Roxb.) Corner	Moraceae	4	5491
Macropanax dispermus (Blume) Kuntze	Araliaceae	1	5189
Madhuca sp.	Sapotaceae	4	5737
Maesa indica Wallich	Myrsinaceae	4	6025
Maesa montana A. DC.	Myrsinaceae	3, 4	5396, 5483, 5748, 6017, 6097
Maesa permollis Kurz	Myrsinaceae	4	5885, 5901
Maesa ramentacea Wallich ex Roxb.	Myrsinaceae	4	5586, 5611
Mallotus paniculatus (Lam.) Muell. Arg.	Euphorbiaceae	4	5833, 5856
Mallotus sp.	Euphorbiaceae	4	5624
Mammea siamensis (Miq.) T. Anderson	Clusiaceae	1	
Mangifera indica L.	Anacardiaceae	1, 11	
Manihot esculenta Crantz	Euphorbiaceae	1, 11	3822
Maoutia puya Wedd.	Urticaceae	2, 4	6073
Markhamia stipulata (Wallich) Seemann ex Schumann	Bignoniaceae	1, 4, 8	4015
Marsdenia tinctoria R. Br.	Asclepiadaceae	6	5395
Marsilea crenata C. Presl	Marsileaceae	1, 4	5410
Melastoma normale D. Don var. *normale*	Melastomaceae	1, 4	5489
Melastoma imbricatum Wallich ex Triana	Melastomaceae	1	5208
Melia azedarach L.	Meliaceae	2, 4, 10	
Melientha suavis Pierre	Opiliaceae	4	
Meliosma simplicifolia (Roxb.) Walp.	Sabiaceae	1, 4	B108
Melocalamus compactiflorus (Kurz) Benth.	Poaceae	1, 4, 10	
Mentha arvensis L.	Lamiaceae	1, 4, 11	5256
Mentha × *villosa* Hudson	Lamiaceae	11	
Mesona sp.	Lamiaceae	1	
Michelia champaca L.	Magnoliaceae	10	
Microcos paniculata L.	Tiliaceae	1, 7	6046
Microglossa pyrifolia (Lam.) Kuntze	Asteraceae	4	5974
Microlepia herbacea Ching & C. Chr. ex Tard. & C. Chr.	Dennstaedtiaceae	4	B65
Microlepia speluncae (L.) T. Moore	Dennstaedtiaceae	4	5501
Microtropis pallens Pierre	Celastraceae	4	B69
Miliusa thorelii Finet & Gagnepain	Annonaceae	4	5966
Miliusa velutina (Dunal) Hook. f. & Thomson	Annonaceae	4	5522
Millettia extensa Benth. ex Baker	Fabaceae	4	5585
Millettia pachycarpa Benth.	Fabaceae	4	5332
Millingtonia hortensis L.	Bignoniaceae	1, 4, 10	5519, 5818

Binomial	Family	Uses	Vouchers
Mimosa invisa Martius ex Colla	Mimosaceae	10	5171
Mimosa pudica L.	Mimosaceae	3, 4	5443, 5842, 5963, 6047, 6163
Mirabilis jalapa L.	Nyctaginaceae	4	Pake
Mitragyna hirsuta Havil.	Rubiaceae	4	5517, 5990
Mitragyna speciosa Korth.	Rubiaceae	4	
Mitragyna spp.	Rubiaceae	5	
Molineria capitulata (Lour.) Herbert	Hypoxidaceae	3, 4, 10	5219
Momordica charantia L.	Cucurbitaceae	1, 11	
Monochoria hastata (L.) Solms	Pontederiaceae	4	
Monochoria vaginalis (Burm. f.) Presl ex Kunth	Pontederiaceae	1, 4	
Morinda angustifolia Roxb.	Rubiaceae	4, 6	5237, 5729
Morinda citrifolia L.	Rubiaceae	1, 4	
Morinda tomentosa Heyne ex Roth	Rubiaceae	4	6023
Morinda sp.	Rubiaceae	4	5660, 5830
Moringa pterygosperma C. F. Gaertner	Moringaceae	1	
Morus alba L.	Moraceae	1, 2, 4	B19
Morus australis Poiret	Moraceae	1	6181
Morus macroura Miq.	Moraceae	4	6070
Mucuna brevipes Craib	Fabaceae	2	5401
Mucuna macrocarpa Wallich	Fabaceae	1, 2, 7, 9	
Mucuna pruriens (L.) DC.	Fabaceae	4	5315, 5423
Mukia javanica (Miq.) C. Jeffrey	Cucurbitaceae	3, 4	6011
Muntingia calabura L.	Elaeocarpaceae	1	5662
Musa acuminata Colla	Musaceae	1, 2, 3, 4, 9	5531
Musa × *paradisiaca* L.	Musaceae	1, 11	
Mussaenda kerrii Craib	Rubiaceae	4	5831, 5853
Mussaenda parva Wallich ex G. Don	Rubiaceae	4, 9, 10	5176, 5398
Mussaenda pubescens Aiton f.	Rubiaceae	4	Pake 75
Mycetia gracilis Craib	Rubiaceae	4	5576
Myriophyllum sp.	Haloragaceae	1	
Nasturtium officinale R. Br.	Brassicaceae	1	
Nauclea orientalis L.	Rubiaceae	4	5747
Nauclea sp.	Rubiaceae	4	6002
Nelsonia canescens (Lam.) Sprengel	Acanthaceae	4	B72
Nephelium hypoleucum Kurz	Sapindaceae	1	
Nephrolepis cordifolia (L.) Presl	Davalliaceae	4	Pake
Nephrolepis falcata (Cav.) C. Chr.	Davalliaceae	1, 4	
Neptunia oleracea Lour.	Mimosaceae	4	4017
Nicotiana tabacum L.	Solanaceae	7, 11	
Nothaphoebe umbelliflora Blume	Lauraceae	4, 6	5484
Nyctanthes arbor-tristis L.	Verbenaceae	6	
Ocimum basilicum L.	Lamiaceae	1, 4	Pake
Ocimum canum Sims	Lamiaceae	1, 4	
Ocimum gratissimum L.	Lamiaceae	4	5449
Ocimum sanctum L.	Lamiaceae	1, 4	
Oenanthe javanica (Blume) DC.	Apiaceae	1, 4	5390, 5957

Binomial	Family	Uses	Vouchers
Oncosperma horridum Scheffer	Arecaceae	4	
Ophioglossum costatum R. Br.	Ophioglossaceae	4	Pake 59
Ophiorrhiza hispidula Wallich ex G. Don	Rubiaceae	4	5705, 5935
Opuntia dillenii (Ker Gawler) Haw.	Cactaceae	3, 4	6036
Oroxylum indicum (L.) Vent.	Bignoniaceae	1, 4	5195, 5427, 6135, 6164, 6173
Orthosiphon aristatus (Blume) Miq.	Lamiaceae	4	5644, 5813
Oryza sativa L.	Poaceae	1, 3, 11	EFA
Osbeckia stellata Ham. ex Ker Gawler	Melastomaceae	1, 4, 10	5489, 5554, 5610, 5775, 5912
Ottelia alismoides (L.) Pers.	Hydrocharitaceae	1	
Oxalis corniculata L.	Oxalidaceae	4	6107
Oxytenanthera albo-ciliata Munro	Poaceae	1, 10	5308, 5319, 5552
Pachyrhizus erosus (L.) Urban	Fabaceae	1, 11	
Paederia pallida Craib	Rubiaceae	4	5437
Paederia pilifera Hook. f.	Rubiaceae	4	5598
Paederia scandens (Lour.) Merr.	Rubiaceae	4	5657
Paederia tomentosa Blume	Rubiaceae	4	5693
Paederia wallichii Hook. f.	Rubiaceae	4	5375, 5415, 5567
Pandanus furcatus Roxb.	Pandanaceae	3, 4, 10	
Panicum sp.	Poaceae	4	6050
Papaver somniferum L.	Papaveraceae	1, 3, 4, 10, 11	4176
Parinari anamensis Hance	Rosaceae	1, 4, 5, 8	
Passiflora edulis Sims	Passifloraceae	1, 11	
Passiflora foetida L.	Passifloraceae	1	
Passiflora quadrangularis L.	Passifloraceae	1	
Pavetta petiolaris Wallich ex Craib	Rubiaceae	4	5928
Pedilanthus tithymaloides (L.) Poit.	Euphorbiaceae	4, 9	5537, 6085
Peliosanthes grandifolia Ridley	Liliaceae	4	5574
Peliosanthes teta Andrews	Liliaceae	4	5580
Peliosanthes violacea Wallich	Liliaceae	4	5606
Pennisetum polystachion (L.) Schultes	Poaceae	4	5724
Peperomia pellucida Kunth	Piperaceae	1	
Peristrophe lanceolaria (Roxb.) Nees	Acanthaceae	4	B51
Persea americana Miller	Lauraceae	1, 11	
Petroselinum crispum (Miller) Nyman ex A.W. Hill	Apiaceae	1, 3, 11	
Peucedanum sp.	Apiaceae	4	5225
Phaseolus lunatus L.	Fabaceae	1, 11	
Phaseolus vulgaris L.	Fabaceae	1, 11	
Phlogacanthus curviflorus (Wallich) Nees var. *curviflorus*	Acanthaceae	4, 6	5726, 5822, 5335, 5708, 5844, 5880, 6086, 6098

Binomial	Family	Uses	Vouchers
Phoebe lanceolata (Wallich ex Nees) Nees	Lauraceae	3, 4	5374, 6066
Phoenix acaulis Roxb.	Arecaceae	1, 10	
Phragmites karka (Retz.) Trin. ex Steudel	Poaceae	4	5542
Phrynium capitatum Willd.	Marantaceae	1, 4, 5, 10	5168, 5664, 5824, 5861, 5862
Phyllanthus acidus (L.) Skeels	Euphorbiaceae	1	
Phyllanthus emblica L.	Euphorbiaceae	1, 2, 4, 6, 7	5196, 5417, 5473, 5591
Phyllanthus sp.	Euphorbiaceae	4	5547
Phyllostachys sp.	Poaceae	1, 10	5680
Physalis angulata L.	Solanaceae	4	5526
Phytolacca acinosa Roxb.	Phytolaccaceae	4	Pake 63
Phytolacca sp.	Phytolaccaceae	4	5541
Pinanga sp.	Arecaceae	5	
Pinus kesiya Royle ex Garden	Pinaceae	5	4036, 4113
Pinus merkusii Junghuhn & Vriese	Pinaceae	4, 5, 8	4112
Piper agyrophyllum Miq.	Piperaceae	4	5476
Piper betle L.	Piperaceae	4, 7	5246, 5258
Piper boehmeriaefolium (Miq.) C. DC.	Piperaceae	1	5494
Piper pedicellatum C. DC.	Piperaceae	3, 4	5774
Piper pellucidum L.	Piperaceae	4	
Piper sarmentosum Roxb.	Piperaceae	1, 4	
Piper sp.	Piperaceae	4	5717
Pisum sativum L.	Fabaceae	1, 11	
Pithecellobium dulce (Roxb.) Benth.	Mimosaceae	1, 4	
Plantago major L.	Plantaginaceae	4	5220, 5281, 5373, 5525, 5560, 5706, 6103
Platycerum wallichii Hook.	Polypodiaceae	4	
Plectranthus hispidus Benth.	Lamiaceae	4	6015b
Plectranthus parviflorus Willd.	Lamiaceae	1	5799
Plectranthus ternifolius D. Don	Lamiaceae	1	3917, 3984, 4110
Plumbago indica L.	Plumbaginaceae	4	5918
Plumbago zeylanica L.	Plumbaginaceae	4	5621, 6100
Plumeria rubra L.	Apocynaceae	3, 4, 9	5757
Polyalthia cerasoides Benth. & Hook. f. ex Hook. f.	Annonaceae	4	5904
Polyalthia simiarum (Ham. ex Hook. f. & Thwaites) Benth. & Hook. f.	Annonaceae	1, 5	5376
Polygala glomerata Lour.	Polygalaceae	4	5696, 5710, 5975
Polygonum barbatum L.	Polygonaceae	4	6051
Polygonum chinense L.	Polygonaceae	1, 4	5465
Polygonum glabrum Willd.	Polygonaceae	1, 4	5386

Binomial	Family	Uses	Vouchers
Polygonum sp.	Polygonaceae	4	5561, 5798
Polyscias fruticosa (L.) Harms	Araliaceae	4	5609
Pothos cathcartii Schott	Araceae	4	5573, 5741
Premna fulva Craib	Verbenaceae	4	B219
Premna nana Collett & Hemsley	Verbenaceae	4	6022, 6116
Protium serratum (Wallich ex Colebr.) Engler	Burseraceae	1	6063
Prunus cerasoides D. Don	Rosaceae	4	5288
Prunus domestica L.	Rosaceae	1, 11	
Prunus persica (L.) Batsch	Rosaceae	1, 3, 4, 11	6054, 6082
Pseuderanthemum andersonii (Masters) Lindau	Acanthaceae	4, 6	5673, 5731
Psidium guajava L.	Myrtaceae	1, 4, 11	5672, 6055
Psophocarpus tetragonolobus (L.) DC.	Fabaceae	1	
Psychotria monticola Kurz var. monticola	Rubiaceae	4	5793
Psychotria morindoides Hutch.	Rubiaceae	4	5469
Psychotria ophioxyloides Wallich	Rubiaceae	4	B260
Pteridium aquilinum (L.) Kuhn	Dennstaedtiaceae	1, 4	5211, 5287
Pteris biaurita L.	Adiantaceae	3, 4	5400, 5485
Pteris semipinnata L.	Adiantaceae	4	Pake 75
Pteris venusta Kunze	Adiantaceae	4	B91
Pteris sp.	Adiantaceae	4	5578
Pterolobium macropterum Kurz	Caesalpinaceae	4	5733
Pterospermum grande Craib	Sterculiaceae	10	
Pueraria phaseoloides (Roxb.) Benth.	Fabaceae	2, 4	6094
Pueraria rigens Craib	Fabaceae	4	5215
Punica granatum L.	Punicaceae	1, 4	5539
Pyrus pyrifolia (Burm. f.) Nakai	Rosaceae	1, 11	
Quercus aliena Blume	Fagaceae	5, 8	
Quercus incana Roxb.	Fagaceae	5, 8	
Quercus kerrii Craib	Fagaceae	5, 8	4228
Quercus semiserrata Roxb.	Fagaceae	5, 8	
Quisqualis indica L.	Combretaceae	4, 9	5649
Radermachera glandulosa (Blume) Miq.	Bignoniaceae	4	B205
Radermachera ignea (Kurz) Steenis	Bignoniaceae	1, 4, 10	5645
Ranunculus cantoniensis DC.	Ranunculaceae	4	Pake 77
Raphanus sativus L.	Brassicaceae	1, 11	
Raphistemma pulchellum Wallich	Asclepiadaceae	4	5761
Rauvolfia sp.	Apocynaceae	4	5712, 5730
Rauwenhoffia siamensis Scheffer	Annonaceae	4	
Rhaphidophora sp.	Araceae	4, 9	5787
Rhus chinensis Miller	Anacardiaceae	1, 3, 4, 10	5198, 5369, 5584
Rhus succedanea L.	Anacardiaceae	1, 4	B195
Rhynchotechum obovatum (Griffith) B. L. Burt	Gesneriaceae	2, 7, 10	5334
Ricinus communis L.	Euphorbiaceae	4, 10, 11	5289, 5767

Binomial	Family	Uses	Vouchers
Rosa helenae Rehder & E. Wilson	Rosaceae	1	5392
Rourea minor (Gaertner) Leenh.	Connaraceae	1, 4	5913
Rubus blepharoneurus Cardot	Rosaceae	1, 3, 4, 7	5207, 5338, 5847, 5914, 6038
Rubus dielsianus Focke	Rosaceae	1, 4	5365
Rubus ellipticus Smith	Rosaceae	1, 4	B96
Saccharum officinarum L.	Poaceae	1, 11	
Saccharum spontaneum L.	Poaceae	10	5202
Salix tetrasperma Roxb.	Salicaceae	3, 4	B143
Salvia coccinea Juss. ex Murray	Lamiaceae	1, 9	
Samanea saman (Jacq.) Merr.	Mimosaceae	4	
Sambucus javanica Reinw. ex Blume	Caprifoliaceae	1, 4, 9, 10	5520, 5671, 5727, 5746, 5768, 5827, 5889, 5947, 6010, 6026, 6147
Sansevieria trifasciata Prain	Agavaceae	3, 4, 9	5508, 6095
Sapium discolor (Champ. ex Benth.) Muell. Arg.	Euphorbiaceae	3	Pake
Saprosma sp.	Rubiaceae	4	5943
Saraca declinata (Jack) Miq.	Caesalpinaceae	4	
Saraca indica L.	Caesalpinaceae	3	
Saurauia nepalensis DC.	Actinidiaceae	4	B270
Saurauia roxburghii Wallich	Actinidiaceae	1, 3, 4	5191, 5272, 5582, 5867, 5890, 5919
Sauropus androgynus (L.) Merr.	Euphorbiaceae	1, 4	
Sauropus quadrangularis (Willd.) Muell. Arg.	Euphorbiaceae	4	B212
Schefflera alongensis R. Viguier	Araliaceae	4	5938
Schefflera clarkeana Craib	Araliaceae	4	5450, 5973
Schefflera sp.	Araliaceae	4	5721
Schima wallichii (DC.) Korth	Theaceae	3, 4, 5, 6, 10	3814, 4093, 5325, 5472, 5549, 5599, 6029, 6032
Scirpus grossus L.	Cyperaceae	4	
Scleria levis Retz.	Cyperaceae	4	5759
Scleria terrestris (L.) Fassett	Cyperaceae	3, 4	5499, 6170
Scleria sp.	Cyperaceae	4	5777, 6058
Scleropyrum wallichianum Arn.	Santalaceae	4	5740
Scoparia dulcis L.	Scrophulariaceae	4	5524, 5994
Scurrula ferruginea (Jack) Danser	Loranthaceae	4	5345
Scutellaria indica L.	Lamiaceae	4	Pake
Sechium edule (Jacq.) Sw.	Cucurbitaceae	1	
Securinega virosa (Roxb. ex Willd.) Baillon	Euphorbiaceae	1, 4, 10	5516, 6146

Binomial	Family	Uses	Vouchers
Selaginella helferi Warb.	Selaginellaceae	1, 4	5329, 5884, 5936
Selaginella minutifolia Spring	Selaginellaceae	4	5581
Selaginella pubescens (Wallich ex Hook. & Grev.) Spring	Selaginellaceae	1	
Selaginella repanda (Desv.) Spring	Selaginellaceae	4	5627
Selaginella roxburghii (Hook. & Grev.) Spring	Selaginellaceae	4	B251
Sesamum indicum L.	Pedaliaceae	1, 10, 11	6112
Sesbania grandiflora (L.) Poiret	Fabaceae	1, 6	
Sesbania javanica Miq.	Fabaceae	10	
Sesbania sesban (L.) Merrr.	Fabaceae	11	
Seseli siamicum Craib	Apiaceae	4	5997
Setaria italica (L.) P. Beauv.	Poaceae	1, 11	
Shorea obtusa Wallich	Dipterocarpaceae	4, 5, 7	5425
Shorea roxburghii G. Don	Dipterocarpaceae	5, 7	
Shorea siamensis Miq.	Dipterocarpaceae	5, 7	
Shuteria vestita Wight & Arn.	Fabaceae	4	6104
Sida acuta Burm. f.	Malvaceae	4	5286, 5692
Sida rhombifolia L. var. *rhombifolia*	Malvaceae	4, 10	5190, 5808, 5908, 6065
Siegesbeckia orientalis L.	Asteraceae	4	B4
Smilax corbularia Kunth subsp. *corbularia*	Smilacaceae	1, 4, 10	5937
Smilax lanceaefolia Roxb.	Smilacaceae	4	5355
Smilax ovalifolia Roxb.	Smilacaceae	2, 3, 4, 9	5419, 5453, 5600, 5758, 6020, 6167
Smilax perfoliata Lour.	Smilacaceae	3	6090
Smilax verticalis Gagnepain	Smilacaceae	1	
Solanum erianthum D. Don	Solanaceae	1, 4, 10	5257, 5295, 5563, 5570, 5625, 5639, 5841, 5855, 6149
Solanum melongena L.	Solanaceae	1, 4, 11	3688
Solanum nigrum L.	Solanaceae	1, 4	5755
Solanum nodiflorum Jacq.	Solanaceae	4	5222
Solanum spirale Roxb.	Solanaceae	4	5795
Solanum torvum Sw.	Solanaceae	1, 4	5193, 5789, 6043
Solanum tuberosum L.	Solanaceae	1, 11	
Solanum sp.	Solanaceae	4	5687, 6014
Solanum sp.	Solanaceae	1	5481
Solena heterophylla Lour.	Cucurbitaceae	4	
Sonchus oleraceus L.	Asteraceae	4	6004
Sorghum bicolor (L.) Moench	Poaceae	1, 3, 9, 11	
Spatholobus parviflorus (Roxb.) Kuntze	Fabaceae	2, 4, 5	5418, 5434, 5553, 5996, 6182

Binomial	Family	Uses	Vouchers
Spatholobus sp.	Fabaceae	4	B174
Sphenomeris chinensis (L.) Maxon	Dennstaedtiaceae	4	Pake
Spilanthes oleracea L.	Asteraceae	1	5411
Spilanthes paniculata Wallich ex DC.	Asteraceae	1, 4, 10	5596, 5652, 5846
Spinacia oleracea L.	Chenopodiaceae	1, 11	
Spondias cytherea Sonn.	Anacardiaceae	1, 4	
Spondias lakonensis Pierre	Anacardiaceae	4	
Spondias pinnata (Koen. & L. f.) Kurz	Anacardiaceae	1, 4, 5	5451
Sporobolus diander (Retz.) P. Beauv.	Poaceae	4	5654
Stahlianthus involucratus (King) Craib ex Loes.	Zingiberaceae	4	Pake 91, 92
Staurogyne lanceolata (Hassk.) Kuntze	Acanthaceae	4	B73
Stemona collinsae Craib	Stemonaceae	4	
Stemona burkillii Prain	Stemonaceae	4	B187
Stemona tuberosa Lour.	Stemonaceae	4	
Stemona sp.	Stemonaceae	4	5699
Stephania brevipes Craib	Menispermaceae	4	5569, 6064
Stephania glabra (Roxb.) Miers	Menispermaceae	4	6106
Stephania japonica (Thunb.) Miers	Menispermaceae	4	
Sterculia foetida L.	Sterculiaceae	1, 2	
Sterculia lanceolata Cav.	Sterculiaceae	4	5493, 5871
Sterculia pexa Pierre	Sterculiaceae	2	5185
Stereospermum colais (Buch.-Ham. ex Dillwyn) Mabb.	Bignoniaceae	4	5999
Stereospermum neuranthum Kurz	Bignoniaceae	1, 4, 7	5456
Streblus asper Lour.	Moraceae	7	5764
Streptocaulon juventas (Lour.) Merr.	Asclepiadaceae	4	5690
Strobilanthes anfractuosa C. B. Clarke	Acanthaceae	4, 6	6030
Strobilanthes lanceifolius T. Anderson	Acanthaceae	6	5615, 6087, 6179
Strobilanthes pentstemonoides (Nees) T. Anderson	Acanthaceae	4, 6	5801
Strychnos nux-vomica L.	Loganiaceae	1, 4	5440, 6126
Styrax benzoides Craib	Styracaceae	4, 10	5298
Symphorema involucratum Roxb.	Verbenaceae	4	5566, 5988
Syzygium cumini (L.) Skeels	Myrtaceae	1, 2, 4	5197
Tacca chantrieri André	Taccaceae	1, 4	5496, 5860, 6096
Tacca integrifolia Ker Gawler	Taccaceae	4	B39
Tacca sp.	Taccaceae	4	5829, 5886, 5898
Tagetes erecta L.	Asteraceae	9, 11	
Tagetes patula L.	Asteraceae	3, 9, 11	
Talinum triangulare Willd.	Portulacaceae	4	5805, 6067
Tamarindus indica L.	Caesalpinaceae	1, 4, 11	3787, 6045
Tectaria polymorpha (Wallich ex Hook.) Copel.	Aspleniaceae	4	5964
Tectona grandis L.	Verbenaceae	4, 5	3788

Binomial	Family	Uses	Vouchers
Telosma pallida (Roxb.) Craib	Asclepiadaceae	1, 7	
Terminalia bellirica (Gertner) Roxb.	Combretaceae	5, 6, 7	
Terminalia mucronata Craib & Hutch.	Combretaceae	5	
Tetrastigma lanceolarum (Roxb.) Planchon	Vitaceae	4	5823
Teucrium viscidium Blume	Lamiaceae	4	Pake 97
Themeda arundinacea (Roxb.) Ridley	Poaceae	1, 4	5593
Thevetia peruviana (Pers.) Schumann	Apocynaceae	9	4162
Thunbergia grandiflora (Roxb. ex Rottl.) Roxb. var. *grandiflora*	Acanthaceae	4	5366, 5588, 5821, 5850
Thunbergia laurifolia Lindley	Acanthaceae	4	5454
Thunbergia similis Craib	Acanthaceae	1, 4, 10	5650
Thyrsostachys oliveri Gamble	Poaceae	1, 5, 7, 10	
Thyrsostachys siamensis (Kurz ex Munro) Gamble	Poaceae	1, 2, 4, 10	5311, 5680
Thysanolaena latifolia (Roxb. ex Horn.) Honda	Poaceae	1, 3, 5, 10	5181
Tiliacora triandra (Colebr.) Diels	Menispermaceae	4	5839
Tinomiscium petiolare Hook. f. & Thomson	Menispermaceae	4	5373, 5487, 5897, 5965, 5970
Tinospora sinensis (Lour.) Merr.	Menispermaceae	4	5840, 5969
Tithonia diversifolia (Hemsley) A. Gray	Asteraceae	4, 9	5290
Toddalia asiatica (L.) Lam.	Rutaceae	1, 3, 4	5336
Torenia siamensis Yamaz.	Scrophulariaceae	4	5668, 5814
Tradescantia sp.	Commelinaceae	4	5955
Trapa bicornis Osbeck	Trapaceae	4	
Trema cannabina Lour.	Ulmaceae	4	5903
Trema orientalis (L.) Blume	Ulmaceae	4	B135
Trevisia palmata (Roxb.) Vis.	Araliaceae	1, 4, 9	5330, 5571, 5825
Trichosanthes anguina L.	Cucurbitaceae	1	
Trichosanthes rubriflos Thornber ex Cayeux	Cucurbitaceae	3	5216
Trichosanthes tricuspidata Lour.	Cucurbitaceae	4	5719, 5859
Triticum aestivum L.	Poaceae	1, 11	
Triumfetta rhomboidea Jacq.	Tiliaceae	4	Pake 100
Turpinia pomifera (Roxb.) Wallich ex DC.	Staphyleaceae	1, 4	5615a, 5616, 5946
Turpinia sphaerocarpa Hassk.	Staphyleaceae	4	5709
Typhonium trilobatum (L.) Schott	Araceae	1, 4	5806
Uncaria macrophylla Wallich	Rubiaceae	4	T272
Uncaria scandens (Smith) Hutch.	Rubiaceae	4	B165
Uncaria sp.	Rubiaceae	4	5907
Uraria cordifolia Wallich	Fabaceae	4	5982
Urena lobata L.	Malvaceae	2, 4	5204, 5406, 5594, 5619, 5934

Binomial	Family	Uses	Vouchers
Uvaria sp.	Annonaceae	4	5780, 5905
Vaccinium sprengelii (G. Don) Sleumer ex Rehder	Ericaceae	1	5426
Vanda sp.	Orchidaceae	4	
Ventilago calyculata Tul.	Rhamnaceae	4	5515, 5835
Verbena officinalis L.	Verbenaceae	4	5701, 6108
Vernonia parishii Hook. f.	Asteraceae	4	5372
Viburnum inopinatum Craib	Caprifoliaceae	4, 9	5421
Vicia faba L.	Fabaceae	1, 11	
Vigna radiata (L.) R. Wilczek	Fabaceae	1, 11	
Vigna umbellata (Thunb.) Ohwi & Ohashi	Fabaceae	1, 10, 11	
Vigna unguiculata (L.) Walp.	Fabaceae	1, 4, 11	6048
Viscum articulatum Burm. f.	Viscaceae	4	Pake 103
Vitex canescens Kurz	Verbenaceae	4	5655
Vitex peduncularis Wallich ex Schauer	Verbenaceae	4	5760
Vitex trifolia L. var. *trifolia*	Verbenaceae	3	B97
Vitex vestita Wallich ex Schauer	Verbenaceae	4	5942
Wallichia caryotoides Roxb.	Arecaceae	1, 2	5507
Wedelia chinensis (Osbeck) Merr.	Asteraceae	4	Pake 104
Xanthium sibiricum Patrin	Asteraceae	4	Pake 105
Xanthosoma violaceum Schott	Araceae	1, 11	
Xylia xylocarpa (Roxb.) Taubert	Mimosaceae	4, 5, 10	
Yucca gloriosa L.	Agavaceae	4	
Zanthoxylum limonella (Dennst.) Alston	Rutaceae	1	5213
Zea mays L.	Poaceae	1, 3, 11	
Zebrina pendula Schnitzl.	Commelinaceae	4, 9	5670
Zingiber cassumunar Roxb.	Zingiberaceae	4	5429, 5430, 5514
Zingiber officinale Roscoe	Zingiberaceae	1, 3, 4, 11	5504, 5579
Zingiber rubens Roxb.	Zingiberaceae	4	5686
Zingiber sp.	Zingiberaceae	4	5613, 5666, 5675, 5676, 5781, 5882
Zingiber sp.	Zingiberaceae	1, 4	5244, 5683, 5715
Zinnia elegans Jacq.	Asteraceae	9	
Ziziphus mauritiana Lam.	Rhamnaceae	1	3832
Ziziphus oenoplia (L.) Miller	Rhamnaceae	4	5910
Zygostelma benthamii Baillon	Asclepiadaceae	4	5340, 5622

APPENDIX 2

Medicinal Plants Used by the Hill Tribes

Key to tribe
1 = Akha
2 = Hmong
3 = Karen
4 = Lahu
5 = Lisu
6 = Mien

Ailment
1 = cough
2 = body pain; internal pain; cancer, analgesic
3 = back pain
4 = stomach ache; indigestion; food poisoning, heart burn
5 = head ache
6 = chest pain
7 = fever
8 = itching/rash; dermal problems, warts
9 = diarrhea
10 = inflammation; swelling; lumps; infection
11 = sore throat, voice
12 = dizziness; fainting
13 = poor appetite; tonic
14 = weakness; tonic; geriatric

15 = diuretic/urinary problems; bladder stones
16 = constipation; purgative; laxative
17 = tooth ache; gum infection; sore mouth
18 = ear ache; hearing problems
19 = numbness; paralysis
20 = eye pain; eye problems
21 = breast milk stimulant
22 = post delivery
23 = blood purifier; bad blood
24 = carminative-gas in stomach
25 = heart problems, stroke
26 = antihelminthic, worms
27 = liver problems, jaundice; hepatitis
28 = dysentery
29 = malaria
30 = tonsillitis
31 = antiseptic; antidote; antibiotic
32 = massage

33 = nosebleed; runny nose; nasal congestion
34 = broken bone; painful bones
35 = ulcer
36 = cuts, wounds; splinters, thorns, stop bleeding
37 = burns
38 = enema
39 = strength; tonic; general health
40 = protect spirit
41 = expectorant
42 = common cold
43 = seizure; spasms, fits, epilepsy, convulsions
44 = stop children crying
45 = sedative/insomnia
46 = stimulant; potency; aphrodisiac, fertility
47 = measles
48 = opium detox withdrawal
49 = breathing problems; lung problems
50 = bleeding; hemorrhaging; internal bleeding

225

Ailment *continued*
51 = antiabortion or antimiscarriage
52 = fish poison
53 = menstrual cramps; cycle
54 = cholera
55 = blood pressure; circulation
56 = fungal infections; dandruff
57 = lice, scalp problems; scabies
58 = hot inside
59 = make sweat; make warm
60 = pus; boils, pimples
61 = help baby delivery
62 = tuberculosis
63 = anemia, paleness, poor circulation
64 = snakebite; centipede bite; caterpillar stings; scorpion stings; insect bites

65 = complexion
66 = vomiting; nausea
67 = chapped lips, cold sore
68 = kidney; kidney stones
69 = polio
70 = steam bath
71 = for animals
72 = chicken pox
73 = sprains & bruises; sore muscles; sore joints; sore feet
74 = VD; genital problems
75 = pneumonia
76 = emetic (induce vomiting)
77 = spleen
78 = abortifacient; antiaphrodisiac
79 = shock; coma; unconsciousness
80 = moles
81 = diabetes
82 = hemorrhoids

83 = rabies; dog bite
84 = asthma
85 = hypertension
86 = leeches

Part used
1 = roots
2 = stem or rhizome
3 = bark
4 = leaves
5 = seeds
6 = whole plant
7 = flowers
8 = fruits
9 = other

Method of use
1 = oral infusion (decoction)
2 = poultice, juice, or latex
3 = steam bath
4 = inhalant
5 = bathe topically
6 = chewed or eaten
7 = other

Latin binomial	Tribe	Ailment	Part used	Method of use
Abelmoschus moschatus	2	36	1	2
Acacia concinna	2, 3	4, 22	4	1, 5
Acacia megaladena var. *megaladena*	1, 5	7, 39	4, 5	1, 5
Acalypha hispida	1	28	1	1
Achyranthes aspera	2, 6	22, 34, 53	4, 6	1, 2
Achyrospermum wallichianum	1	17	4	6
Acorus gramineus	1	4	1	6
Adenanthera pavonina	1, 4	7, 79	1, 5	1, 6
Aeginetia indica	3	10, 22	7	2
Aesculus assamica	3	21, 60	8	2
Afzelia xylocarpa	1, 6	4, 64	5, 6	1, 2
Aganosma marginata	3	26	4	6
Aganosma sp.	3	22, 39, 53	2, 4	1
Agastache rugosa	2	22	7	7
Agave sisalana	1	39	4	1
Ageratum conyzoides	1, 2, 4	4, 5, 7, 36, 39, 64	1, 4	1, 2, 6
Aglaonema costatum	6	42	6	1
Aglaonema sp.	1	19, 64	2, 4, 6	2, 5
Agrimonia nepalensis var. *nepalensis*	1	15, 68	1	1
Albizia chinensis	1	10, 22, 60	6	2, 6

Latin binomial	Tribe	Ailment	Part used	Method of use
Alocasia cucullata	2, 6	2, 11, 60	1, 4, 6	1, 2
Alocasia macrorrhiza	1, 4	4, 11, 71	2	2, 6
Aloe vera	1, 2, 3, 4, 6	7, 37, 43, 83	4	2
Alpinia bracteata	2, 6	9, 15	6	1
Alpinia malaccensis	1	4, 18	2	1
Alpinia zerumbet	1	18	4	2
Alpinia sp.	1	64	2	2
Alstonia scholaris	1, 3, 4	4, 8, 29, 36, 60, 74	3, 4	1, 2, 5
Altenanthera sp.	2, 6	22, 39	6	1
Amalocalyx microlobus	6	34	6	1, 5
Amomum megalocheilos	2, 6	25, 39	1, 3	1
Amomum siamense	1	2, 70	1, 4	2, 3
Amomum sp.	1, 3	4, 22	2	1
Amorphophallus campanulatus	2	17	4	2
Ananas comosus	1	10	4, 8	2
Annona reticulata	6	25	6	1
Anthurium andraeanum	1	64	4	2
Antidesma acidum	3	22	4	1, 5
Antidesma ghaesembilla	2	63	4, 8	5
Antidesma sp.	5	8	1, 4	1, 5
Antidesma sp.	6	16	1	1
Aporusa wallichii	1, 6	4, 5, 12, 24	3, 4	1
Araiostegia pulchra	1	36	4	2
Aralia chinensis	2, 6	15	2, 4, 7	1
Aralia sp.	2	15, 16	4	1
Araucaria bidwillii	4	45	4	1
Archidendron clypearia	1, 4	7, 17, 20, 37	4	2, 5
Ardisia crenata	2	14	?	1
Arenga pinnata	2	2	4	5
Argyreia capitiformis	1	37	6	2
Argyreia wallichii	1, 6	2, 39	6	1
Argyreia sp.	1	2, 21	3, 6	1, 2
Artabotrys sp.	1	36	6	2
Artemisia atrovirens	1, 3, 4, 5	2, 5, 7, 22, 34, 70, 72, 73	1, 4, 6	1, 2, 3, 5, 6
Artemisia austroyunnanensis	3, 4	4, 79	4, 6	2, 4
Artemisia dubia	1, 4	4, 5, 8, 33, 70	4, 6	2, 3, 4
Artocarpus hypargyreus	3	2	3	1
Asparagus filicinus	2	3	6	1
Averrhoa carambola	2, 4, 6	4, 9, 10, 15	4, 6, 8	1, 5, 6
Azadirachta indica	3	8	3	5
Baccaurea ramiflora	1	1, 4, 9, 28, 39	1, 8	1
Baliospermum montanum	3	66	1, 4	1
Bambusa arundinacea	6	8	2	1

Latin binomial	Tribe	Ailment	Part used	Method of use
Bambusa tulda	6	2	2	2
Bambusa vulgaris	6	2	1	1
Barleria strigosa	3	39	1	1
Basella alba	1, 6	1, 3, 22, 36	1, 2, 4, 6, 8	1, 2
Bauhinia acuminata	6	39	4	1, 6
Bauhinia purpurea	2, 6	63, 68	4, 6	1, 5
Bauhinia sp.	1	31	2	1
Berchemia floribunda	1	37	4	2
Betula alnoides	4	2, 14, 66	1, 3	1, 5, 6
Bidens pilosa	2, 3, 5	2, 3, 7, 36, 42	1, 4, 6	1, 2
Bischofia javanica	2, 5	11, 28	2, 3, 4	1, 2
Blumea balsamifera	1, 2, 3, 4, 5, 6	1, 2, 4, 7, 15, 22, 29, 39, 50	1, 4, 6	1, 2, 5, 7
Blumea fistulosa	3	9, 28	6	1
Blumea lacera	1, 3	1, 10	1, 6	1, 5
Blumea lanceolaria	4	7	4	1
Blumea membranacea	4, 5	1, 4, 7, 15, 34, 76	1, 6	1
Boerhavia chinensis	1	14, 22	1	1
Bougainvillea spectabilis	1	39	1	1
Brassaia actinophylla	2	4	1	1
Brassaiopsis ficifolia	4	7, 29, 70	6	3
Broussonetia papyrifera	1, 6	64, 82	1, 2, 4	1, 2
Brucea mollis	1, 2	8, 26, 56, 73	4, 6	2, 5
Bryophyllum pinnatum	1, 2, 3, 6	10, 25, 34, 37	4, 6	2
Buddleja asiatica	1, 4	4, 8, 10, 14, 34, 52, 56	4, 6	1, 2, 5
Byttneria pilosa	1	56	6	5
Caladium × *hortulanum*	1	19	2, 4	2, 5
Callicarpa arborea **var.** *arborea*	1	36	2, 3, 4	1, 2, 5
Calotropis gigantea	4	7	3	5
Canarium subulatum	2	39	6	5
Canavalia virosa	1	7	5	1, 5
Canthium parvifolium	1, 4	8, 10, 15, 36, 64, 74	1, 4, 6, 8	1, 2, 5
Capsella bursa-pastoris	2	64	6	2
Capsicum annuum	5	19	1	1, 5
Carex sp.	4	8, 15	6	1, 5
Careya arborea	1, 3	9, 20, 22, 37	2, 3, 4	1, 2
Carica papaya	4	2	?	2
Caryota mitis	6	39	1	1
Cassia fistula	4	12, 14, 70	2, 4	3
Cassia garrettiana	1	22	6	1
Cassia hirsuta	5	22	6	1
Cassia occidentalis	1, 3, 5	4, 15, 29, 31, 39	1, 6	1, 5

Latin binomial	Tribe	Ailment	Part used	Method of use
Cassia timoriensis	1	8, 81	3, 7	5
Cassytha filiformis	2	34	6	1
Cayratia trifolia	3	60	4	2
Celosia argentea	1, 2, 5	2, 4, 8, 26, 47, 53, 74	1, 7	1, 2
Celtis tetrandra	4	14, 59, 63, 70	3	1, 3
Centella asiatica	2, 3, 4, 6	4, 8, 9, 24, 27, 39, 47, 58	4, 6	1, 2, 5, 6
Centipeda minima	3	42	7	4
Chloranthus elatior	1, 2, 3, 4, 5	1, 2, 3, 4, 7, 8, 14, 15, 22, 29, 31, 39, 50, 63, 70	1, 2, 4, 6	1, 2, 3, 5, 6
Chloranthus nervosus	5	29	1	1
Chloranthus sp.	5	4	1	1
Chlorophytum intermedium	2	22	6	1
Chrysanthemum × *morifolium*	1	10	6	2
Cinnamomum camphora	2	22	8	6
Cinnamomum tamala	4	22, 70	4	3
Cinnamomum sp.	1	83	2, 4	2
Cinnamomum sp.	1	36	4	2
Cinnamomum sp.	1	4	4	1
Cissampelos hispida	1, 2, 3, 5	7, 26, 35, 46	1, 2	1
Cissus cf. *discolor*	1	8	4	2
Cissus hastata	2, 3, 5	34, 36, 64	1, 4, 6	2
Cissus repanda	2, 6	19, 22, 34, 70	1, 2, 4	1, 2, 3
Cissus repens	2, 6	2, 8, 39	6	5
Citrus maxima	2	29	4	2
Claoxylon indicum	2	18	2	2
Clausena excavata	1, 2, 4, 5	8, 10, 29, 31, 36, 37, 64, 71	4, 6	1, 2, 3, 5
Clausena longipes	3	22, 63	2, 4	1, 5
Cleidion spiciflorum	3	7, 29	1, 2, 4	1, 5
Cleistanthus hirsutulus	4	9, 31, 66	1, 3	1, 2
Clematis siamensis	1	36	4	2
Clematis subpetala	1	1, 49	1	1
Clerodendrum colebrookeanum	1, 4, 6	2, 4, 8, 10, 22, 31, 36, 50, 57, 62	1, 3, 4	1, 2, 5
Clerodendrum fragrans	5	22, 58	1, 4	1, 2, 5
Clerodendrum japonicum	2	1	1	1
Clerodendrum paniculatum	1, 2, 3, 4, 5	4, 14, 22, 37	1, 4	1, 2

Latin binomial	Tribe	Ailment	Part used	Method of use
Clerodendrum serratum	1, 2, 3, 4, 5, 6	1, 4, 7, 9, 10, 22, 26, 29, 31, 35, 36, 46, 61, 68, 84	1, 4, 6, 7	1, 2, 5
Clerodendrum sp.	5	22, 39	4	5
Clitoria macrophylla	1	36	6	2
Cnestis palala	1	5, 13	1	1
Coccinia grandis	6	39	6	1
Codonopsis javanica	1, 4	21	6	1
Coelogyne trinervis	2	34	6	2
Colocasia esculenta	4, 6	9, 11	1, 4	1, 2
Colocasia sp.	1, 4	4, 8	4	1, 2
Combretum deciduum	3	9	1	1
Commelina diffusa	1, 2	8, 18, 36, 67	2, 4, 6	2
Congea tomentosa	1, 2	1, 2, 15, 36, 64, 68	4, 6	1, 2
Connarus semidecandrus	2	4	4	6
Conyza sumatrensis	2	8, 17	4	2
Costus speciosus	1, 2, 4, 5, 6	3, 15, 18, 19, 39, 43, 46	2, 4, 6	1, 2, 5, 6
Crateva magna	1, 2, 4, 6	12, 14, 17, 34	2, 4	1, 2, 5
Cratoxylum cochinchinense	4	14, 63	1	1
Cratoxylum formosum subsp. *pruniflorum*	3, 4, 6	2, 12, 17, 31, 36, 50	4, 6	1, 2, 6
Crinum asiaticum	1, 2	10, 14, 34, 60	1, 2	1, 2
Crotalaria dubia	1	4, 39	6	6
Crotalaria pallida	2, 4	15, 46	1	1
Crotalaria verrucosa	3	39, 45	6	1
Crotalaria sp.	5	15	1	1
Croton crassifolius	2	4	1	1
Croton oblongifolius	1, 2, 3, 4, 6	4, 13, 22, 34, 36, 64, 71	3, 4	1, 2, 5
Croton robustus	1, 4	22, 29, 36	3, 4	1, 2
Crypsinus cruciformis	3	8, 57	2	2
Cucurbita maxima	2	78	8	6
Cucurbita pepo	4	2	6	2
Curcuma aeruginosa	3	4, 64	2	1, 5
Curcuma domestica	2, 3	4, 8, 24, 35	2	1, 2, 6
Curcuma parviflora	1	36	4	2
Curcuma zedoaria	2, 3	28, 36, 79	2	1, 5
Curcuma sp.	2, 3	2, 4, 35, 36, 64	2	1, 2, 6
Cuscuta australis	6	3	6	1

Latin binomial	Tribe	Ailment	Part used	Method of use
Cuscuta reflexa	2	7, 29	6	2
Cyathula prostrata	6	22, 53	6	1
Cycas revoluta	2	36	9	2
Cycas siamensis	3, 6	25	4	1
Cydista aequinoctialis	1	34, 73	6	2
Cymbopogon citratus	2, 4, 5	34, 39, 41, 70, 73	1, 6	1, 2, 3
Dalbergia cf. *paniculata*	1	10	6	2
Datura metel	2	17, 73	4	1, 2
Debregeasia velutina	2	5	4	1
Dendrobium sp.	2	34	4	2
Dendrocalamus hamiltonii	6	4	6	1
Dendrocalamus membranaceus	2, 6	27, 63	4, 6	1
Derris elliptica	1, 4, 5, 6	8, 52	4	5
Desmodium gangeticum	1, 4	31	1	1
Desmodium heterocarpon	2, 6	9, 14, 25, 28	1, 2, 6	1
Desmodium laxiflorum	6	39, 55	1, 4	1, 5
Desmodium longipes	1	39, 43	1	1, 5
Desmodium oblongum	5	58	1	1
Desmodium renifolium	4, 6	22, 39, 55	1, 4	1, 5
Desmodium triangulare	5	1	2, 4	1
Desmodium triquetrum	1, 2, 3, 4	4, 9, 14, 15, 16, 27	1, 6	1, 5, 6
Desmos sootepense	1	34, 43	1, 4	1, 2
Dianella ensifolia	4	41	1	1
Dicliptera roxburghiana	2, 6	4, 7, 22, 39	4, 6	1
Dichrocephala integrifolia	2	4, 31	6	1, 2
Dicranopteris linearis	1, 4	8, 45	4	1, 5
Dillenia indica	2, 6	4	3, 4, 8	1, 6
Dioscorea alata	1, 6	4, 15, 36	1, 4, 6	1, 2
Dioscorea bulbifera	2	80	1	2
Dioscorea hispida	1, 2, 3, 4	4, 10, 15, 27, 64	1, 2, 3, 6, 8	1, 2
Dioscorea pentaphylla	2	56	1, 4	5
Diplazium esculentum	2	39	2, 4	1, 5
Disporum calcaratum	2, 5	13, 36	1, 2, 4	1, 2, 6
Dolichandrone serrulata	4	71	6	6
Dolichos lablab	6	11	6	1
Dombeya wallichii	1	4	?	1
Dracaena angustifolia	6	1	1	1
Dracaena fragrans	1, 3	39	1	1
Dracaena loureiri	1	22	?	5
Dregea volubilis	2, 3, 4, 6	8, 21, 37, 57	1, 4, 6	5, 6
Drymaria cordata	1	7, 36, 56	6	2
Dryopteris cochleata	2	20, 36, 63	2, 4, 6	1, 2, 5
Drypetes sp.	5	73	1	1, 5
Duabanga grandiflora	1	4	5	1
Dumasia leiocarpa	1	34, 39	4	1, 5

Latin binomial	Tribe	Ailment	Part used	Method of use
Dysoxylum andamanicum	1	4, 9, 28	3	1
Elaeagnus conferta	1	8, 57	4	5
Elephantopus scaber	1, 2, 3, 4	1, 15, 21, 41, 42, 59, 67	1, 6	1, 2
Eleusine indica	1, 2, 6	51, 73, 79	6	2
Eleutherine bulbosa	1, 2, 3, 5	2, 4, 12, 15, 28, 36, 53	2 2	1, 2, 5, 6 5
Elsholtzia blanda	1	10, 36	6	5
Embelia stricta	4	21	1	1
Embelia sp.	6	21, 53, 61	2	1
Entada rheedii	1	2, 4, 10	5	2
Equisetum debile	2, 5	15, 22, 34	2	1, 2, 5
Eranthemum tetragonum	1	36	6	2
Eryngium foetidum	2, 4, 6	16, 29, 64	4, 6	1, 2
Erythrina subumbrans	6	34	4	2
Erythrina variegata	4	34	3	5
Etlingera littoralis	6	4, 24	1	1
Eucalyptus canaldulensis	4	2, 42, 73	4	2
Eucalyptus citriodora	4	33, 42	4	2
Euodia glomerata	2, 3, 4, 6	7, 10, 29, 42, 43, 71	1, 2, 4, 6	1, 2, 5, 6
Euodia triphylla	1	71	1	1
Euodia sp.	5	22, 70	6	3, 5
Euonymus sootepensis	2	37	2, 4	5
Eupatorium adenophorum	1, 4	8, 10, 19, 36, 62, 73	1, 2, 4, 6	1, 2, 5
Eupatorium odoratum	1, 2, 3, 4, 5, 6	2, 3, 8, 15, 29, 31, 34, 36, 43	1, 4, 6	1, 2, 5, 6
Eupatorium stoechadorum	1	18, 34	6	2
Euphorbia antiquorum	1, 2, 4, 6	21, 37, 77, 80	2	1, 2, 5
Euphorbia heterophylla	3, 5	16, 39, 45	4, 6	1, 6
Euphorbia hirta	2, 3	21, 35, 36	6	1, 6
Euphorbia sp.	3	36	4	2
Eurya acuminata var. *cerasifolia*	4	4, 8	6	2
Eurya acuminata var. *wallichiana*	4	10	6	2
Eurysolen gracilis	1	36	4	2
Fagopyrum dibotrys	1	34	6	2
Fagopyrum esculentum	1	71	4	1, 2, 6
Ficus chartacea	2	8	4	5
Ficus fistulosa	4	15, 74	1	1
Ficus hispida	3, 4	21, 42	1, 2	1, 2
Ficus racemosa	3	2, 10	2	2
Ficus semicordata var. *semicordata*	1, 2, 4, 6	2, 9, 22, 39	3, 4, 6	1, 5
Ficus sp.	4	10, 70	2, 4	1, 3
Flemingia macrophylla	4	4	4	1

Latin binomial	Tribe	Ailment	Part used	Method of use
Flemingia paniculata	1	20	4	5
Flemingia sootepensis	5	58	1	1
Foeniculum vulgare	2	42, 72	4	6
Garuga pinnata	1	10, 60	3	2, 5
Gelsemium elegans	2	74	6	?
Gendarussa vulgaris	2	15	1	1
Girardinia heterophylla	2, 6	11, 19	1, 4	1, 5
Gladiolus × *hortulanus*	2	31	2	1
Glochidion eriocarpum	2	74	1	1
Glochidion kerrii	1	4, 31	1	1
Gmelina arborea	5	8	3	5
Gnetum montanum	1	36, 55, 60	4	5
Gnetum sp.	5	7, 22, 29, 39	4	5
Gomphostemma wallichii	5	37	4	5
Gonocaryum lobbianum	3	4, 22	3	1, 5
Gossypium barbadense	5	60	2	2
Grewia hirsuta	1	8, 56	4	2
Gymnema inodorum	2	10	4	5
Gynostemma pedatum	4	34, 73	6	2
Gynura pseudochina	6	22	6	6
Harrisonia perforata	3	8, 17	1, 8	5, 6
Hedychium coronarium	3, 4	1, 70, 73	2, 4	3, 6
Hedyotis elegans	4	43	1, 4	1, 2
Hedyotis capitellata var. *pubescens*	2	19, 70	6	3, 5
Hedyotis tenelliflora	2	17	6	6
Helichrysum bracteatum	1	39	6	1
Heliciopsis terminans	5	22, 39	4	5
Helicteres elongata	3, 4	1, 29	1	1
Helicteres hirsuta	1	36	2, 4	2
Helicteres lanceolata	1	1, 39	1	1
Helicteres plebeja	1, 3	1, 2, 4, 39, 41	1, 6	1, 6
Hemigraphis glaucescens	1	10	4	2
Holarrhena pubescens	3	4, 7	2, 4	1
Homalium ceylanicum	4	2, 10, 26	1, 2, 3	1
Homalomena occulta	2	14, 27	2, 4	1
Homalomena sp.	2	60	2	2
Homalomena sp.	4	64	2, 4	2
Houttuynia cordata	1, 2, 4	7, 29, 39, 75	1, 4, 6	1, 5, 6
Hydrocotyle javanica	1	36	6	2
Hymenodictyon excelsum	3	35	4	1
Hypericum japonicum	2	10	6	2
Impatiens balsamina	1, 2	10, 14, 61	6	1
Imperata cylindrica	1, 2, 3, 5	15, 36, 65, 68, 74	1, 2, 6	1, 5, 6
Indigofera squalida	5	15	1	1
Inula cappa	2	14, 36, 61	1, 4	1, 2
Ipomoea batatas	1	34	2	2

Latin binomial	Tribe	Ailment	Part used	Method of use
Iresine herbstii	1, 2	4, 14, 22, 34, 35, 36	4, 6	1, 2
Iris collettii	4	15	2	1
Ixora cibdela	1	2, 26	1	1
Ixora finlaysoniana	1	1, 39	1	1
Jacaranda obtusifolia	1	22	?	?
Jasminum nervosum	6	74	1, 4	1
Jatropha curcas	1, 2, 3, 6	4, 16, 17, 35, 36, 37	2, 3, 4	1, 2, 5
Justicia glomerulata	6	46	6	1
Justicia quadrifaria	2	4	6	1
Kaempferia rotunda	3, 4	4, 16, 66	2	1, 6
Kaempferia sp.	2	50	2	2
Kaempferia sp.	1, 2, 4, 6	12, 25, 36, 50, 64	2	1, 2
Kalanchoe laciniata	2	14, 15, 46	4	1
Kalanchoe sp.	1, 2, 6	34, 37, 61	4, 6	1, 2
Lagerstroemia calyculata	4	10, 22	4	2
Laportea interrupta	1, 4, 6	8, 9, 25, 45, 56	1, 4	1, 2
Lasia spinosa	2, 6	4	6	1
Lasianthus sp.	2, 6	7, 74	4, 6	1, 5
Leea indica	1, 2, 4, 6	9, 15, 22, 28, 34, 36, 50, 73	1, 2, 4, 6	1, 2, 5
Leea rubra	4, 5	8, 63	1, 2, 4	1, 2
Leea sp.	1, 3	9, 36, 54	1, 4	1, 2
Lepisathes rubiginosa	6	3	1	1
Lespedeza parviflora	4	4	1	1
Leucaena leucocephala	1	7	8	5
Limnophila rugosa	1, 2, 4, 5	4, 8, 16, 21, 22, 35	2, 4, 6	1, 2, 6
Lithocarpus polystachyus	1	31	?	1
Litsea glutinosa	3, 4	34, 35, 56, 57, 60	3, 4	2, 5
Litsea monopetala	1	4	?	1
Lobelia angulata	1, 4	8, 64	?	2
Lonicera macrantha	2	29	?	2
Lophatherum gracile	2	1	2	1
Lophopetalum sp.	6	8	6	2
Lygodium flexuosum	1, 2, 4, 5, 6	3, 11, 15, 19, 28, 50, 64, 71, 82	1, 2, 4, 6	1
Lygodium japonicum	2	14, 15	2	1
Macaranga denticulata	6	4	4	6
Machilus parviflorus	1	34, 64	4	2
Machilus sp.	1	36	4	2
Maclura amboinensis	1	2	9	5

Latin binomial	Tribe	Ailment	Part used	Method of use
Maclura aff. *fruticosa*	4	39	1	1
Madhuca sp.	3	1, 7, 11	2, 3	1, 5, 6
Maesa indica	5	8	1, 4	1, 5
Maesa montana	1, 2, 5	2, 4, 12, 31	1, 3, 6	1
Maesa permollis	2, 4	16, 21	1, 4	1
Maesa ramentacea	6	37	6	5
Mallotus paniculatus	6	15	6	1
Mallotus sp.	6	73	4	5
Maoutia puya	2	7, 43	1	1
Markhamia stipulata	1, 3	2, 3, 8, 18, 36, 37	1, 2, 3, 4, 8	2, 5
Marsilea crenata	6	11	6	1
Melastoma normale var. *normale*	4	29	1	1
Melia azedarach	3	8	3	5
Melientha suavis	6	39	4	1
Meliosma simplicifolia	1	8	3	5
Mentha arvensis	6	45	4	1
Microglossa pyrifolia	1, 3	10, 42	6	2, 5
Microlepia herbacea	1	36, 64	4	2
Microlepia speluncae	4	7	2, 4	1
Microtropis pallens	1	10	4	2
Miliusa thorelii	3	2, 78	1, 2	1
Miliusa velutina	3	67	3	6
Millettia extensa	6	8, 56, 57	6	5
Millettia pachycarpa	1	31, 37	6	2
Millingtonia hortensis	1, 2, 3, 4, 6	4, 7, 10, 28, 29, 39	1, 3, 6	1, 5
Mimosa pudica	3, 4, 5	15	1, 6	1
Mirabilis jalapa	2	14, 15, 22, 46	1	1
Mitragyna hirsuta	3	39	2, 4	1, 5
Mitragyna speciosa	2, 6	34, 39	6	1, 2
Molineria capitulata	1	36	2	2
Monochoria hastata	6	39	6	1
Monochoria vaginalis	2, 6	11, 39	6	1
Morinda angustifolia	1	2, 4, 10	1, 2, 4	1, 2, 5
Morinda citrifolia	2, 3	37, 42	4, 8	2, 6
Morinda tomentosa	5	4, 66	1	1
Morinda sp.	1, 6	7, 14, 15	1, 4, 6	1, 2
Morus alba	2	37	4	2
Morus macroura	2	31	2	2
Mucuna pruriens	1	37	4	2
Mukia javanica	5	76	?	1
Musa acuminata	1, 2, 3, 6	4, 7, 9, 15, 16	2, 4, 9	1, 5, 6
Mussaenda kerrii	6	7, 26, 58	1, 2, 4	1
Mussaenda parva	1	2, 34, 37, 39	1, 6	1, 2
Mussaenda pubescens	2	18, 29	4	1, 2

Latin binomial	Tribe	Ailment	Part used	Method of use
Mycetia gracilis	2	18	6	1, 5
Nauclea orientalis	3, 5	2, 31, 36	1, 2, 4	1, 2
Nauclea sp.	5	22, 66	1	1
Nelsonia canescens	1	8, 17	4	2
Nephrolepis cordifolia	2	1	2	1
Nephrolepis falcata	2	2, 19, 39	2, 6	1, 5
Neptunia oleracea	1, 4	64, 71	6	1, 2
Nothaphoebe umbelliflora	1	34, 36	3, 4	2
Ocimum basilicum	2	20, 56	4	2, 5
Ocimum canum	3	24	6	6
Ocimum gratissimum	2, 3, 6	8, 10, 18, 27, 35, 42	1, 3, 4, 6	1, 2
Ocimum sanctum	1	4, 20	5, 6	1, 7
Oenanthe javanica	1, 2	2, 22, 39	6	2, 6
Oncosperma horridum	2	4	6	1
Ophioglossum costatum	2	3, 46	2	1, 6
Ophiorrhiza hispidula	2, 4	16, 39	1 6	1
Opuntia dillenii	2	4, 24, 27, 77	2	1, 2, 6
Oroxylum indicum	1, 2, 5, 6	2, 4, 8, 9, 10, 15, 29, 34, 36, 37, 61, 71, 77, 78	1, 2, 3, 4, 6	1, 2, 5, 6
Orthosiphon aristatus	2, 5, 6	4, 15, 73	4, 6	1, 2
Osbeckia stellata	1, 2, 4, 5, 6	4, 9, 12, 41, 46, 66	1, 2, 3, 4, 6	1, 5
Oxalis corniculata	2	1, 11, 41	6	1
Oxytenanthera albo-ciliata	2	8	2	5
Paederia pallida	3	27	6	1, 5
Paederia pilifera	6	7, 22	6	1
Paederia scandens	5	22	6	1
Paederia wallichii	1, 2, 4	19, 24, 29, 35, 60, 64, 70, 71	4, 6	1, 2, 3, 5, 6
Paederia tomentosa	3	28	6	1
Pandanus furcatus	2, 4	7, 15	4, 8	1
Panicum sp.	1, 2	4, 35, 71	1, 4	1, 2
Papaver somniferum	1, 2, 3, 4, 5, 6	2, 9	8	4, 6
Parinari anamensis	6	10	6	5
Pavetta petiolaris	4	14	1	1
Pedilanthus tithymaloides	2, 4	34, 73	6	2
Peliosanthes grandifolia	1	1, 39	1	1
Peliosanthes teta	2	1, 42	6	1
Peliosanthes violacea	6	18, 22	2	1, 5
Pennisetum polystachion	3	15	1	1
Peristrophe lanceolaria	1	10	4	2
Peucedanum sp.	1	18, 34, 77	4, 6	1, 2, 5

Latin binomial	Tribe	Ailment	Part used	Method of use
Phlogacanthus curviflorus	1, 2, 3, 6	2, 7, 8, 10, 27, 34, 57, 82	1, 4, 6	1, 2, 5
Phoebe lanceolata	2	60	4	2
Phragmites karka	2	34	2, 4	2
Phrynium capitatum	2, 4, 6	3, 4, 10, 14, 15, 39	2, 4, 6	1, 2, 5, 6
Phyllanthus emblica	2, 4, 6	4, 9, 17	1, 3, 4, 8	1, 6
Phyllanthus sp.	2	9	3	1
Physalis angulata	2, 3	8, 36, 60, 73	4	2
Phytolacca acinosa	2	79	?	1
Phytolacca sp.	2	63	2, 4	6
Pinus merkusii	1, 4	8, 15	4, 9	1, 5
Piper agyrophyllum	4	64	1	2
Piper betle	1, 3, 4	8, 33, 53	2, 4	1, 2, 6
Piper pedicellatum	1	40	1	1
Piper pellucidum	4	53	2	1
Piper sarmentosum	2, 6	6, 7, 11	4, 6	1, 5
Piper sp.	3	24, 26, 70	6	3
Pithecellobium dulce	6	4	6	1
Plantago major	1, 2, 3, 4, 6	1, 4, 7, 11, 15, 34, 36, 49, 50, 73	4, 6	1, 2
Platycerum wallichii	1	18	4	5
Plectranthus hispidus	5	37	4	2
Plumbago indica	4	17, 39, 53, 78	1	1, 6
Plumbago zeylanica	1, 2, 6	1, 3, 15, 29, 36, 39, 61, 66	1, 2, 4	1, 2, 5
Plumeria rubra	4	4	2, 4	1, 2
Polyalthia cerasoides	4	29, 82	1, 4	1, 2
Polygala glomerata	2, 3	13	6	1
Polygonum barbatum	2	29	4	2
Polygonum chinense	1, 2, 4	8, 26, 31, 36, 37, 60	4, 6	1, 2
Polygonum glabrum	1	36	6	5
Polygonum sp.	2, 6	15, 31, 49	1, 2, 4, 6	1
Polyscias fruticosa	6	15	6	1
Pothos cathcartii	2, 3	1, 2, 50	6	1
Premna fulva	1	2	6	5
Premna nana	5	4, 66	1	1
Protium serratum	2	4	1	1
Prunus cerasoides	4	1, 33, 73	3	2
Prunus persica	2, 4	10, 29	4	2, 5
Pseuderanthemum andersonii	2, 3	8, 14, 82	2, 4	1, 2, 5
Psidium guajava	1, 2, 3	4, 9, 17, 29, 39	1, 2, 4	1, 2, 6

Latin binomial	Tribe	Ailment	Part used	Method of use
Psychotria monticola **var.** *monticola*	6	15, 53	1	1
Psychotria morindoides	4	4, 5, 7	1	1
Psychotria ophioxyloides	1	1	1	1
Pteridium aquilinum	1, 2, 4	36, 64	4	2
Pteris biaurita	1, 4	2, 7, 29, 70	4	3
Pteris semipinnata	2	7	4	1
Pteris venusta	1, 2	16, 35, 36, 64	2, 4	1, 2
Pteris sp.	2	50	4	2
Pterolobium macropterum	3	22	2	1
Pueraria phaseoloides	1	15, 68	2, 8	1
Pueraria rigens	4	34	6	2
Punica granatum	2	9	8	1
Quisqualis indica	5	22	6	1
Radermachera glandulosa	1	39, 71	1, 4	1, 6
Radermachera ignea	1, 5	9, 36	3, 4	1, 2
Ranunculus cantoniensis	2	14, 15, 20, 46, 73	1	1
Raphistemma pulchellum	4	21	1	6
Rauvolfia sp.	3	7, 29	1	1
Rauwenhoffia siamensis	1	36	?	1, 5
Rhaphidophora sp.	1	10, 14	4	1
Rhus chinensis	1, 2	1, 4, 7, 8, 22, 36	1, 4	1, 2, 5
Rhus succedanea	1	10, 26, 57, 71	6	1, 2, 5
Ricinus communis	1, 2, 3, 4, 5, 6	2, 4, 8, 10, 16, 18, 22, 34, 68, 70	4, 5, 6	1, 2, 3, 5
Rourea minor	1, 4	1, 9, 31, 66	1, 3	1
Rubus blepharoneurus	1, 4	4, 9, 62, 66	1, 3	1
Rubus dielsianus	1	10	6	2
Rubus ellipticus	1	4, 28	1	1
Salix tetrasperma	1	36	3, 4	2
Samanea saman	4	17	?	1
Sambucus javanica	1, 2, 3, 5, 6	2, 10, 15, 22, 29, 34, 36, 39, 46, 63, 70, 73	1, 2, 4	1, 2, 3, 5
Sansevieria trifasciata	2, 4	2, 34, 64	6	2
Saprosma sp.	5	22	6	5
Saraca declinata	1	39	?	1
Saurauia nepalensis	1	36	3, 4	2
Saurauia roxburghii	2, 4	7, 10, 15, 29, 34, 60	1, 3, 4	1, 5
Sauropus androgynus	3, 4	2, 22, 39	4, 6	5, 6
Sauropus quadrangularis	1	36	1, 2, 4, 8	1, 2
Schefflera alongensis	4	34	4	2

Latin binomial	Tribe	Ailment	Part used	Method of use
Schefflera clarkeana	3	2, 22, 63	1, 2, 6	1, 5
Schefflera sp.	3	3	2	1
Schima wallichii	1, 2, 6	4, 9, 17, 28, 77	2, 4	1, 2, 6
Scirpus grossus	2, 3	4, 15, 68	1	1, 6
Scleria levis	4	4	1	1, 6
Scleria terrestris	1, 4	7	1, 4	1
Scleria sp.	1, 2	1, 41, 46	1, 2	1
Scleropyrum wallichianum	3	2	2	1
Scoparia dulcis	2, 3	4, 11, 29, 31, 35, 36, 37	1, 4, 6	1, 2, 5, 6
Scurrula ferruginea	1	2, 58	1, 4	1
Scutellaria indica	2	35	6	2
Securinega virosa	3	22	2, 4	5
Selaginella helferi	1, 2, 4	15, 22, 36, 37, 64	2, 4, 6	1, 2
Selaginella minutifolia	2	15	6	1
Selaginella repanda	2, 6	10, 11	6	1, 5
Selaginella roxburghii	1	34	6	2
Seseli siamicum	5	13, 63	6	5
Shorea obtusa	1	8	4	5
Shuteria vestita	2	37	2, 4	2
Sida acuta	3, 4	15, 16, 60	1, 4	1, 2
Sida rhombifolia var. *rhombifolia*	2, 4, 6	15, 25, 36, 51, 53, 74	1, 4, 6	1, 2
Siegesbeckia orientalis	1	34	2, 4	2
Smilax corbularia subsp. *corbularia*	4	14	1	1
Smilax lanceaefolia	1	2, 8, 36	4	2
Smilax ovalifolia	2, 3, 4, 5, 6	2, 3, 4, 8, 15, 28, 35, 39, 50, 73, 74	1, 4, 6	1, 2
Solanum erianthum	1, 2, 6	1, 2, 3, 7, 14, 22, 29, 36, 39, 73, 74	1, 6, 9	1, 2
Solanum melongena	1	9	8	1, 6
Solanum nigrum	1, 4	14, 86	1, 8	1, 2
Solanum nodiflorum	1	34	4	2
Solanum spirale	6	1, 6, 49	3, 4	1, 6
Solanum torvum	1	19	2, 4	2, 5
Solanum sp.	5	21	1	1
Solanum sp.	3, 5	2, 16, 21, 26	1, 3, 7	1, 6
Solena heterophylla	4, 6	4, 17	6, 8	1, 6
Sonchus oleraceus	5	39	1, 7	1
Spatholobus parviflorus	1, 2, 3	4, 8, 22, 46, 53, 63	2, 4	1, 2, 5

Latin binomial	Tribe	Ailment	Part used	Method of use
Spatholobus sp.	1	23	9	1
Sphenomeris chinensis	2	10	1, 4	2
Spilanthes paniculata	6	17	4	2
Spondias cytherea	2, 6	4, 8	3, 4	1, 5
Spondias lakonensis	2	10	6	5
Spondias pinnata	3	9	4	6
Sporobolus diander	1, 5	15, 72	1	1, 2
Stahlianthus involucratus	2	58, 73	1	1
Staurogyne lanceolata	1	1, 39	1, 4	1, 6
Stemona collinsae	1, 2	4, 14	1	1, 2, 6
Stemona burkillii	1	57	4, 8	5
Stemona tuberosa	2, 6	15, 69	1, 6	1, 5
Stemona sp.	2	4, 16	1, 6	1
Stephania brevipes	1, 2	8, 15, 33, 49	1, 4, 6	1, 2, 5
Stephania glabra	2	34	6	2
Stephania japonica	2, 3, 6	4, 8, 15, 36	2, 4, 6	1, 2, 5
Sterculia lanceolata	1, 4	2, 9, 21	1, 4	1
Stereospermum colais	5	34, 73	2, 3, 4	2
Stereospermum neuranthum	3	8	3	5
Streptocaulon juventas	1, 3	9, 14	6	1
Strobilanthes anfractuosa	1	2, 11, 73	2, 6	1
Strobilanthes lanceifolius	4	7	?	?
Strobilanthes pentstemonoides	6	58	4	2
Strychnos nux-vomica	3	24	1	1
Styrax benzoides	4	7	?	1
Symphorema involucratum	2, 3	13, 34, 36, 63	6	1, 2, 5
Syzygium cumini	4	21, 64, 66	1, 6	1, 2
Tacca chantrieri	2, 4	2, 4, 39	1, 2, 4	1, 6
Tacca integrifolia	1	1, 2, 36, 39	1, 2, 6	1, 2
Tacca sp.	2, 4, 6	7, 8, 34, 53, 73	1, 4, 6	1, 2, 5
Talinum triangulare	2, 6	13, 22	6	1
Tamarindus indica	1, 2	8, 36	4	2
Tectaria polymorpha	3	10, 36	2, 4	2, 5
Tectona grandis	4	7	3	5
Tetrastigma lanceolarum	6	64	4	5
Teucrium viscidium	2	79	?	1
Themeda arundinacea	6	8	6	5
Thunbergia grandiflora var. *grandiflora*	1, 6	8, 10, 15, 34, 36, 39	1, 4, 6	1, 2, 5, 6
Thunbergia laurifolia	3	64	4	2, 5, 6
Thunbergia similis	5	73	1	1
Thyrsostachys siamensis	6	4	6	1
Tiliacora triandra	6	22, 73	6	1
Tinomiscium petiolare	1, 3, 4	10, 34, 36	1, 2, 4, 6	1, 2, 5
Tinospora sinensis	1, 2, 3, 6	4, 19, 34, 39	2, 6	1, 2, 5

Latin binomial	Tribe	Ailment	Part used	Method of use
Tithonia diversifolia	4	4	4	1, 2
Toddalia asiatica	1	10	6	2
Torenia siamensis	2, 6	4, 22, 36, 37	6	1, 2
Tradescantia sp.	2	61	6	1
Trapa bicornis	1, 2	4, 8, 35, 77	4	1, 5
Trema cannabina	4	2, 7, 29	1	1
Trema orientalis	1	4, 66	1, 3	6
Trevisia palmata	2, 6	15, 16	6	1
Trichosanthes tricuspidata	3	29, 70	6	1, 3
Triumfetta rhomboidea	2	4, 7	?	1
Turpinia pomifera	5, 6	1, 2	2, 4	1
Turpinia sphaerocarpa	2	9, 28	1, 3	1
Typhonium trilobatum	2, 4, 6	2, 5, 8, 76	1, 2, 4, 6	1, 2, 5
Uncaria macrophylla	1	37	6	2
Uncaria scandens	1	37	2	2
Uncaria sp.	4	15	1	1
Uraria cordifolia	3	22	4	5
Urena lobata	1, 2, 4, 6	15, 31, 36, 51, 74	1, 6	1, 2
Uvaria sp.	1, 4	22, 26, 35	1, 3	1
Vanda sp.	1	37	?	5
Ventilago calyculata	3, 6	3, 13, 23, 55	6	1
Verbena officinalis	1, 2	3, 4, 31, 56, 57, 71	1, 4, 6	1, 2, 5
Vernonia parishii	1	15	?	1
Viburnum inopinatum	1	4	?	1
Vigna unguiculata	2	4, 35	1	1
Viscum articulatum	2	29	6	1
Vitex canescens	5	22	6	1
Vitex peduncularis	4	58	1, 4	1, 2
Vitex vestita	5	22	6	5
Wedelia chinensis	2	17	4	6
Xanthium sibiricum	2	15	6	1
Xylia xylocarpa	1	52	3	?
Yucca gloriosa	1	10, 70	4	3
Zebrina pendula	2	2	6	1
Zingiber cassumunar	3, 4	22, 24	2	1, 5
Zingiber officinale	2	3	2	1
Zingiber rubens	3	4	2	2, 6
Zingiber sp.	1, 2, 3, 6	2, 3, 4, 9, 21, 24, 36	2, 4	1, 6
Ziziphus oenoplia	4	15, 74	1	1
Zygostelma benthamii	1, 6	14, 15, 29, 39	1, 6	1

APPENDIX 3

Units of Measure

1 mile = 1.6 kilometers
1 rai = 0.16 hectare
1 rai = 40 × 40 meters
6.25 rai = 1 hectare
1 hectare = 2.47 acres
1 rai = 0.4 acre
1 *viss* (*joi*) = 1.6 kilograms
1 *tang* (basket) = 20 liters
1 *tang* = 15 kilograms
6 *tang* = 1 sack
1 tin (*biip*) = 20 liters
1 tin = 12 kilograms

APPENDIX 4

Hill Tribe Villages Visited, 1976–91

Akha

Pa Kluay Akha (Ba Go Akha)
Ba Kha Akha
Huay Mae Liam
Li Pha
Maw La Baw Soe (Mae Suai Akha)
A Gu
A Hai
Huoe Eurn
Paya Phai Kao

Hmong

Cheng Meng
Huey Man Jan
Pha Kia Nai
Ban Huay Nam Chang
Mae Tho
Ban Tung Sai
Mae Sa Mai
Hoey Manao
Mae Sa village
Khun Klang
Buak Jan
Pa Kluoy

Karen

Den Hom
Samaeng
Mae Sa Lang
Mae Cha Ta
Olota
Mae Pae Kee
Ping Kong
Pa Kia Nok
Pa Kia Nai
Mae Yang Ha
Ta Ho Ta
Mae Ka Pu Luang
Moung Phaem
Mae Tia
Ruam Mit
Klo Mee Glo
Huey Nam Yen
Boh Kaew
Som Poy
Pha Kia
Pha Pum
Pong Klang Nam
Luang Muang
Huay Buhling
Pha Klouy

Lahu

Goshen
La Bah
Mu Ban
Suan Dok
Ban Da Luang
Ban Doi Mok
Mae Poon Luang
Pak Thang
Pa Kluay
Lang Muang
Ba Jo
Nong Khieu

Lisu

Huay Kong
Ban Khun Jae
Nong Khaem
Mae Moo
Pa Daeng
Lang Muang
Ban Mae Tang
Nong Pha Chan
Ba Bo Ngar

Mien

Ban Pha Daeng
Ban Pang Pu Lou
Ban Huay Nam Yin
Huay Gaew
Pha Daeng (Mae San)
Ban Yao Huai Nam Sai (Mae Yao)
Pha Deua

Bibliography

Ajopho, M. 1989. The disappearing Akha medical knowledge. *Life on the Mountain* 1:34–37.
_____ . 1990. The empire with no boundaries. *Life on the Mountain* 2(3):24–27.
Anderson, E. F. 1986a. Ethnobotany of hill tribes of North Thailand. I. Medicinal plants of Akha. *Economic Botany* 40:38–53.
_____ . 1986b. Ethnobotany of hill tribes of North Thailand. II. Lahu medicinal plants. *Economic Botany* 40:442–450.
Applied Scientific Research Corporation of Thailand. 1966. *An Initial List of Thai Medicinal Plants.* Bangkok: Technological Research Institute, Tribal Research Centre.
Ashton, P. S. 1981. Forest conditions in the tropics of Asia and the Far East. In *Where Have All the Flowers Gone? Deforestation in the Third World.* Eds. V. H. Sutlive, N. Altshuler, and M. D. Zamora. Studies in Third World Societies. Williamsburg, VA: Williamsburg Publishing of College of William and Mary. 169–179.
Austin, R., and K. Veda. 1970. *Bamboo.* New York: Weatherhill.
Backus, M., ed. 1884. *Siam and Laos, as Seen by Our American Missionaries.* Philadelphia: Presbyterian Board of Publication.
Banijbhatana, D. 1956. Teak forests of Thailand. In *Proceedings of the 4th World Forestry Congress.* Dehra Dun 1954, 3[1958]. 299–311.
Bänziger, H. 1988. How wildlife is helping to save Doi Suthep: Buddhist sanctuary and national park of Thailand. In *Proceedings of the Symposium Systematic Botany—A Key Science for Tropical Research and Documentation, Sweden, 14–17 Sept. 1987.* Ed. I. Hedberg. Almqvist & Wiksell International. Stockholm.
A Barefoot Doctor's Manual. 1985. New York: Gramercy.
Barker, R., R. W. Herdt, and B. Rose. 1985. *The Rice Economy of Asia.* Washington, D.C.: Resources for the Future.
Beeching, J. 1975. *The Chinese Opium Wars.* New York: Harcourt Brace Jovanovich.
Belanger, F. W. 1989. *Drugs, the U.S., and Khun Sa.* Bangkok: Editions Duang Kamol.
Benedict, P. K. 1972. *Sino-Tibetian: A Conspectus.* Cambridge: Cambridge University Press.
Bernatzik, H. A. 1970. *Akha and Miao: Problems of Applied Ethnography in Farther India.* Trans. A. Nagler. New Haven: Human Relations Area Files.
Bhandhachat, P., G. L. Robert, M. Roongruangsee, and S. Sarobol. 1987. *A Study of Attitudes of Hilltribes Towards Thai-Norwegian Church Aid; Highland Development Project.* PRDI Report #21. Chiang Mai: Center for Research and Development, Payap University.
Bhikkhu, D., and V. Bhikkhu. 1977. *The Meo.* 3rd ed. Bangkok: Suksa Samphau.
Bhruksasri, W. 1989a. Government policy: highland ethnic minorities. In *Hill Tribes Today.* Eds. J. McKinnon and B. Vienne. Bangkok: White Lotus-Orstom. 5–31.
_____ . 1989b. The principle hill tribe problem: tribal deforestation and its solution. *Life on the Mountain* 1(2):8–15.

Biology Design Team, Institute for Promotion of Teaching Science and Technology. 1982. *Some Common Flora and Fauna in Thailand*. Bangkok: Institute for Promotion of Teaching Science and Technology.

Black, R. 1985. *Three-Language Check Lists of Plant Names (Thai-English-Scientific)*. Chiang Mai: Highland Plant Protection Programme, Royal Project.

Boonkert, S. 1985. *Some Bamboos in Thailand* (in Thai). 7th series. Bangkok: Faculty of Forestry, Kasetsart University.

Boucaud, A., and L. Boucaud. 1988. *Burma's Golden Triangle: On the Trail of the Opium Warlords*. Bangkok: Asia Books.

Boyes, J., and S. Piraban. 1989. *A Life Apart*. Chiang Mai: Jareuk Publications.

_____ . 1990. *Hmong Voices*. 2nd ed. Chiang Mai: Trasvin Publications.

Bragg, K. 1989a. *Akha Ethnobotany: Forest Resources of a Mountain People*. Manuscript.

_____ . 1989b. Traditional medicine and use of plants. *Life on the Mountain* 1:2–5.

_____ . 1989c. The Akha: folk botany and forest traditions. In *A Northern Miscellany*. Ed. G. Walton. Chiang Mai: Jareuk Publications.

Brun, V., and T. Schumacher. 1987. *Traditional Herbal Medicine in Northern Thailand*. Berkeley: University of California Press.

Campbell, M., N. Pongnoi, and C. Voraphitak. 1981. *From the Hands of the Hills*. 2nd ed. Hong Kong: Media Transasia.

Capistrano, A. D., and G. G. Marten. 1986. Agriculture in Southeast Asia. In *Traditional Agriculture in Southeast Asia: A Human Ecology Perspective*. Ed. G. G. Marten. Boulder, CO: Westview Press. 6–19.

Chaipigusit, P. 1989. Anarchists of the highlands? A critical review of a stereotype applied to the Lisu. In *Hill Tribes Today*. Eds. J. McKinnon and B. Vienne. Bangkok: White Lotus-Orstom. 173–190.

Chandler, R. F. 1979. *Rice in the Tropics: A Guide to the Development of National Programs*. Boulder, CO: Westview Press.

Chaturabhand, P. 1988. *People of the Hills*. 3rd ed. Bangkok: Editions Duang Kamol.

Cheke, A. S., W. Nanakorn, and C. Yankoses. 1979. Dormancy and dispersal of seeds in secondary forests species under the canopy of a primary tropical rain forest in Northern Thailand. *Biotropica* 11:88–95.

Christiany, L. 1986. Shifting cultivation and tropical soils: patterns, problems, and possible improvements. In *Traditional Agriculture in Southeast Asia: A Human Ecology Perspective*. Ed. G. G. Marten. Boulder, CO: Westview Press. 226–240.

Clapham, W. B., Jr. 1973. *Natural Ecosystems*. New York: Macmillan.

Clawson, D. L. 1985. Harvest security and intraspecific diversity in traditional tropical agriculture. *Economic Botany* 39:56–67.

Cobley, L. S. 1976. *An Introduction to the Botany of Tropical Crops*. 2nd ed. London: Longman.

Cohen, E. 1983. Hill tribe tourism. In *Highlanders of Thailand*. Eds. J. McKinnon and W. Bhruksasri. Kuala Lumpur: Oxford University Press. 307–325.

Collinson, A. S. 1977. *Introduction to World Vegetation*. London: George Allen & Uniwin.

Conklin, H. C. 1954. An ethnoecological approach to shifting agriculture. *Transactions of the New York Academy of Science Ser. 2*, 17:133–142.

Conrad, Y. 1989. Lisu identity in northern Thailand. In *Hill Tribes Today*. Eds. J. McKinnon and B. Vienne. Bangkok: White Lotus-Orstom. 191–221.

Cooper, R. 1984. *Resource Scarcity and the Hmong Response*. Singapore: Singapore University Press.

Cronquist, A. 1981. *An Integrated System of Classification of Flowering Plants*. New York: Columbia University Press.

Cutler, K. 1988. *Reforestation of Land Used for Swidden Agriculture in Northern Thailand*. B. A. Thesis, Whitman College, Walla Walla, WA.

Darnton-Hill, I., N. Hassan, R. Karim, and M. R. Duthie. 1988. *Tables of Nutrient Composition of Bangladesh Foods*. Bangladesh: Helen Keller International.

Dastur, J. F. 1964. *Useful Plants of India and Pakistan*. Bombay: D. B. Taraporevala.

Davies, J. R., and T. Wu. 1990. *A Trekkers Guide to: The Hill Tribes of Northern Thailand*. 2nd ed. Salisbury, UK: Footloose Books.

Dessaint, A. Y. 1980. *Minorities of Southwest China.* New Haven: Human Relations Area Files.

Dessaint, W. Y., and A. Y. Dessaint. 1982. Economic systems and ethnic relations in Northern Thailand. *Contributions to South-East Asian Ethnography* 1:72–85.

Duke, J. A. 1973. Utilization of *Papaver. Economic Botany* 27:390–400.

――――. 1974. Notes on Meo and Yao poppy cultivation. *Phytologia* 28(1):5–8.

――――. 1985. *CRC Handbook of Medicinal Herbs.* Boca Raton: CRC Press.

Duke, J. A., and E. S. Ayensu. n.d. *Medicinal Plants of China.* 2 vols. Algonac, MI: Reference Publications.

Durrenberger, E. P. 1979. Rice production in a Lisu village. *Journal of Southeast Asian Studies* 10(1):139–145.

Edwards, M. V. 1950. Burma forest types. *Indian Forest Records* [New Series] 7:137–173.

Elliott, S., J. F. Maxwell, and O. P. Beaver. 1989. A transect survey of monsoon forest in Doi Suthep-Pui National Park. *Natural History Bulletin of the Siam Society* 37:137–171.

Embree, J. F., and L. D. Dotson. 1950. *Bibliography of the Peoples and Cultures of Mainland Southeast Asia.* New Haven: Yale University Press.

Embree, J. F., and W. L. Thomas, Jr. 1950. *Ethnic Groups of Northern Southeast Asia.* New Haven: Yale University Southeast Asian Studies. Mimeo.

An Englishman's Siamese Journals: 1890–1893. 1895. Bangkok: Siam Media International Books.

Ewins, P., and D. Bazely. 1989. Jungle law in Thailand's forests. *New Scientist* (18 November): 42–45.

Farnsworth, N. R. 1977. The current importance of plants as a source of drugs. In *Crop Resources.* Ed. D. S. Seigler. New York: Academic Press. 61–73.

――――. 1988. Screening plants for new medicines. In *Biodiversity.* Ed. E. O. Wilson. Washington, D.C.: National Academy Press. 83–97.

Farrelly, D. 1984. *The Book of Bamboo.* San Francisco: Sierra Club.

Fay, P. W. 1975. *The Opium War, 1840–1842.* Chapel Hill: University of North Carolina Press.

Feeny, D. 1988. Agricultural expansion and forest depletion in Thailand, 1900–1975. In *World Forest Deforestation in the Twentieth Century.* Eds. J. F. Richards and R. P. Tucker. Durham: Duke University Press. 112–143, 281–287.

Food and Agriculture Organization of the United Nations. 1981. *Tropical Forest Resources Assessment Project: Forest Resources of Tropical Asia.* Technical Report #3, Rome.

――――. 1983. *Plants and Plant Products of Economic Importance.* Terminology Bulletin 25/1, Rome.

Food and Agriculture Organization of the United Nations and the United Nations Economic Commission for Asia and the Far East. 1961. *Timber Trends and Prospects in the Asia-Pacific Region.* Geneva.

Freeman, M. 1989. *Hilltribes of Thailand.* Bangkok: Asia Books.

Gamble, J. S. 1896. *Annals of the Royal Botanic Garden.* Vol. 7, *The Bambuseae of British India.* Calcutta: Bengal Secretariat Press.

Geddes, W. R. 1973. The opium problem in Northern Thailand. In *Studies of Contemporary Thailand.* Eds. R. Ho and E. C. Chapman. Canberra: Australian National University, Research School of Pacific Studies, Department of Human Geography. 213–234.

――――. 1976. *Migrants of the Mountains; The Cultural Ecology of the Blue Miao (Hmong Njua) of Thailand.* Oxford: Clarendon Press.

Goldrick, R. 1989. *Chiang Mai and the Hill Tribes.* Bangkok: Sangdad.

Golley, R. F., ed. 1975. *Tropical Ecological Systems.* New York: Springer-Verlag.

Golomb, L. 1976. The origin, spread and persistence of glutinous rice as a staple crop in mainland Southeast Asia. *Journal of Southeast Asian Studies* 7:1–15.

Grandstaff, T. B. 1980. *Shifting Cultivation in Northern Thailand.* Resource Systems Theory and Methodology Series, #3. Tokyo: United Nations University.

Grew, K. 1991. *Fiber and Dye Plants Used by the Hilltribes of Northern Thailand.* B. A. Thesis, Whitman College, Walla Walla, WA.

Grunfeld, F. V. 1982. *Wayfarers of the Thai Forest. The Akha.* Peoples of the Wild Series. Amsterdam: Time-Life Books.

Hanks, L. M. 1972. *Rice and Man: Agricultural Ecology in Southeast Asia.* Chicago: Aldine-Atherton.

Harada, J., Y. Paisooksantivatana, and S. Zungsontiporn. 1987. *Weeds in the Highlands of Northern Thailand.* Project Manual #3. Bangkok: National Weed Science Research Institute, Department of Agriculture.

Hawkes, J. G. 1983. *The Diversity of Crop Plants.* Cambridge: Harvard University Press.

Held, S. E. 1978. *Weaving: A Handbook of the Fiber Arts.* 2nd ed. New York: Holt, Rinehart and Winston.

Higham, C. 1979. The economic basis of prehistoric Thailand. *American Scientist* 67:670–679.

Higham, C. F. W. 1984. Prehistoric rice cultivation in Southeast Asia. *Scientific American* 250:138–146.

Hill, A. F. 1952. *Economic Botany.* New York: McGraw-Hill.

Hilltribe Research Institute, Chiang Mai University. 1986. *Survey of Socio-Economic Conditions in the Project Areas of the Thai-Norwegian Church and Highland Development Project.* Technical Document Series #3, Payap University Research Center, Chiang Mai.

Hinton, P. 1973. Population dynamics and dispersal trends among the Karen of Northern Thailand. In *Studies of Contemporary Thailand.* Eds. R. Ho and E. C. Chapman. Canberra: Australian National University Research School of Pacific Studies Department of Human Geography. 235–251.

———. 1983. Why the Karen do not grow opium: competition and contradiction in the highlands of North Thailand. *Ethnology* 22:1–16.

Hiranramdej, S. P. 1986. *Dictionary of Thai Medicinal Plant Names.* Vol. 1. Chiang Mai: Faculty of Pharmacy, Chiang Mai University.

Hoare, P. W. C. 1985. The movement of Lahu hill people towards a lowland life style in North Thailand: a study of three villages. *Contributions to South-East Asian Ethnography* 4:75–117.

Holdridge, L. R. 1967. *Life Zone Ecology.* Rev. ed. San Jose, Costa Rica: Tropical Science Center.

Holdridge, L. R., W. C. Grenke, W. H. Hatheway, T. Liang, and J. A. Tosi, Jr. 1971. *Forest Environments in Tropical Life Zones. A Pilot Study.* Oxford: Pergamon Press.

Holttum, R. E. 1958. The bamboos of the Malay Peninsula. *The Gardens' Bulletin* (Singapore) 16:1–135.

Howes, F. N. 1974. *A Dictionary of Useful and Everyday Plants and Their Common Names.* Cambridge: Cambridge University Press.

Hsu, H., Y. Chen, S. Shen, C. Hsu, C. Chen, and H. Chang. 1986. *Oriental Materia Medica.* Long Beach, CA: Oriental Healing Arts Institute.

Hutterer, K. L. 1984. Ecology and evolution of agriculture in Southeast Asia. In *An Introduction to Human Ecology Research on Agricultural Systems in Southeast Asia.* Ed. T. A. Rambo. Laguna, Philippines: University of the Philippines at Los Baños. 75–97.

Jaafar, S. J. 1975. The Meo people: an introduction. In *Farmers in the Hills: Ethnographic Notes on the Upland Peoples of North Thailand.* Ed. A. R. Walker. Georgetown: Penerbit Universiti Sains Malaysia. 61–72.

Jaafar, S. J., and A. R. Walker. 1975. The Akha people: an introduction. In *Farmers in the Hills: Ethnographic Notes on the Upland Peoples of North Thailand.* Ed. A. R. Walker. Georgetown: Penerbit Universiti Sains Malaysia. 169–181.

Jacobs, M. 1962. Reliquiae Kerrianae. *Blumea* 11:427–493.

Jacquat, C. 1990. *Plants from the Markets of Thailand.* Bangkok: Editions Duang Kamol.

Janick, J., R. W. Schery, F. Woods, and V. W. Ruttan. 1981. *Plant Science: An Introduction to World Crops.* 3rd ed. San Francisco: W. H. Freeman.

Janzen, D. H. 1976. Why bamboos wait so long to flower. *Annual Review of Ecologic Systems* 7:347–391.

Jones, A. D. 1979. Medical practice and tribal communities. In *Health and Disease in Tribal*

Societies. Eds. K. Elliott and J. Whelan. Ciba Foundation Symposium 49 (new series). Amsterdam: Exerpta Medica. 243–267.

Joshi, A. R., and J. M. Edington. 1990. The use of medicinal plants by two village communities in the central development region of Nepal. *Economic Botany* 44:71–83.

Judd, L. C. 1964. *Dry Rice Agriculture in Northern Thailand.* Data Paper #52, Southeast Asia Program Department of Asian Studies, Cornell University, Ithaca, NY.

_____. 1969. The agricultural economy of the hills and adjacent areas: the hill Thai. In *Tribesmen and Peasants in North Thailand.* Ed. P. Hinton. Proceedings of the first symposium of the Tribal Research Centre, Chiang Mai, 1967. Bangkok: Sompong Press. 86–99.

Kammerer, C. A. 1988. Of labels and laws: Thailand's resettlement and repatriation policies. *C S Quarterly* 12:7–12.

_____. 1989. Territorial imperatives: Akha ethnic identity and Thailand's national integration. In *Hill Tribes Today.* Eds. J. McKinnon and B. Vienne. Bangkok: White Lotus-Orstom. 259–301.

Keen, F. G. B. 1963. *Land Development and Settlement of Hill Tribes in the Uplands of Tek Province.* Bangkok: Department of Public Welfare.

_____. 1978. Ecological relationships in a Hmong (Meo) economy. In *Farmers in the Forest.* Eds. P. Kunstadter, E. C. Chapman, and S. Sabhasri. Honolulu: East-West Center. 210–221.

_____. 1983. Land use. In *Highlanders of Thailand.* Eds. J. McKinnon, and W. Bhruksasri. Kuala Lumpur: Oxford University Press. 293–306.

Kerala Forest Research Institute. 1985. *Dipterocarps of South Asia.* RAPA Monograph 4/85, Food and Agriculture Organization Regional Office for Asia and the Pacific, Bangkok.

Kerr, A. F. G. 1911. I. Contributions to the flora of Siam. I. Sketch of the vegetation of Chiengmai. *Bulletin of Miscellaneous Information, Royal Botanical Garden* 1:1–6.

Kesmanee, C. 1985. *The Impact of Modernization on the Cultures of Ethnic Groups in Northern Thailand: A Case Study of the Hmong.* Chiang Mai: Tribal Research Institute.

_____. 1987. *Hilltribe Resettlement Policy: Ways Out of the Labyrinth. A Case Study of Kamphaengphet Province.* Chiang Mai: Tribal Research Institute.

_____. 1988. Hilltribe relocation policy in Thailand. *C S Quarterly* 12:2–6.

_____. 1989. The poisoning effect of a lovers triangle: highlanders, opium and extension crops, a policy overdue for review. In *Hill Tribes Today.* Eds. J. McKinnon and B. Vienne. Bangkok: White Lotus-Orstom. 61–102.

Keyes, C. F., ed. 1979. *Ethnic Adaptation and Identity. The Karen on the Thai Frontier with Burma.* Philadelphia: Institute for the Study of Human Issues.

Khankeaw, S., and P. Lewis. 1979. *Lahu/Akha Survey 1979.* Chiang Mai: Family Planning International Assistance.

Kleinman, A. 1978. Concepts and a model for the comparison of medical systems as cultural systems. *Social Science and Medicine* 12:85–93.

Küchler, A. W., and J. O. Sawyer, Jr. 1967. A study of the vegetation near Chiengmai, Thailand. *Transactions of the Kansas Academy of Science* 70:281–348.

Kunstadter, P. 1967. Introduction. In *Southeast Asian Tribes, Minorities, and Nations.* Ed. P. Kunstadter. Vol. 1. Princeton, NJ: Princeton University Press. 369–400.

_____. 1978a. Ecological modification and adaptation: an ethnobotanical view of Lua' swiddeners in northwestern Thailand. In *The Nature and Status of Ethnobotany.* Ed. R. Ford. Ann Arbor: University of Michigan Press. 169–200.

_____. 1978b. Subsistence agricultural economies of Lua' and Karen hill farmers, Mae Sariang District, northwestern Thailand. In *Farmers in the Forest.* Eds. P. Kunstadter, E. C. Chapman, and S. Sabhasri. Honolulu: East-West Center. 74–133.

_____. 1983. Highland populations in Northern Thailand. In *Highlanders of Thailand.* Eds. J. McKinnon and W. Bhruksasri. Kuala Lumpur: Oxford University Press. 15–45.

Kunstadter, P., and E. C. Chapman. 1978. Problems of shifting cultivation and economic development in Northern Thailand. In *Farmers in the Forest.* Eds. P. Kunstadter, E. C. Chapman, and S. Sabhasri. Honolulu: East-West Center. 3–23.

Kunstadter, P., E. C. Chapman, and S. Sabhasri. 1978. *Farmers in the Forest.* Honolulu: East-West Center.

Kunstadter, P., S. Sabhasri, and T. Smitinand. 1978. Flora of a forest fallow farming environment in northwestern Thailand. *Journal of the National Research Council of Thailand* 10:1–45.

Leach, R. E. 1949. Some aspects of dry rice cultivation in North Burma and British Borneo. *Advances of Science* 6:26–28.

LeBar, F., G. C. Hickey, and J. K. Musgrave. 1964. *Ethnic Groups of Mainland Southeast Asia.* New Haven, CT: Human Relations Area Files Press.

Lewis, P. 1968. *Akha-English Dictionary.* Data Paper #7, Southeast Asia Program, Department of Asian Studies, Cornell University, Ithaca, NY.

_____ . 1968–70. *Ethnographic Notes on the Akhas of Burma.* New Haven: Human Relations Area Files.

_____ . 1970. *Introducing the Hill Tribes of Thailand.* Chiang Mai: Faculty of Social Sciences, Chiang Mai University. Mimeo.

_____ . 1982. Basic themes in Akha culture. *Contributions to Southeast Asian Ethnography* 1:86–101.

_____ . 1986. *Lahu-English-Thai Dictionary.* Chiang Mai: Thailand Lahu Baptist Convention.

_____ . 1989. *Akha-English-Thai Dictionary.* Chiang Rai: Project for Akha (DAPA).

Lewis, P., and E. Lewis. 1984. *Peoples of the Golden Triangle.* London: Thames and Hudson.

Lewis, W. H., and L. V. Avioli. 1991. Leaves of *Ehretia cymosa* (Boraginaceae) used to heal fractures in Ghana increase bone remodeling. *Economic Botany* 45(2):281–282.

Lewis, W. H., and M. P. F. Elvin-Lewis. 1977. *Medical Botany.* New York: John Wiley.

Li, Hi-Lin. 1970. The origin of cultivated plants in southeast Asia. *Economic Botany* 24:3–19.

Liberty Hyde Baily Hortorium. 1976. *Hortus Third: A Concise Dictionary of Plants Cultivated in the United States and Canada.* 3rd ed. New York: Macmillan.

Lin, Wei-chih. 1968. *The Bamboos of Thailand (Siam).* Special Bulletin of Taiwan Forestry Research Institute, Tapei.

Litzenberger, S. C., ed. 1974. *Guide for Field Crops in the Tropics and the Subtropics.* Washington, D.C.: Office of Agriculture, Technical Assistance Bureau, Agency for International Development.

Longman, K. A., and J. Jeník. 1974. *Tropical Forest and Its Environment.* Tropical Ecology Series. Norfolk: Longman.

Lötschert, W., and G. Beese. 1983. *Collins Guide to Tropical Plants.* London: Collins.

Lugo, A. E. 1988. Estimating reductions in the diversity of tropical forest species. In *Biodiversity.* Ed. E. O. Wilson. Washington, D.C.: National Academy Press. 58–70.

Mabberley, D. J. 1987. *The Plant Book.* Cambridge: Cambridge University Press.

McClure, F. A. 1957. Bamboo in the economy of Oriental peoples. *Annual Report of the Smithsonian Institute.* 391–412.

_____ . 1966. *The Bamboos: A Fresh Perspective.* Cambridge: Harvard University Press.

McCoy, A. W. 1972. *The Politics of Heroin in Southeast Asia.* New York: Harper & Row.

McKenzie, J. L., and N. J. Chrisman. 1977. Healing herbs, gods, and magic. *Nursing Outlook* 25(5):326–329.

McKinnon, J. 1989. Structural assimilation and the consensus: clearing grounds on which to rearrange our thoughts. In *Hill Tribes Today.* Eds. J. McKinnon and B. Vienne. Bangkok: White Lotus-Orstom. 303–359.

McKinnon, J., and W. Bhruksasri. 1983. *Highlanders of Thailand.* Kuala Lumpur: Oxford University Press.

McMakin, P. D. 1988. *A Field Guide to the Flowering Plants of Thailand.* Bangkok: White Lotus.

Mahaphol, S. 1954. *Teak in Thailand.* Bangkok: Royal Forest Department.

Malla, S. B. 1982. *Medicinal Plants of Nepal.* Bangkok: Food and Agriculture Organization Regional Office for Asia and the Pacific.

Maneeprasert, R. 1989. "Women and children first?" A review of the current nutritional status in the highlands. In *Hill Tribes Today.* Eds. J. McKinnon and B. Vienne. Bangkok: White Lotus-Orstom. 143–158.

Mann, R. S. 1990. *Highland Development and the Thai Government*. Mimeo.

Marden, L. 1980. Bamboo, the giant grass. *National Geographic* 158:502–529.

Marlowe, D. H. 1969. Upland-lowland relationships: the case of the S'kaw Karen of central upland western Chiang Mai. In *Tribesmen and Peasants in North Thailand*. Ed. P. Hinton. Proceedings of the first symposium of the Tribal Research Centre, Chiang Mai, 1967. Bangkok: Sompong Press. 53–68.

Marshall, H. I. 1922. The Karen people of Burma: a study in anthropology and ethnology. *The Ohio State University Bulletin* 26:1–328.

Martin, F. W., ed. 1984. *CRC Handbook of Tropical Food Crops*. Boca Raton: CRC Press.

Matisoff, J. A. 1983. Linguistic diversity and language contact. In *Highlanders of Thailand*. Eds. J. McKinnon and W. Bhruksasri. Kuala Lumpur: Oxford University Press. 56–86.

Maxwell, J. F. 1988. The vegetation of Doi Suthep-Pui National Park, Chiang Mai Province, Thailand. *Tiger Paper* (Oct-Dec):6–14.

Meikle, R. D. 1980. *Draft Index of Author Abbreviations Compiled at the Herbarium, Royal Botanic Gardens, Kew*. Kew: Royal Botanic Gardens.

Merlin, M. D. 1984. *On the Trail of the Ancient Opium Poppy*. Rutherford, NJ: Fairleigh Dickinson University Press.

Miles, D. J. 1969. Shifting cultivation: threats and prospects. In *Tribesmen and Peasants of North Thailand*. Ed. P. Hinton. Proceedings of the first symposium of the Tribal Research Centre, Chiang Mai, 1967. Bangkok: Sompong Press. 93–99.

————. 1973. Some demographic implications of regional commerce: the case of North Thailand's Yao minority. In *Studies of Contemporary Thailand*. Eds. R. Ho and E. C. Chapman. Canberra: Australian National University, Research School of Pacific Studies, Department of Human Geography. 253–272.

Mottin, J. 1980. *History of the Hmong*. Bangkok: Odeon Store.

Mountain People's Culture and Development Project. 1989. Policy: deforestation and relocation for highlanders of Thailand. *Life on the Mountain* 1:42–47.

Myers, N. 1980. *Conversion of Tropical Moist Forests*. Washington, D.C.: National Academy of Sciences.

————. 1981. Deforestation in the tropics: who gains, who loses? In *Where Have All the Flowers Gone: Deforestation in the Third World*. Eds. V. H. Sutlive, N. Altshuler, and M. D. Zamora. Williamsburg, VA: Williamsburg Publishing of College of William and Mary. 1–21.

————. 1988. Tropical forests and their species, going, going . . . ? In *Biodiversity*. Ed. E. O. Wilson. Washington, D.C.: National Academy Press. 28–35.

National Academy of Sciences. 1975. *Herbal Pharmacology in the People's Republic of China*. Washington, D.C.: National Academy of Sciences.

National Institute for the Control of Pharmaceutical and Biological Products. 1987. *Colour Atlas of Chinese Traditional Drugs*. Vol. 1. Beijing: Science Press.

National Research Council (U.S.) Committee on Selected Biological Programs in the Tropics. 1982. *Ecological Aspects of Development in the Humid Tropics*. Washington, D.C.: National Academy Press.

Nawigamune, W., ed. 1989. *Chiang Mai and the Hill Tribes*. Bangkok: Sangdad.

Nepal Ministry of Forests and Soil Conservation. Department of Medicinal Plants. 1982. *Medicinal Plants of Nepal*. 3rd ed. Katmandu, Nepal.

Noda, K., M. Teerawatsakul, C. Prakongvongs, and L. Chaiwiratnukul. 1986. *Major Weeds in Thailand*. Bangkok: National Weed Science Research Institute, Department of Agriculture.

Norman, M. J. T. 1978. Energy imputs and outputs of subsistence cropping systems in the tropics. *Agro-Ecosystems* 4:355–366.

Nuttonson, M. Y. 1963. *The Physical Environment and Agriculture of Thailand*. Washington, D.C.: American Institute of Crop Ecology.

Ogawa, H., K. Yoda, and T. Kira. 1961. A preliminary survey on the vegetation of Thailand. In *Nature and Life in Southeast Asia*. Eds. T. Kira and T. Umesao. Vol. 1. Kyoto: Fauna and Flora Research Society. 21–157.

Oughton, G. 1969. Hill-tribe Agriculture in Northern Thailand. Chiang Mai: Tribal Research Center. Mimeo.

Pake, C. E. 1986. *Herbal Medicines used by Hmong Refugees in Thailand*. M. P.H. Thesis, University of Minnesota.

Pei, S. 1985. Preliminary study of ethnobotany in Xishuang Banna, People's Republic of China. *Journal of Ethnopharmacology* 13:121–137.

———. 1987. Medicinal plants in tropical areas of China. In *Medicinal and Poisonous Plants of the Tropics*. Comp. A. J. M. Leeuwenberg. Proceedings of symposium 5–35, 14th International Botanical Congress, Berlin. Wageningen: Purdoc. 16–35.

Pelzer, K. J. 1978. Swidden cultivation in Southeast Asia: historical, ecological, and economic perspectives. In *Farmers in the Forest*. Eds. P. Kunstadter, E. C. Chapman, and S. Sabhasri. Honolulu: East-West Center. 271–286.

Pendleton, R. L. 1962. *Thailand: Aspects of Landscape and Life*. New York: Duell, Sloan & Pearce.

Perry, L. M. 1980. *Medicinal Plants of East and Southeast Asia: Attributed Properties and Uses*. Cambridge: MIT Press.

Peters, W. J., and L. F. Neuenschwander. 1988. *Slash and Burn*. Moscow: University of Idaho Press.

Phengklai, C., and S. Khamsai. 1985. Some non-timber species of Thailand. *Thailand Forestry Bulletin* 15:108–148.

Phillips, J. F. O., W. R. Geddes, and R. J. Merrill. 1967. *Report of the United Nations Survey Team on the Economic and Social Needs of the Opium Producing Areas in Thailand*. Bangkok: Government House Printing Office.

Phothiart, P. 1989. Karen: when the wind blows. In *Hill Tribes Today*. Eds. J. McKinnon and B. Vienne. Bangkok: White Lotus-Orstom. 369–392.

Pinratana, A. 1974–86. *Flowers in Thailand*. 12 vols. Bangkok: Viratham Press.

Plotkin, M. J. 1988. The outlook for new agricultural and industrial products from the tropics. In *Biodiversity*. Ed. E. O. Wilson. Washington, D.C.: National Academy Press. 106–116.

Pongpangan, S., and S. Poobrasert. 1971. *Edible and Poisonous Plants in Thai Forests*. Bangkok: Science Society of Thailand.

Ponglux, D., S. Wongseripipatana, T. Phadungcharoen, N. Rungrungsri, and K. Likhitwitayawiud. 1987. *Medicinal Plants*. Bangkok: International Congress on Natural Products.

Pope, G. G. 1989. Bamboo and human evolution. *Natural History* (October):48–56.

Punyarajun, S. 1981. *Medicinal Plants as Economic Replacement Crop for Opium Poppy in Northern Thailand*. USDA Contract No. 12-1-4-0605-181. Chiang Mai: United States Department of Agriculture.

Purseglove, J. W. 1968. *Tropical Crops: Dicotyledons*. Harlow: Longman Scientific and Technical.

Rambo, T. A. 1984. No free lunch: a reexamination of the energetic efficiency of swidden agriculture. In *An Introduction to Human Ecology Research on Agricultural Systems in Southeast Asia*. Eds. T. A. Rambo and P. E. Sajise. Laguna: University of the Philippines at Los Baños. 154–163.

Rashid, M. R., and P. H. Walker. 1975a. The Karen people: an introduction. In *Farmers in the Hills: Ethnographic Notes on the Upland Peoples of North Thailand*. Ed. A. R. Walker. Georgetown: Penerbit Universiti Sains Malaysia. 87–95.

———. 1975b. The Lisu people: an introduction. In *Farmers in the Hills: Ethnographic Notes on the Upland Peoples of North Thailand*. Ed. A. R. Walker. Georgetown: Penerbit Universiti Sains Malaysia. 157–164.

Raven, P. H. 1988. Our diminishing tropical forests. In *Biodiversity*. Ed. E. O. Wilson. Washington, D.C.: National Academy Press. 119–122.

Renard, R. D., P. Bhandhachat, G. L. Robert, M. Roongruangsee, S. Sarabol, and N. Prachadetsuwat. 1988. *Changes in the Northern Thai Hills: An Examination of the Impact of Hill Tribe Development Work 1957–1987*. Research Report #42. Chiang Mai: Research and Development Center, Payap University.

Renard, R. D., R. Prachadetsuwat, and S. Moe. 1991. *Some Notes on the Karen and Their Music.* Chiang Mai: Center for Arts and Culture, Payap University.

Reutrakul, V., C. Sagwansupyakorn, and others. 1980. *Research on Identification and Production of Diosgenin Produced Plants for Opium Poppy Substitute in the Highlands of Northern Thailand.* Semi-Annual Report (June-December 1980). Bangkok: Kasetsart University.

Richards, P. W. 1966. *The Tropical Rain Forest.* Cambridge: Cambridge University Press.

_____ . 1973. The tropical rain forest. *Scientific American* 229:58–67.

Robbins, R. G., and T. Smitinand. 1966. A botanical ascent of Doi Inthanond. *The Natural History Bulletin of the Siam Society* 21:205–227.

Robert, G. L. 1987. *Let Them Eat Trees.* Chiang Mai: Research and Development Center, Payap University.

_____ , ed. 1988. *Approaches to Improving Nutrition of the Hilltribes.* Chiang Mai: Food and Agriculture Organization and Payap University.

Robert, G. L., R. Batzinger, and R. D. Renard. 1986. *Census of Problems of Highlands in Thai-Norwegian Church Aid Highland Development project Areas. An Analysis, with Suggested Priorities in Development.* Research Report #5. Chiang Mai: Payap Research Center, Payap University.

Robert, G. L., K. Bond, P. Bhandhachat, B. Boonjaiphet, S. Sinth, T. Sawat, and P. Kid-arn. 1988. *Evaluation of the Health Card Project. Chiang Mai Province.* Research Report #34. Chiang Mai: Research and Development Center, Payap University.

Robert, G. L., T. Kunkhajonphan, S. Sarobol, N. Knoll, and K. Phochanachai. 1990. *Agricultural Labor Survey 1989.* Research Report #69. Chiang Mai: Payap University Research and Development Institute.

Roux, H. and T. V. Chu. 1954. Qualques Minorités Ethniques du Nord-Indochine. *France-Asie* 10:135–419.

Royal Forest Department. 1948. *Siamese Plant Names. Part 1. Botanical Names-Local Names.* Bangkok: Royal Forest Department.

_____ . 1950. *Types and Distribution of Forests in Thailand.* Bangkok: Royal Forest Department.

_____ . 1962. *Types of Forests of Thailand.* Bangkok: Royal Forest Department.

Ruaysungneun, S., and N. Purivirojkul. 1980. *Bamboos* (in Thai). Bangkok: Royal Forest Department. Mimeo.

Rural Advancement Fund International. 1989. Biotechnology and medicinal plants. *RAFI Communique* (March):1–9.

Salzer, W. 1987. *The TG-HPP Approach Towards Sustainable Agriculture and Soil and Water Conservation in the Hills of Northern Thailand.* Internal Paper 80, Thai-German Highland Development Programme, Chiang Mai.

Samapuddhi, K. 1957. *Some Plant Foods in the Forest of Thailand.* Bangkok: Royal Forest Department.

Sawyer, J. O., Jr., and C. Chermsirivathana. 1969. A flora of Doi Suthep, Doi Pui, Chiang Mai, North Thailand. *The Natural History Bulletin of the Siam Society* 23:99–132.

Schimper, A. F. W. 1903. *Plant-Geography Upon a Physiological Basis.* Trans. W. R. Fisher, P. Groom, and I. B. Balfour. Oxford: Oxford University Press.

Setamanit, S. 1987. Thailand. In *Environmental Management in Southeast Asia.* Ed. L. S. Chia. Singapore: Faculty of Science, National University of Singapore.

Simpson, B. B., and M. Conner-Ogorzaly. 1986. *Economic Botany: Plants in Our World.* New York: McGraw-Hill.

Smil, V. 1985. China's food. *Scientific American* 253:116–124.

Smitinand, T. 1957. *Materials Used For Thatching in Thailand.* Procedures of the ninth Pacific Science Congress, vol. 4. 248–249.

_____ . 1966. The vegetation of Doi Chiangdao, a limestone massive in Chiang Mai, Northern Thailand. *Natural History Bulletin of the Siam Society* 21:94–218.

_____ , ed. 1977. *Plants of Chao Yai National Park.* Bangkok: New Thammada Press.

_____ . 1980. *Thai Plant Names (Botanical Names-Vernacular Names).* Bangkok: Royal Forest Department.

Smitinand, T., and C. Chaianant. n.d. *Classification of Thai Bamboos* (in Thai). Bangkok: Royal Forest Department. Mimeo.

Smitinand, T., and K. Larsen, eds. 1970–87. *Flora of Thailand.* 5 vols. Bangkok: Applied Scientific Research Corporation of Thailand.

Smitinand, T., and S. Ramyarangsi. 1980. Thailand. In *Bamboo Research in Asia.* Eds. G. Lessard and A. Chouinard. Ottowa: International Development Research Centre.

Smitinand, T., S. Sabhasri, and P. Kunstadter. 1978. The environment in Northern Thailand. In *Farmers in the Forest.* Eds. P. Kunstadter, E. C. Chapman, and S. Sabhasri. Honolulu: East-West Center. 24–40.

Smitinand, T., T. Santisuk, and C. Phengklai. 1980. *The Manual of Dipterocarpaceae of Mainland South-east Asia.* Bangkok: Royal Forest Department.

Soderstrom, T. R. 1979. Ecology and phytosociology of bamboo vegetation. 1. Distribution and environment of the Bambusoideae. In *Ecology of Grasslands and Bamboolands in the World.* Ed. M. Numata. Jena: Gustav Fischer. 223–236.

———. 1981. Some evolutionary trends in the Bambusoideae (Poaceae). *Annals of the Missouri Botanical Garden* 68:15–47.

Soderstrom, T. R., and C. E. Calderón. 1979. A commentary on the bamboos (Poaceae: Bambusoideae). *Biotropica* 11:161–172.

Soderstrom, T. R., and S. M. Young. 1983. A guide to collecting bamboos. *Annals of the Missouri Botanical Garden* 70:128–136.

Sono, P. 1974. *Merchantable Timbers of Thailand.* Bangkok: Royal Forest Department.

Spencer, J. E. 1966. *Shifting Cultivation in Southeastern Asia.* Berkeley: University of California Press.

Srimongkol, K., and G. G. Marten. 1986. Traditional agriculture in northern Thailand. In *Traditional Agriculture in Southeast Asia: A Human Ecology Perspective.* Ed. G. G. Marten. Boulder, CO: Westview Press. 85–102.

Srisavasdi, B. C. 1967. *The Hill Tribes of Siam; Photographic Book.* Bangkok: Odeon Store.

Srisawas, N., and M. Suwan. 1985. Thailand-swidden cultivation in the north (White Meo). In *Swidden Cultivation in Asia.* Vol. 2. Ed. UNESCO. Bangkok: UNESCO Regional Office for Education in Asia and the Pacific. 296–349.

Sukthumrong, A., J. Akavipat, V. Varunyanondh, R. Tasanee, and S. Kasemsap. 1979. *Research on Cultivated Crops and Wild Plants for Dye Production in the Highlands of Northern Thailand.* Bangkok: Highland Agriculture Project, Kasetsart University.

Sutthi, C. 1989a. Appendices II, III, & IV. In *Hill Tribes Today.* Eds. J. McKinnon and B. Vienne. Bangkok: White Lotus-Orstom. 427–470.

———. 1989b. Highland agriculture: from better to worse. In *Hill Tribes Today.* Eds. J. McKinnon and B. Vienne. Bangkok: White Lotus-Orstom. 107–142.

Suvarnasuddhi, K. 1950. *Some Commercial Timbers of Thailand.* Bangkok: Royal Forest Department.

Suvatabandhu, K. 1960. *Vegetation of Thailand and Its Correlation with Climate and Soil Type.* Proceedings of the symposium on Humid Tropics Vegetation. 170–175.

Tan, C. B. 1975. The Yao people: an introduction. In *Farmers in the Hills: Ethnographic Notes on the Upland Peoples of North Thailand.* Ed. A. R. Walker. Georgetown: Penerbit Universiti Sains Malaysia. 21–31.

Tapp, N. 1986. *The Hmong of Thailand: Opium People of the Golden Triangle.* Indigenous Peoples and Development Series, Report #4, Anti-Slavery Society, London.

Task Force on Hill Tribes and Minority Groups, Office of Special Activites, Ministry of Education. 1987. *Survey of Hill Tribes & Minority Groups in Northern Thailand.* Bangkok: Ministry of Education and United States Agency for International Development.

Technical Service Club. 1987. *The Hill Tribes of Thailand.* 2nd ed. Chiang Mai: Tribal Research Institute.

Thai-Norway Highland Development Program (TN-HDP). 1988. *Thai-Norway Church Aid Highland Development Project.* Chiang Mai: Chang-Puak Press.

Thein, A. 1969. *Illustrated Medical Dictionary.* 2nd ed. Rangoon: Mingala. 4 vols. (In Burmese)

Thongtham, M. L. C., and P. Theeravuthichai. 1981. *Developmental Research on Economic Ferns as Cash Crop for the Hill Tribes of Northern Thailand.* Bangkok: Highland Agriculture Project, Kasetsart University.

Tiyawalee, D. 1979. *Multiple Cropping for Highland.* Report to Agriculture Research Service, United States Department of Agriculture. Chiang Mai: Faculty of Agriculture, Chiang Mai University.

Tiyawalee, D., V. Pattaro, M. Samnaneechai, P. Wivatvongvana, and V. Hengsawad. 1978. *Legumes for Highland.* Report to Agriculture Research Service, United States Department of Agriculture. Chiang Mai: Faculty of Agriculture, Chiang Mai University.

Tonanon, N. 1974. *A Short Introduction for Some Common Timbers of Thailand.* Bangkok: Royal Forest Department.

Trease, G. E., and W. C. Evans. 1983. *Pharmacognosy.* 12th ed. London: Bailière Tindall.

Tribal Integrated Health Project. 1979. *Lahu/Akha Survey: 1979.* Chiang Mai: Family Planning International Assistance.

Tribal Research Centre. 1976a. *Directory of Tribal Villages in Thailand. (13) Changwat Chiang Rai.* Bangkok: Ministry of the Interior Tribal Data Project, Hill Tribes Welfare Division, Public Welfare Department.

———. 1976b. *Directory of Tribal Villages in Thailand. (14) Changwat Chiang Rai.* Bangkok: Ministry of the Interior Tribal Data Project, Hill Tribes Welfare Division, Public Welfare Department.

Tribal Research Institute. 1989. Tribal Population Summary. In *Hill Tribes Today.* Eds. J. McKinnon and B. Vienne. Bangkok: White Lotus-Orstom. Appendix I.

Tung, T. M. 1980. *Indochinese Patients: Cultural Aspects of Medical and Psychiatric Care of Indochinese Refugees.* Washington, D.C.: Action for South East Asians.

U.S. Congress, Office of Technology Assessment. 1984. *Technologies to Sustain Tropical Forest Resources.* OTA-F-214, March 1984, Washington, D. C.

U.S. Department of the Army. 1970. *Minority Groups in Thailand.* Ethnographic Study Series. Department of the Army Pamphlet #555–107, Washington, D. C.

Vidal, J. 1962. Noms Vernaculaires de Plantes (Lao, Meo, Kha) en usage au Laos. *Bulletin de L'Ecole Francaise d'Extrème-Orient* 49:1–197.

Vidal, J., and J. Lemoine. Contribution a l'Ethnobotanique des Hmong du Laos. *Journal d'Agriculture Tropicale et de Botanique Appliquée* 17:1–59.

Vienne, B. 1989. Facing development in the highlands: a challenge for Thai society. In *Hill Tribes Today.* Eds. J. McKinnon and B. Vienne. Bangkok: White Lotus-Orstom. 33–60.

Viriyapanyakul, V. 1990. Opium de-addiction and Akha concerns for the future. *Life on the Mountain* 2(3):5–9.

Viseshakul, D., V. N. Songla, and S. Punsa. 1982. Rice consumption of hilltribe people of Northern Thailand. *Journal of the Medical Association of Thailand* 65:133–138.

von Geusau, L. A. 1979. *Research into and Validation of the Traditional Medical System of the Akha in Northern Thailand.* New York. Manuscript.

———. 1983. Dialectics of Akhazaz: the interiorizations of a perennial minority group. In *Highlanders of Thailand.* Eds. J. McKinnon and W. Bhruksasri. Kuala Lumpur: Oxford University Press. 243–277.

———. 1990. Opium de-addiction programmes: a learning experience 1981–89. *Life on the Mountain* 2(3):10–22.

Vryheid, R. D., S. Wongcharoen, and G. L. Robert. 1987. *Nutrition Education Campaign and Survey. TG-HDP Nam Lang Project Area, Pangmapha Subdistrict, Mae Hongson Province.* Internal Paper 92, Thai-German Highland Development Programme, Chiang Mai.

Wakeman, F. 1975. *The Fall of Imperial China.* New York: Free Press.

Walker, A. R. 1974. The Lahu of the Yunnan-Indochina borderlands: an introduction. *Folk* 16–17:329–344.

———. 1975. The Lahu people: an introduction. In *Farmers in the Hills: Ethnographic Notes on the Upland Peoples of North Thailand.* Ed. A. R. Walker. Georgetown: Penerbit Universiti Sains Malaysia. 111–126.

———. 1983. The Lahu people: an introduction. In *Highlanders of Thailand.* Eds. J. McKinnon and W. Bhruksasri. Kuala Lumpur: Oxford University Press. 227–237.

Walker, A. R., and S. Wongprasert. 1968. *The Preparation of the Opium Swiddens and the Sowing of Poppy Seeds.* Red Lahu Project Field Report #10, Tribal Research Center, Chiang Mai.

Walter, H. 1971. *Ecology of Tropical and Subtropical Vegetation.* Trans. D. Mueller-Dombois. Edinburgh: Oliver & Boyd.

Wang, D., and S. Shen. 1987. *Bamboos of China.* Portland, OR: Timber Press.

Wanleelag, N. 1980. *Investigation of Insecticidal Plants and Their Production in Highland Area.* Bangkok: Highland Agriculture Project, Kasetsart University.

Webster, C. C., and P. N. Wilson. 1980. *Agriculture in the Tropics.* 2nd ed. White Plains: Longman.

Webster, P. J. 1981. Monsoons. *Scientific American* 245:109–118.

Weil, A. T. 1981. Botanical vs. chemical drugs: pros and cons. In *Folk Medicine and Herbal Healing.* Eds. G. G. Meyer, K. Blum, and J. G. Cull. Springfield: Charles C. Thomas. 287–312.

Westermeyer, J. 1982. *Poppies, Pipes, and People. Opium and Its Use in Laos.* Berkeley: University of California Press.

White, J. C. 1982a. Natural history investigations at Ban Chiang: the study of natural resources and their use today aids reconstruction of early village farming in prehistory. *Expedition* 24:25–32.

———. 1982b. Prehistoric environment and subsistence in Northeast Thailand. *South-East Asian Studies Newsletter* 9.

White, P. T. 1985. The poppy. *National Geographic* 167:142–189.

Whitmore, T. C. 1984. *Tropical Rain Forests of the Far East.* 2nd ed. Oxford: Clarendon Press.

Whittaker, R. H. 1975. *Communities and Ecosystems.* 2nd ed. New York: Macmillian.

Widjaja, E. A. 1987. A revision of Malesian *Gigantochloa* (Poaceae–Bambusoideae). *Reinwardtia* 10:291–380.

Wiens, H. J. 1954. *China's March Toward the Tropics.* Hamden, CT: Shoe String Press.

Williams, L. 1967. *Forests of Southeast Asia, Puerto Rico, and Texas.* Washington, D.C.: United States Department of Agriculture, Agricultural Research Service.

Wilson, E. O. 1988. The current state of biological diversity. In *Biodiversity.* Ed. E. O. Wilson. Washington, D.C.: National Academy Press. 3–18.

Wong, H. 1984. Peoples of China's far provinces. *National Geographic* 165:283–333.

Wongsprasert, S. 1989. Opiate of the people? A case study of Lahu opium addicts. In *Hill Tribes Today.* Eds. J. McKinnon and B. Vienne. Bangkok: White Lotus-Orstom. 159–172.

Xiong, L., J. Xiong, and N. L. Xiong. 1984. *English-Hmong-English Dictionary.* Bangkok: Pandora.

Young, G. 1974. *The Hill Tribes of Northern Thailand.* 5th ed. Bangkok: Siam Society.

Zeven, A. C., and P. M. Zhukovsky. 1975. *Dictionary of Cultivated Plants and Their Centres of Diversity.* Wageningen: Centre for Agricultural Publishing and Documentation.

General Index

Index to Scientific Names

Index to Common Plant Names